Hilary Bauerman

Text-Book of Systematic Mineralogy

Elibron Classics
www.elibron.com

Elibron Classics series.

ISBN 0-543-90528-4 (paperback)
ISBN 0-543-90527-6 (hardcover)

This Elibron Classics Replica Edition is an unabridged facsimile
of the edition published in 1892 by Longmans, Green, and Co., London and
New York.

TEXT-BOOKS OF SCIENCE

ADAPTED FOR THE USE OF

ARTISANS AND STUDENTS IN PUBLIC AND SCIENCE SCHOOLS

SYSTEMATIC MINERALOGY

TEXT-BOOK

OF

SYSTEMATIC MINERALOGY

BY

HILARY BAUERMAN, F.G.S.

ASSOCIATE OF THE ROYAL SCHOOL OF MINES
PROFESSOR OF METALLURGY, ROYAL ARTILLERY COLLEGE

THIRD EDITION

LONDON
LONGMANS, GREEN, AND CO.
AND NEW YORK : 15 EAST 16th STREET
1892

PRINTED BY
SPOTTISWOODE AND CO., NEW-STREET SQUARE
LONDON

PREFACE.

In preparing the present volume, two main objects have been kept in view: first, that it should form a useful guide to students desirous of acquiring a general knowledge of the subject; and secondly, that it should serve as an elementary introduction to larger text-books, such as those of Dana, Miller, Descloizeaux, and Schrauf, an acquaintance with which is essential to those who wish to familiarise themselves with the higher branches of the subject. For this purpose, the treatment adopted has been as general as possible, the descriptions of the crystalline forms dealing only with their symmetry and general geometrical properties, without entering into the question of the practical calculation and determination of individual examples, which would have increased its bulk beyond admissible proportions. In this part of the text, the methods followed have been mainly those of Groth's admirable treatise on 'Physical Crystallography,' except that the plan there adopted of considering the physical structure of crystals before their geometrical properties has been abandoned in favour of the less logical, though more familiar, one of giving precedence to the latter. The optical properties of crystals have been considered at somewhat greater length than is usual in rudimentary books, on account of the great and increasing use made of this branch of investigation.

Upon similar utilitarian considerations a mixed system of notation has been adopted in the crystallographic part, the forms being designated in the text by their symbols according to Naumann, while the notation of their faces is by indices on Miller's system. As a matter of personal preference, the latter system would have been adopted exclusively; but, having regard to the fact that the former is used much more extensively than any other system, both in text-books and in original memoirs, familiarity with its use is very desirable to students.

In the hexagonal system, the Bravais-Miller notation by indices on four axes has been adopted, as showing most clearly the relation between it and the tetragonal system.

In the chemical portion of the volume the classification followed is that of the second edition of Rammelsberg's 'Handbuch der Mineral-Chemie,' as being the standard modern authority upon the chemistry of minerals.

The systematic part having been extended somewhat more than was originally intended, it has been found impossible to include physiography, or general descriptive mineralogy, in the same volume, without deviating too widely from the plan of the series. This will therefore be issued as a companion volume.

In the preparation of the work, valuable assistance and advice has been received from many friends. In gratefully acknowledging these services, the writer has to mention particularly those rendered by the editor of the series, Mr. Merrifield, who has made several important additions to the text, Mr. R. T. Glazebrook, of Trinity College, Cambridge, and Mr. F. W. Rudler, who has passed the later sheets through the press during the writer's absence abroad.

LONDON: *January* 10, 1881.

CONTENTS.

SYSTEMATIC MINERALOGY.

CHAPTER I.

PRELIMINARY.

MINERALOGY is the science that treats of the substances known as Minerals—that is, the constituents of the earth considered as they actually occur in nature.

The constitution of the solid earth, excluding all consideration of its inhabitants—that is, of the animals and plants living in the atmosphere—may be regarded in many different ways.

In the most general view the earth is a spheroid, about five times as heavy as an equal volume of water. Geology, with somewhat more detail, considers the accessible portion a shell of some ten to twelve miles thick, as made up of about the same number of different kinds of rock, the inaccessible interior portion being probably not very dissimilar in composition; while Chemistry is concerned mainly with the ultimate elementary constituents of the mass, and the properties of these elements as derived from the study of their combinations artificially formed. The position of Mineralogy is intermediate between those of geology and chemistry. With the former it considers the structure of the solid mass of the earth, but in greater detail, resolving the rock masses into a larger number of more exactly defined constituents or minerals; while, with the latter, it considers the elementary constitution of such substances, restricting its study, however, to such chemical compounds as are actually found in nature. This individual, natural existence is essential to the idea of a mineral, as distinguishing it from a chemical salt or other artificial preparation. Whether the latter is or is not represented in nature can only be determined by experience, but, speaking generally, it may be said that only the most stable and least soluble compounds, or precisely those that are most difficultly obtainable in the laboratory, are represented in nature, and therefore the chemistry of minerals, though essentially fragmentary, is of no small importance in the general body of chemical knowledge.

The qualities essential to the distinction of minerals among themselves are of three kinds—namely, *form*, *structure*, and *chemical composition*—all of which must be investigated and determined before the specific independence of any mineral can be regarded as properly established. The first of these considers the external form of the substance by methods which are essentially those of descriptive solid geometry, qualified by certain special principles—those of symmetry and numerical rationality—generalised from the whole body of such observations. This part of the subject is known as *Geometrical* or *Morphological Crystallography*. The second quality, that of structure, considers the substance as made up of similar material molecules, whose arrangement is indicated by the physical properties, such as density, cohesion, colour, &c.; or generally, by their elastic resistance to forces tending to disturb their molecular equilibrium, which may or may not vary in different directions. These investigations are essentially part of the work of experimental physicists, but their results, when combined with those of the geometrical crystallographer, collectively form the branch of *Physical Crystallography*. Lastly, the investigation of the third and most important character, that of elementary composition, is the work of ***Mineral Chemistry***; and the combination of this with the knowledge derived from the study of form gives rise to the important principle of *Isomorphism*, or the relation of form to constitution, upon which the most natural and satisfactory systems of classification are founded.

In addition to these three principal heads, information upon subsidiary matters is requisite for the attainment of a complete knowledge of any mineral. These have reference to its natural habitat, such as association with other minerals, geographical and geological distribution, and evidences of possible changes from the condition of original formation. This last point especially has an important bearing upon the speculative matter of the origin and mode of formation of minerals, which is, or should be, the province of geology proper, as the basis of any reasonable speculation upon the structure of the earth in its largest sense. This branch of the subject is, however, generally spoken of as *Chemical Geology*.

The whole body of knowledge comprised under these several heads, when classified in an orderly manner, forms *Descriptive Mineralogy*, or *Physiography*.

The number of elements known to the chemist is now between sixty and seventy, all of which are concerned in the production of minerals, though in widely different proportions. Some eight or ten are found in the free or uncombined state, and with six or seven hundred combinations of two or more, make up the roll of natural minerals or mineral species.

The term *species* is applied to any substance whose form, structure, and composition are definite, constant, and peculiar to itself, and therefore serve to distinguish it from all other species. By constancy of form and composition in the above definition is not meant that there must be absolute identity in these particulars between different examples of the same substance, but that the variations in either shall be subject to known laws. Thus, carbonate of calcium in the species Calcite appears in several hundreds of different forms, all of which may, by the application of crystallographic laws, be shown to be derivatives of a single geometrical form—the rhombohedron—and again its composition may differ sensibly in different specimens, but these differences are explainable by the law of isomorphism, which shows that one dyad metal may be substituted for another without altering the general molecular constitution.

The exact limits to be given to species is to a great extent matter of opinion. Upon purely chemical grounds, all substances combining the same type of molecular constitution with analogous forms may be considered as varieties of a single species, without reference to the nature of the elements composing them; but this is too wide a definition to be of much practical use to the mineralogist. It is therefore customary with such a class of similarly constituted substances to classify them according to their contained metals, giving a different name to each, and to call the whole group by the name of the most prominent species. On the other hand, a variety of any substance marked by some constant peculiarity, whether of form, colour, or other apparent property, may often receive a particular name with advantage, even when the distinguishing difference is too slight to allow of the separation on the grounds of systematic form or composition. The ultimate guide is in all cases the convenience of the observer; and if the practice of giving names without a previous complete determination of the composition and physical properties be avoided, it is generally better to form new specific names rather than unduly widen the boundaries of older ones.

The proportion in which different minerals enter into the composition of the crust of the earth varies very considerably, as does also the size of the individual masses of any one. Thus we may find the same substance in particles of microscopic minuteness in some places, and in others in masses measurable by cubic feet or yards, and weighing up to hundreds of tons, or even forming mountain masses, without sensible admixture of other substances. Experience, however, shows that the characteristic properties of any mineral, and especially that of form, are best developed with individuals of a moderate size, as when very minute they become invisible and incapable of exact measurement, and when very large the characteristic form is not, as a rule, apparent. Such undefined mineral masses are generally spoken of as Rocks. Quartzite and statuary marbles, for instance, are aggregates of particles of quartz and calcite into masses of a slaty or granular texture in which their proper forms are entirely lost; while, on the other hand, the crystalline lava called basalt is made up of individuals of the species Labradorite, Augite, Olivine, and Magnetite, all perfectly well defined in form and physical characters, but so minute as to be indistinguishable to the unaided eye, the general effect being that of a uniform, opaque, black substance, separating into masses whose shapes bear no obvious relation to those of their constituent minerals. The distinction between rocks and minerals is, however, one of geological convenience only, and in mineralogy is almost without significance; the nature of a mineral mass being defined according to its constituents, either as of a single species or an aggregate of two or more, without reference to the size or perfection of the individual components. It often happens that groups of two or more minerals, distinctly separated, pass in the same mass by inappreciable gradations into aggregates in which the individuals are indistinguishable by ordinary means, so that it is difficult to say where either condition begins or ends. This difficulty is further increased by the use of the microscope, which very commonly resolves substances apparently uniform into aggregates of dissimilar ones, and there is no reason to suppose that the individuality of the constituents ceases when the microscope is no longer able to reveal them. The same class of observation, however, shows that foreign substances are so commonly included even in the most perfectly developed mineral individuals or crystals, that the condition of homogeneity required by the ordinary definition of a mineral as a homogeneous inorganic substance is seldom, if ever, realised, and therefore this definition can only be accepted as an approximation requiring considerable qualification in use.

The complete discussion of all the subjects indicated in the preceding pages, or even of any one of them, being beyond the scope of an elementary book of limited extent, the space at command will be devoted to a sketch of the principles upon which the methods of determining form, structure, and other elements of classification in minerals are based, without entering into the details of the methods of observing or reduction of observations, for which matters the student is referred to the larger special works as given in the list at the end of the volume. A physiographic sketch of the more important species, classified according to Berzelius' and Rammelsberg's method, forms the subject of a companion volume.

CHAPTER II.

GENERAL PRINCIPLES OF FORM.

WITH the exception of water, mercury, and some hydrocarbons which are liquids at ordinary temperatures, minerals are solids, and occur in masses which in some cases are of irregular, and in others of regular, shape. The first of these are called *amorphous*, and the second *crystalline*, substances, and any individual mass of the latter kind is a *crystal*. The branch of mineralogy that treats of the study of such form is called Crystallography.

The term 'crystal,' derived from the Greek κρύσταλλος, which was applied to ice and transparent quartz or rock-crystal, the latter having been supposed to be produced from water by extreme cold in mountain regions, is applied to natural and artificial substances which, in solidifying, from a state whether of solution or fusion, assume definite polyhedral forms, which are constant for the same substance. The leading property of crystals, as distinguished from mere geometrical solids, is the invariability of the angles between corresponding faces in different individuals of the same substance. There is usually a very marked *symmetry* to be noticed in the arrangement of their plane faces and edges, and occasionally of their points also, although this latter symmetry is not essential, crystallographic symmetry being one of direction and not of position, so that two parallel planes or two parallel lines are not distinguished from one another, and on that account the invariability of the angles is a paramount consideration. The character of the symmetry varies in different groups of crystals, and forms the basis of their classification into systems.

If we take any polyhedron and place it upon a looking-glass, the object and its image together constitute a symmetrical figure, of which the reflecting surface is the plane of symmetry; and if in any given solid we can find a plane such that, if we were to cut it in half by that plane, and to place against the section a mirror, the reflected image would exactly reproduce the other half (identically, and not reversed, as objects generally are by reflection), the solid is said to be symmetrical about that plane.[1] This is a

[1] This may be shown with a model of a cube, painted white, and a plate of red glass, held perpendicularly to one of its faces, when a white image of the part of the face in front will be seen by reflected, and a red one of the hinder part by transmitted, light. If the direction of the plate be parallel to that of a plane of symmetry the two images will

more particular supposition than need be made in crystallography, in which two parallel planes are not distinguished; nevertheless it is convenient to use it for purposes of description, in order to escape a vagueness which would otherwise make the description unintelligible. For instance, although figs. 1 and 2 have exactly the same crystallographic symmetry, it will be more convenient to consider and describe the first, and to regard the second as another example of the same form. They are both symmetrical, as

FIG. 1.

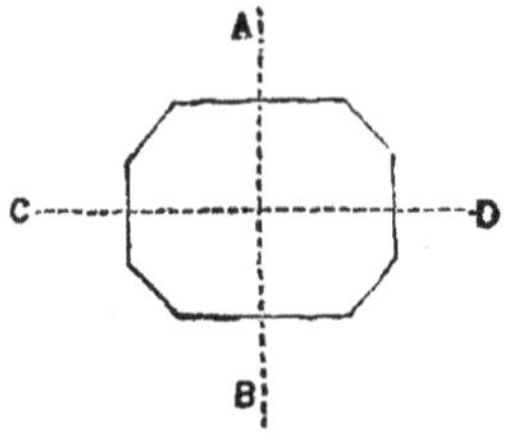

FIG. 2.

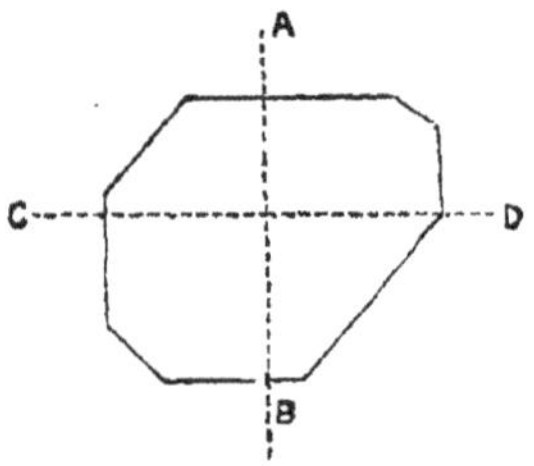

regards direction, to the lines A B, C D, but the symmetry of position of fig. 1 renders it a much more definite thing to talk about and to apply linear measure to. Only it must not be forgotten that this symmetry of position is neither essential nor inherent, but is merely adopted as an aid in forming definite ideas.

Symmetry about a line in plane figures corresponds to symmetry about a plane in space. In the latter, symmetry about lines or axes is also observed in many cases, but the discussion of this may be deferred until another characteristic feature of crystals, and that the most important one, as forming the basis of all exact crystallography—namely, the principle of rationality—has been noticed.

The term 'Rationality' will be best understood by using

apparently coincide; but in any other position they will deviate to a greater or less extent. The particular positions of symmetry for the cube are shown in figs. 11, 12, 13.

plane space of two dimensions, instead of actual space of three dimensions, for an illustration. Without considering the exact meaning of axes of reference, let it be assumed that O A, O B, fig. 3, represent two such axes, to which two planes belonging to one crystal, represented by P Q, P′ Q′, are referred. Then the principle of rationality requires that if P Q″ be drawn through P parallel to P′ Q′, the ratio of O Q to O Q″ shall always be rational, or, as it may be more generally stated,

FIG. 3.

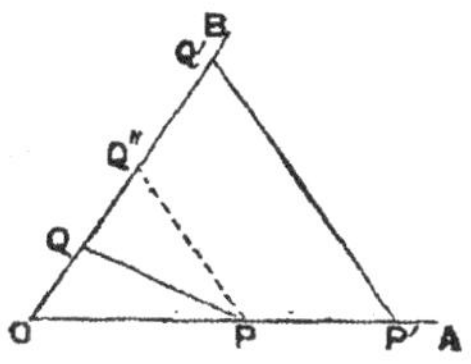

$$\frac{OP}{OQ} : \frac{OP'}{OQ'} = \text{a simple rational fraction.}$$

Usually, the relation is one of very simple numbers, such as 2 : 1, 1 : 2, 2 : 3, 4 : 5, &c., while it can never be that of an incommensurable surd, such as $\sqrt{2}$ or $\sqrt{5}$, to unity. This law is an empirical one—that is, it expresses the results of observations without explaining their cause—but there are no known exceptions. Its geometrical consequences are—

1. The exclusion of all but the simpler types of symmetry about an axis, namely, binary, quaternary, ternary, and senary.

2. The exclusion, from possible crystalline forms, of the Platonic or regular geometrical solids of higher order than the cube or octahedron.

The regular dodecahedron and icosahedron involve pentagonal symmetry, and they bring the irrational value $\sqrt{5}$ into the axial relations. There are numerous examples among natural crystals, notably in iron pyrites, of forms approximating in shape to these—that is, they are contained by twelve five-sided or twenty three-sided faces—but these are never regular pentagons, or all equilateral triangles. When applied to three dimensions, as in actual crystals, the principle of rationality requires that if planes of different kinds occur in the same crystal they must be so related that their intercepts[1] upon the axes of reference are in rational proportion to one another. That is to say, if one plane meets the three axes in the points *p q r*, and the other in the points P Q R, the relation expressed by

$$\frac{\text{O}p}{\text{OP}} : \frac{\text{O}q}{\text{OQ}} : \frac{\text{O}r}{\text{OR}}$$

will be that of quantities having rational ratios to one another, and usually these will be found to be in low numbers.

Axes of symmetry. When a polyhedron is turned about a line so selected that after passing through an aliquot part of a whole revolution its position in space as a whole has not been changed, and there is no apparent difference in shape for the two aspects, it is said to be symmetrical to the line or axis of rotation. The kind of symmetry is denoted by the number of times the positions of symmetry recur in a complete revolution, or between the starting of a marked point on the crystal and its return to the original position. Thus, in a cube, the straight line joining the middle points of two opposite edges is an axis of binary symmetry ; for by turning the solid about such a line through half a revolution, it assumes a position apparently similar to the original one, and the change can only be perceived by observing the alteration in the place of a marked point or face with reference to some external object. The cube has also ternary symmetry about a diagonal line joining opposite points, and quaternary symmetry about a line joining the centres of opposite faces, the original position being apparently restored by rotation through one third of a revolution in the former, and one quarter or one right angle in the latter case. Quaternary also includes the lower condition of binary

[1] An intercept is the distance intercepted or cut off by a plane upon an axis measured from the origin of the latter. Thus in fig. 3, O Q, O Q′, O P, O P′, are intercepts upon the axes O B, O A, whose origin is at O.

symmetry. The only other class of symmetry possible in crystals is senary or hexagonal, corresponding to a rotation of one-sixth of a revolution, such as that of a regular hexagonal prism about its axis; this includes ternary symmetry. Quinary symmetry, such as that of a Platonic or regular icosahedron about a diagonal, or of a Platonic dodecahedron about a line joining the centres of opposite faces, is crystallographically impossible, as it introduces irrational relations. The remarks made on page 8 apply equally in this case: the necessary symmetry being one of direction only, the same symmetry exists about any line parallel to the axis as about the axis itself; but for convenience of description it is best to consider the cases in which there is also symmetry of position, always bearing in mind that this is a mere matter of convenience, and not essential, or affecting the classification.

As has already been shown in the case of the cube, a crystal may be symmetrical about more than one axis. If there is binary or quaternary symmetry about two axes at right angles to one another there is a third axis of the same kind at right angles to both.

Crystals are classified into systems according to the number and character of their axes of symmetry. Six of such systems are possible, all of which are represented by natural minerals. They are as follows, commencing with those of lowest symmetry :—

1. The triclinic system. This is without any axis of symmetry, the faces of any form being only symmetrical to a central point.

2. The oblique system. This has one axis of binary symmetry, and consequently one plane of symmetry.

3. The rhombic system. This has three axes of binary symmetry at right angles to one another.

4. The hexagonal system. This is characterised by one axis of senary or hexagonal symmetry and six of binary symmetry at right angles to the first. It includes the case

in which there is an axis of ternary symmetry, or the rhombohedral system.

5. The tetragonal system. This has one axis of quaternary symmetry at right angles to two of binary symmetry, which are also at right angles to each other.

6. The cubic system. This is specially characterised by three axes of quaternary symmetry at right angles to one another, besides which there are four of ternary and six of binary symmetry, whose positions have been already alluded to on p. 10, and will be more specially noticed subsequently.

When a system has more than one kind of symmetry, it may be distinguished by the number of its axes of the highest kind, or axes of principal symmetry. Upon this distinction is founded the classification of the systems into the following three groups, which are closely related to their physical properties :—

1. *Without a principal axis of symmetry.* This includes the triclinic, oblique, and rhombic systems, the first being without linear symmetry, while in the second and the third the symmetry is all of the same kind, namely, binary.

2. *With one principal axis of symmetry.* This includes the hexagonal and tetragonal systems, the principal symmetry of the first being senary and of the second quaternary.

3. *With three principal axes of symmetry.* This is special to the cubic system, the three axes being those of quaternary symmetry.

When a crystal is contained by all the planes or faces[1] required by the complete symmetry of the system, each one has a counterpart plane parallel to it, so that their total number is always even and not less than six. These are

[1] It is convenient to call the natural surfaces of crystals *faces*, and those produced artificially, or required in geometrical construction, *planes*, e.g. planes of symmetry and cleavage planes.

said to be holohedral (full-faced) forms. There are also certain forms in which only one half or one quarter of the full number of faces are present; these are respectively called hemihedral (half-faced) and tetartohedral (quarter-faced) forms. The selection of these faces may in some instances be made in more ways than one, subject to the condition that the relation of the faces to the axes of symmetry must be the same as in the holohedral form, or each equivalent axis must cut an equal number of faces—namely, one half or one quarter of that which it would do at the same inclination in the full-faced form.

Hemihedral forms are not possible in the triclinic system, that being without plane symmetry; in the oblique system there may be one kind, but it has not been observed in nature; in the rhombic there may be two kinds, but it is doubtful whether more than one actually exists in nature. In the remaining systems, hexagonal, tetragonal, and cubic, the forms are susceptible of hemihedral development in three ways, but it is only in the tetragonal that all three are actually known to exist, and of these one has only been observed in artificial organic compounds.

Tetartohedral crystals are possible in all systems where there are more than two kinds of hemihedrism, that is in the tetragonal, hexagonal, and cubic. As they may be considered as resulting from the successive application of two kinds of hemihedry to holohedral forms, three classes of tetartohedra might be possible in the first two systems, were it not for the circumstance that in either only two out of the three combinations give rise to forms that satisfy the general conditions of symmetry. In the cubic system all three kinds of hemi-hemihedrism produce the same class of form or there is only one kind of tetartohedron. This has not been found in natural minerals, but is characteristic of a group of metallic salts, the nitrates of the lead-barium group. In the tetragonal and hexagonal systems the two possible methods give rise to two different kinds of forms, neither of which has been observed either in natural or artificial crystals in the former system, but in the latter both kinds are known, one of them being specially characteristic of the commonest mineral constituent of the earth's crust, namely, quartz or rock-crystal.

By hemihedral or tetartohedral development a crystal belonging to any system loses a portion of the symmetry which characterises it when possessed of the full number of faces. Thus, in the three kinds of hemihedra possible in the systems with principal axes, one is entirely without plane symmetry, while each of the other two has only one of the two kinds necessary in the full faced forms. It must therefore be remembered that in defining systems by their characteristic symmetry, such definitions only apply to the holohedral forms.

There is an essential distinction between the geometrical and the mineralogical idea of hemihedrism and tetartohedrism, which it may be well to notice as early as possible. In the former only those forms are considered as hemihedra and tetartohedra that are geometrically distinguishable from the holohedral forms, while in the latter the presence of any such form in any crystal of a substance is considered as imparting the same character to all other crystals of the same substance, whether they be geometrically distinguishable from holohedral forms or not.

The same kind of symmetry as is displayed in crystals is furnished by an orderly arrangement of points in space, which has the further analogy with crystals of suggesting rational ratios. These arrangements of points may be extremely simple; indeed all the crystallographic systems may be represented by the points of intersection of three sets of parallel and equidistant planes, as in fig. 4, where we may call $a_1a_4 = a$, $a_1a_2 = b$, and $a_1b_1 = c$, and the angles $a_5a_4b_4 = \alpha$, $a_1a_4b_4 = \beta$, and $a_1a_4a_5 = \gamma$. The whole series of points may be regarded as a sort of net in space, of which the strings represent the lines of intersections of these planes, and the knots or nodes the points of intersection. Then, if the equal distances at which one set of planes intersects the other two be taken as *a*, *b*, and *c* respectively, and the angles between the lines as α, β, and γ, these quantities will be the characteristics of the system, and symmetrical relations of equality between those characteristics will determine a crystallographic system. The planes of the system, corresponding to faces of the crystal, will be planes drawn through any three points of the system. The limitation to rational ratios is at once suggested by the necessity of taking a whole number of intervals between any two points, in order to satisfy the definition of a plane of the system being one passing through three points of the system. As to the particular modes of symmetry, if α β γ are all different, there is no symmetry. This corresponds to the anorthic or triclinic system. If β and γ are right angles, *a* is perpendicular to the plane of *b c*, and is an axis of binary symmetry. We have then the oblique system. If α β γ are all right angles, we have three axes of binary symmetry, corresponding to the rhombic system. If, in addition, $b = c$, *a* is an axis of quaternary symmetry, and we obtain the tetragonal system; while if $a = b = c$ we have the cubic system. If, on the other hand, $\alpha = \beta = \gamma$, without being right angles, and also $a = b = c$, then there is an axis of ternary symmetry equally inclined to *a*, *b*, and *c*. This represents the rhombohedral system of ternary symmetry which has been mentioned as included in the hexagonal system.

FIG. 4.

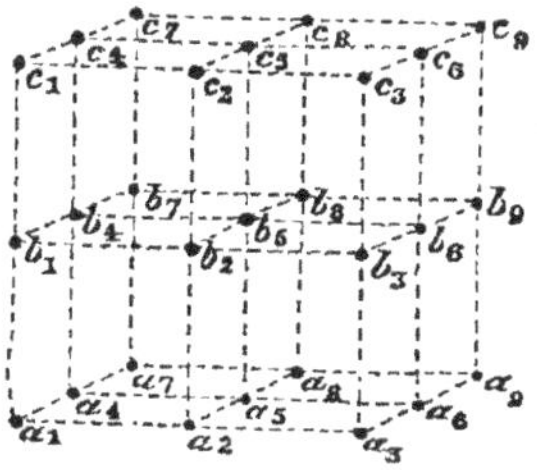

It is to be remarked that while these relations between *a b c*, α β γ, yield the crystallographic systems in the easiest way, they are sufficient, but not necessary. For instance, if α β γ are all right angles, it is not necessary that *a*, *b*, and *c*

should be all equal in order to give a cubic system : it is sufficient that they should be commensurable. For, let $a = 1$, $b = 2$, $c = 3$, then a cube whose side is 6 will include all, and will thus yield cubic symmetry. There may also be particular relations between $a\ b\ c$, $\alpha\ \beta\ \gamma$, of a less simple character, which will afford a similar increase of symmetry. This will be at once seen by joining three points of a cubical system, taken arbitrarily, to a fourth. If we complete the parallelepiped, and take its planes as the basis of a new system of points, the old axes will no longer be edges of the fundamental parallelepiped, but they will not thereby cease to be axes of symmetry, for it will always be possible to find planes of the new system (that is, each passing through three points of it) which shall be at right angles to them. As a particular example, if $a = b = c$, and $\alpha = \beta = \gamma$, we have in general the rhombohedral system ; but if α takes the particular value, $70° \ 31' \ 44''$, so that the dihedral angles are of 60 degrees, this introduces further symmetry, and throws us back on the cubic system.

The conditions on which an arbitrary parallelepipedal system should have a given symmetry are not fully known, and must at any rate be extremely complicated. In place of attempting any such reductions, it will be preferable to work from known axes of symmetry in each system.

Notation of a plane face. In all the systems, the position of a face of a crystal is represented by the points at which it cuts the three selected axes. Since, moreover, the ratios of the intercepts must always be commensurable, this is secured by taking the intercepts as integral multiples (or submultiples) of definite lengths measured along each axis, and called the parameters or units. These lengths are not arbitrary, but depend upon the nature of the substance forming the crystal. If we consider this as composed of a series of molecules in parallelepipedal order, it secures that the faces shall be planes of the system.

Notation of faces of crystals. There are four principal

methods of indicating the faces of crystals by their axial relations in use—those of Weiss, Miller, Naumann, and Lévy; and as all are of equal authority, it will be necessary for the student to become acquainted with the principles of each. The last of these is a modification of the oldest system, that of the Abbé Haüy, which was in the earlier years of this century in common use by mineralogists in most European countries; but during the last quarter of a century it has in many cases been superseded by one or other of the remaining systems, none of which have, however, up to the present time become sufficiently popular to be universally accepted. In the first two methods a face is indicated by a symbol compounded of three signs derived from the ratios of the axial intercepts to the unit parameters—that is, one for each axis—which are written in an invariable order; while in the other two, arbitrary signs, which vary with the systems, are used in addition.

Weiss's notation. In this, the least conventional of all the methods, a unit face A B C is indicated by the length of its intercepts upon the axes of reference in the order shown in fig. 5, as $a : b : c$, where O A $= a$, O B $= b$, O C $= c$. This supposes the parameters to be dissimilar for all three axes, as in the rhombic, oblique, and triclinic systems; but when $a = b$, as in the hexagonal and tetragonal systems, the symbol used is $a : a : c$, and in the cubic system, where all three parameters are alike, it is $a : a : a$.

FIG. 5.

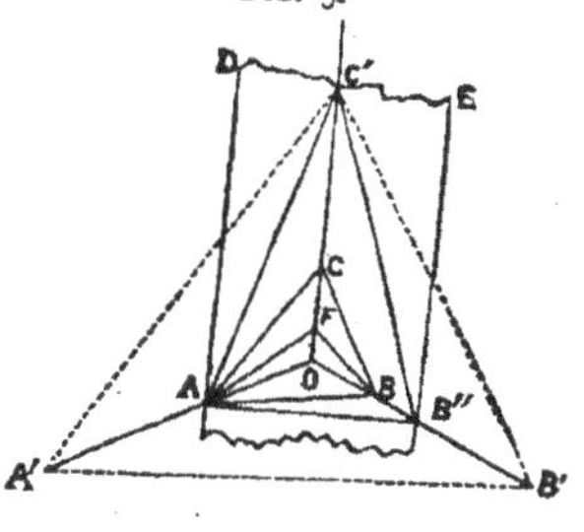

For any face having different intercepts, such as A B'' C', the longer ones are expressed as multiples of the shortest considered as unity; or for the particular case where the intercept O B'' is $1\frac{1}{2}$ times O B, and O C' 3 times O C, $a : \frac{3}{2} b : 3 c$, all the signs being positive; while for a parallel face placed behind, to the left, and below the point O, the same symbol is used, but with negative signs or accented letters thus, $-a : -\frac{3}{2}b : -3c$, or $a' : \frac{3}{2}b' ; 3c'$, the co-efficients being any rational whole number or fraction, including 0 and ∞. For a face such as A F B, where O F $= \frac{1}{2}$ O C, the symbol is $a : b : \frac{1}{2}c$. When one of the intercepts is immeasurably long, its co-efficient becomes ∞, as in the face A B'' D E, which is parallel to the vertical axis C, the symbol is $a : \frac{3}{2}b : \infty c$.

The general expression for any face meeting the three axes at dissimilar distances is $a : nb : mc$, where n and m may be any rational numbers, the first greater, and the second either greater or less, than unity.

This is the least artificial class of notation, as no contractions are employed; but it is on that account rather inconvenient, the symbols being somewhat cumbrous in use. It is employed in the works of the Berlin school of mineralogists, including those of Weiss, Rose, and Rammelsberg.

Miller's notation. If in a series of planes like those of fig. 5, the unit form be supposed to lie outside the others, the intercepts of any enclosed plane upon two of the axes may be expressed as fraction of an unit intercept, which then becomes the longest instead of, as in Weiss's method, the shortest. Thus, in fig. 5, supposing a plane to be drawn through A' B' C' parallel to A B C, the lengths of the intercepts will all be multiplied by three, which does not alter their ratios, so that the new plane will still have the symbol $a : b : c$ as before; but the relation of the plane A B'' C' to the new axes is $\frac{a}{3} : \frac{b}{2} : \frac{c}{1}$, and if the symbol of the unit face be written as $\frac{a}{1} : \frac{b}{1} : \frac{c}{1}$, the denominators of these groups of fractions alone, if written down in the proper order—thus, 1 1.1, 3 2 1—give a new kind of symbol perfectly expressing the axial relations of the two planes. This method is adopted by Miller, the numbers making up the symbols being called *indices*. These are always whole numbers including 0, and in the greater number of instances not exceeding 6. Intercepts on the negative side of any axis are indicated by a minus sign written above the corresponding index number. Thus, the counterpart planes of 1 1 1 and 3 2 1 are $\bar{1}\,\bar{1}\,\bar{1}$ and $\bar{3}\,\bar{2}\,\bar{1}$.

The general expression for a plane intercepting three axes at unequal distances in this system is obtained by the method shown in fig. 6, where, supposing H K L to represent such a plane, and A B C that of the unit form, the axial lengths of the former will be fractions of the latter, as in the preceding figure, or putting $OH = \frac{1}{h} OA$, $OK = \frac{1}{k} OB$, and $OL = \frac{1}{l} OC$, will be represented by

FIG. 6.

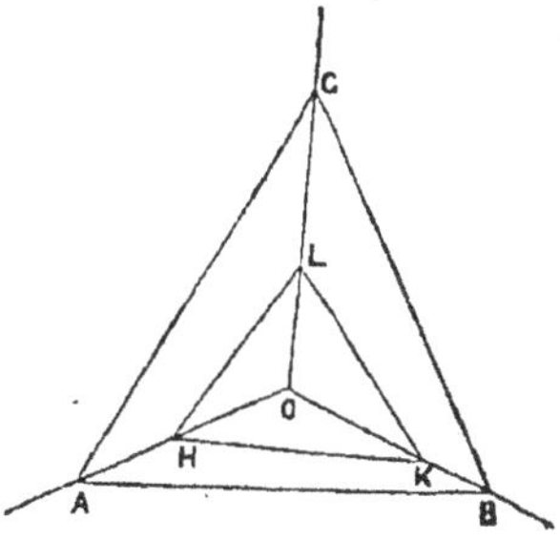

$$OH = \frac{OA}{h},\ OK = \frac{OB}{k},\ \text{and } OL = \frac{OC}{l},$$

and, conversely, the indices will be

$$h = \frac{OA}{OH},\ k = \frac{OB}{OK},\ \text{and } l = \frac{OC}{OL}.$$

These values are in no wise altered by multiplying them by any numerical co-efficient, as the unit parameter will be similarly altered by the same process, and the ratio of the two series will be unchanged. The symbol hkl is therefore the most general expression that can be obtained for a form by this method of notation, where all three indices are dissimilar; and from it the symbols of all other planes in a system may be obtained by assigning particular values to the different indices. Thus, 1 1 1 is a special form in which $h = k = l$; 0 0 1 another where $h = k = 0$ and $l = 1$; and so on with all possible values that can be assigned to them. This system of notation by indices was first used by the Rev. Dr. Whewell, but has been developed and extended by Professor W. H. Miller, of Cambridge, to whom it is substantially due in its present form. It is by far the most elegant of all the methods in use, and will probably be adopted at no very distant date by all mineralogists, although up to the present time its use has been somewhat restricted.

If Miller's symbol of the unit plane (111) corresponding to Weiss's $a : b : c$, be written as $\frac{a}{1} : \frac{b}{1} : \frac{c}{1}$, that of hkl will be $\frac{a}{h} : \frac{b}{k} : \frac{c}{l}$, which ratios, when brought to whole numbers by multiplying each term by hkl, become $kla : hlb : hkc$, or the parameter co-efficient of any axis will be the product of the indices of the other two, and these products, when written in the proper order and reduced to their simplest expression, will give the required symbol. For instance, Weiss's symbol corresponding to $(hkl) = (1\ 2\ 3)$ is

$$\frac{a}{1} : \frac{b}{2} : \frac{c}{3} = 6a : 3b : 2c = 3a : \tfrac{3}{2}b : c.$$

Similarly (346) corresponds to

$$\frac{a}{3} : \frac{b}{4} : \frac{c}{6} = 24a : 18b : 12c = 2a : \tfrac{3}{2}b : c.$$

When one index = 0, the corresponding co-efficient $= \infty$, and the co-efficients of the other two axes are found by interchanging their indices; thus $(3\ 2\ 0) = 2\,a : 3\,b : \infty\,c = a : \frac{3}{2}\,b : \infty\,c$. When two indices or co-efficients = 0 or ∞, the third index or co-efficient is put = 1, which means that the intercept on that axis is finite, and not that it is of any particular unit length; thus $(0\ 0\ 1) = \infty\,a : \infty\,b : c$; $(0\ 1\ 0) = \infty\,a : b : \infty\,c$; and $(1\ 0\ 0) = a : \infty\,b : \infty\,c$.

Weiss's numbers being directly, and Miller's inversely, proportional to the lengths of the intercepts, the highest co-efficient will always correspond to the lowest index, whatever may be the order of the symbols. When h is the highest and l the lowest of three dissimilar indices $h\,k\,l$, the corresponding ratio of the co-efficients in the same order will be $1 : n : m$ where $n > 1$ and $m > n$. Dividing this by $m\,n$ it becomes $\frac{1}{m\,n} : \frac{1}{m} : \frac{1}{n}$, whence are deduced $h = m\,n$, $k = m$, and $l = n$, as the expressions for the conversion of Weiss's into Miller's symbols.

Thus:

$$a : \tfrac{4}{3}\,a : 4\,a = (\tfrac{16}{3}\ 4\ \tfrac{4}{3}) = (16\ 12\ 4) = (4\,3\,1).$$

The above case applies to the cubic system, where the positions of the signs are interchangeable; in the remaining systems n is always > 1, but m is specially restricted to the vertical axis, and may be either greater or less than n or 1.

Thus:

$$a : \tfrac{3}{2}\,b : \tfrac{1}{2}\,c = (3\,2\,6);\ \tfrac{3}{2}\,a : b : 2\,c = (4\,6\,3), \text{\&c.}$$

In arranging the indices or other signs forming the symbol of any face, care must be taken that the axes are always noted in the same order. Unfortunately, there is no uniformity of practice in this matter, either one of the three axes being considered as the first by different authors. In Miller's system the three axes are indicated by the letters $x\ y\ z$, the first extending right and left, the second from front to back, and the third above and below the centre. Weiss calls the axes $a\ b\ c$, their order being a front and back, b right and left, and c top and bottom. The latter system, as being the oldest and most generally known, is adopted in this volume. The position of the first and second indices in the symbol of any face as noted in the figures must therefore be transposed to make them correspond with those of Miller's order.[1]

Naumann's notation. The symbol of any face may be used by implication to indicate the whole of the same kind of faces in a crystal, if the symmetry of the system is known.

[1] The difference expressed in words is as follows: The plane whose parameters are positive on all three axes, is, according to Miller, the right front top face, while in Weiss's order it is the front right top one.

This is apparent in Weiss's notation by the use of index letters for the different axes ; but not in Miller's, where the symbols are similar for all the systems, and do not of themselves show the character of the form. Another method modified from that of Weiss, due to the late Dr. C. F Naumann, indicates both form and symmetry in a single symbol by combining the parameter values with certain arbitrary signs, that differ in each system, representing the form. In the cubic system, where the three axes are all rectangular and the parameters equal, the unit form—Miller's 1 1 1 or Weiss's $a : a : a$—is the regular octahedron. This is represented by the capital letter O, the initial of octahedron. In the other systems, where the forms corresponding to the unit values of the parameters are of the kind known as pyramids, the initial P is used as the unit symbol. The derived forms in any system are indicated by the addition of Weiss's co-efficients, according to the number of axes on which the intercepts vary from unity, but the unit letter is never used more than once in any symbol. The general symbol corresponding to $h\ k\ l$, or $m\,a : b : n\,c$, is m P n in the systems with rectangular axes, but in those with one or more oblique axes accents and other arbitrary signs are added to the characteristic letter to show the symmetry of the system. This notation is more extensively used than any other, mainly from the circumstance that the most popular text-book on Mineralogy is written by its author ;[1] and in a slightly modified form it is also used in Dana's manual and text-books, which are probably the most abundantly circulated books of their class in the English language. The symbols have the advantage of being short and convenient ; and being essentially arbitrary, when their nature is once understood they cannot be mistaken for anything else, and are therefore well suited for descriptive pur-

[1] *Elemente der Mineralogie.* C. F. Naumann. The first edition was published in 1846, and the ninth shortly before the author's death in 1873. A tenth edition, edited by F. Zirkel, appeared in 1877. The same system is followed in the late Professor Nicol's manual.

poses. They are, however, only adapted for indicating forms—that is, the whole system of faces corresponding to any particular set of parameters, and not individual planes —as the parts of the symbols are arranged in an invariable and arbitrary form. The special method of arranging these for different kinds of symmetry will be considered under the description of the different systems.

Lévy's notation. In a system with three axes the solid, contained by faces, parallel to two of the axes, and intersecting the third at some measurable distance, will either be a cube or some parallelepipedon, and in the hexagonal system it will be an hexagonal prism. In such forms, when exactly developed, the edges will be of the same length as the axes to which they are parallel, and the plane angles of the parallelograms forming the faces will have the characteristic angles of the axes. The system originally invented by the Abbé Haüy, and subsequently modified by Lévy and Descloizeaux, uses reference solids of this kind known as 'primitive' forms, which are essentially the same as the molecular networks of Bravais. The faces in such forms are indicated by the capital letters P M T, their points by vowels *a e i o*, and their edges by consonants in an invariable order, from left to right, each primitive form requiring as many letters for its description as it has different parts. The number of these, therefore, is an indication of the degree of symmetry. The symbols of derived faces are compounded of those of the part of the primitive modified, whether an edge or a solid angle, and a series of signs indicating the ratio of the intercepts of the new plane upon the edges, written as exponents. Thus, in the cubic system, the octahedron is written as a^1, which means that it intercepts an equal length upon each of the edges measuring from point *a* where three faces meet. This is the oldest of all the systems of notation, and was at one time almost universally current, but at present it may be considered as restricted to the mineralogists of France, by whom it is generally used. The important works of Descloizeaux and Mallard being written in this system, a knowledge of its principles will be found useful to the student.[1]

Relations of faces. The symbol of any face in a crystal may, by an extension of meaning be considered as typical of the whole *form*—that is, of all faces similarly related to the axes of reference. When this is meant, the symbol is enclosed in brackets, thus—$\{h\,k\,l\}$; but when restricted to an individual face it is put in parentheses, thus—$(h\,k\,l)$. As will be subsequently seen, the possible number of faces in a form varies with the symmetry of the system, the maximum of forty-eight occurring in t e cubic, and the minimum of two—that is the face and its counterpart—in the triclinic system. There are many instances of crystals contained by the faces of only a single form, especially in the cubic system; but it is far more common to find them made up of two or more forms grouped in regular order, such compound crystals being known as combinations. There is no limit to the number of forms that may enter into a combination, subject to the conditions that they all have the same degree of symmetry, and are so arranged that all the faces meet in convex angles. Crystals, in which some of the faces meet in concave or re-entering angles, are not uncommon, but these are never simple, being peculiarly arranged groups of two or more, known as twin crystals. The general solution of the problem of the determination of the direction of the edge or line intersection of two dissimilar faces in terms of their parameters is as follows:—Let H K L, H′ K′ L′ (fig. 7)

FIG. 7

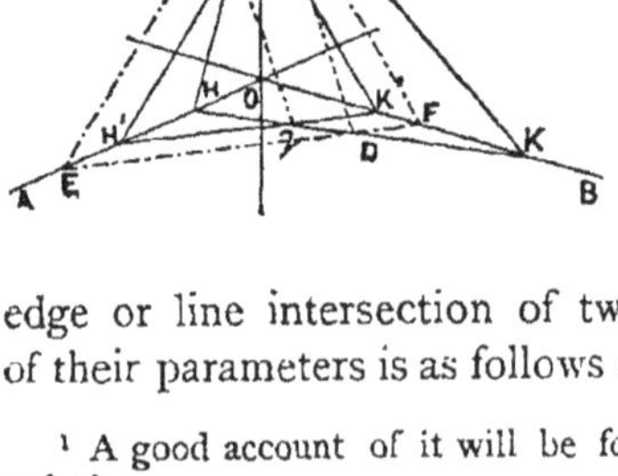

[1] A good account of it will be found in Pisani's *Traité de Minéralogie.*

be two such planes, their intersections H K and H K', with the axial plane, A O B, cross in the point γ, which will therefore be a point in the required edge as common to both faces; β will be a second point of a similar kind in the plane A O C, and α a third in the plane A O B[1] As the direction of the line of intersection is unchanged by moving either face parallel to itself, the face $(h' k' l')$ if shifted to the position E L F gives the new line L D as the required direction. This, however, is equivalent to multiplying the parameters of the face by $\frac{\text{O L}}{\text{O L}'}$. These latter will therefore be for the new position

$$\text{O E} = \text{O H}\ \frac{\text{O L}}{\text{O L}'},\ \text{O F} = \text{O K}'\ \frac{\text{O L}}{\text{O L}'},\ \text{and O L}$$

—that is, the intercept on the third axis, common to both faces, is an original parameter of $(h\ k\ l)$; the position of the point L is therefore determined. To find the point D, draw in fig. 8, D U parallel to O B, and D V parallel to O A, when the problem takes the form of the determination of the co-ordinates of the point D, the lengths O U and O V, in terms of the parameters, as when these are known, the sides of the parallelogram O U D V, and with them the position of D, are determined.

The triangle O K H is similar to U D H, also O F E is similar to U D E, whence follows

$$\text{O K} : \text{U D} = \text{O H} : \text{U H} = \text{O H} : (\text{O H} - \text{O U})$$
$$\text{O F} : \text{U D} = \text{O E} : \text{U E} = \text{O E} : (\text{O E} - \text{O U}).$$

The first of these ratios gives the equation

$$\text{O K} . \text{O H} - \text{O U} . \text{O K} = \text{O H} . \text{U D},$$

and the second,

$$\text{O F} . \text{O E} - \text{O U} . \text{O F} = \text{O E} . \text{U D};$$

[1] This demonstration is given in Groth's *Physikalische Krystallographie.*

whence the two unknown quantities OU and UD are derived as follows:

$$OU = \frac{OE.OK.OH - OF.OE.OH}{OE.OK - OH.OF}$$

$$UD = \frac{OK.OF.OE - OK.OH.OF}{OE.OK - OH.OF}$$

or,

$$OU = OH.OE \frac{OK - OF}{OE.OK - OF.OH}$$

$$UD = OV = OK.OF \frac{OE - OH}{OE.OK - OF.OH}.$$

Substituting the proper values for OE and OF, these become

$$OU = OH.\frac{OL}{OL'}.OH'.\frac{OK - \frac{OL}{OL}.OK'}{\frac{OL}{OL'}.OH'.OK - \frac{OL}{OL'}.OK'.OH}:$$

$$= OH.\frac{OL}{OL'}.OH'.\frac{\frac{OL'}{OL}.OK - OK'}{OH'.OK - OK'.OH}$$

$$OV = OK.\frac{OL}{OL'}.OK'.\frac{\frac{OL}{OL'}.OH' - OH}{\frac{OL}{OL'}.OH'.OK - \frac{OL}{OL'}.OK'.OH}$$

$$= OK.\frac{OL}{OL'}.OK'.\frac{OH' - \frac{OL'}{OL}.OH}{OH'.OK - OK'.OH}$$

$$OU = \frac{OH.OH'}{OL'}.\frac{OK.OL' - OL.OK'}{OK.OH' - OH.OK'}$$

$$OV = \frac{OK.OK'}{OL'}.\frac{OL.OH' - OH.OL'}{OK.OH' - OH.OK'}.$$

—which are the equations for the co-ordinates of the point D in terms of the parameters.

If in fig. 8 a length O W = O L be laid off on the negative side of the axis C, and the parallelepiped O U D V W R Q S constructed, the diagonal O Q will also give the direction of the edge between the two faces being parallel to L D. The multiplication of all the parameters of $(h\,k\,l)$ by any quantity, m, does not affect the direction of the plane or of its intersection with the other plane, so that, if for O H, O K, O L in the preceding formulæ the new values m O H, m O K, and m O L be substituted, we obtain O V′ = m O V, O V′ = m O V, O W = m O W—that is, all the sides of the above parallelepiped will be multiplied by the same quantity, whereby the direction of its diagonal is not altered. The same holds good when the expressions for O U, O V, and O W are multiplied by

FIG. 8.

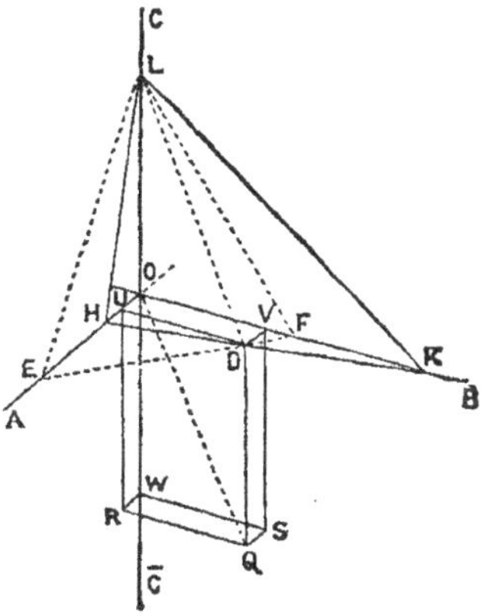

$$\frac{\mathrm{OK}\,.\,\mathrm{OH}' - \mathrm{OH}\,.\,\mathrm{OK}'}{\mathrm{OH}\,.\,\mathrm{OH}'\,.\,\mathrm{OK}\,.\,\mathrm{OK}'\,.\,\mathrm{OL}},$$

whence the following perfectly symmetrical expressions result :

$$\mathrm{OU} = \frac{1}{\mathrm{OL}\,.\,\mathrm{OK}'} - \frac{1}{\mathrm{OK}\,.\,\mathrm{OL}'},$$

$$\mathrm{OV} = \frac{1}{\mathrm{OH}\,.\,\mathrm{OL}'} - \frac{1}{\mathrm{OL}\,.\,\mathrm{OH}'},$$

$$\mathrm{OW} = \frac{1}{\mathrm{OK}\,.\,\mathrm{OH}'} - \frac{1}{\mathrm{OH}\,.\,\mathrm{OK}'}.$$

In the application of these equations to the determination of the direction of intersection of two faces, this intersection is supposed to pass through the origin of the axes, which gives one point; the second is found by laying off upon the three axes the values found for the parameters, in the proper directions, positive or negative, according to their signs, which gives the three sides of the parallelepiped whose diagonal drawn from the origin of the axes is the direction required.

If it is required to express O U, O V, O W, by the indices instead of the parameters of the faces, the values of the latter, expressed by the former, must be introduced into the equation, or for the unit parameters a, b, c,

$$\text{O H} = \frac{a}{h}, \quad \text{O K} = \frac{b}{k}, \quad \text{O L} = \frac{c}{l}.$$

$$\text{O H}' = \frac{a}{h'}, \quad \text{O K}' = \frac{b}{k'}, \quad \text{O L}' = \frac{c}{l'}.$$

If these values are substituted for O U, O V, O W, in the formulæ given above, and each expression is brought to one denomination, we obtain—

$$\text{O U} = \frac{k\,l' - l\,k'}{b\,c} = \frac{u}{b\,c},$$

$$\text{O V} = \frac{l\,h' - h\,l'}{a\,c} = \frac{v}{a\,c},$$

$$\text{O W} = \frac{h\,k' - k\,h'}{a\,b} = \frac{w}{a\,b},$$

when the differences forming the numerators are expressed by the contractions u, v, w. Multiplying these by the product $a\,b\,c$, which does not alter their relative magnitudes, they become

$$a\,u, \quad b\,v, \quad c\,w,$$

where the sides of the parallelepiped whose diagonal is the required direction are represented by magnitudes depending only on the indices. For their determination the following method is given by Miller, which is easily remembered :—

Write down the indices one above the other twice, cut

off the first and last columns, multiply the others crosswise, and subtract one product from the other. Thus:

$$\begin{array}{c|ccccc|c} h & k & \times\; l & \times\; h & \times\; k & & l \\ h' & k' & l' & h' & k' & & l' \\ \hline \end{array}$$

$$kl' - lk',\; lh' - hl',\; hk' - kh'$$
$$= u \qquad = v \qquad = w,$$

which, like the indices, are rational whole numbers.

Zones. Any number of faces parallel to any right line is called a *zone.* Faces belonging to the same zone, and which consequently intersect in lines which are parallel to one another and to every face of the zone, are said to be *tautozonal.* If we suppose the faces of a zone to pass all through one point, they will intersect in a right line passing through that point, and the direction of that line is called the *zone axis.* For instance, in the cube, the front and back and right and left pairs of faces constitute a zone whose edges are all vertical and parallel to the axis c, which is therefore their zone axis; and either of these may be grouped with the third pair of faces, forming zones whose axes are a and b respectively.

Any three faces, P Q R, whose symbols are (efg) (hkl) (pqr), will lie in the same zone when the line of intersection of P and Q is parallel to that of Q and R, or, in other words, when the diagonals of the parallelepipeds found by the method shown in fig. 8 coincide for each pair. This, however, is only possible when the ratios of the three sides are the same in both—that is, when the sides of the first only differ from those of the second by a constant factor. The intersection of P and Q is determined by the parallelepiped, whose sides are—

$$au = a\,(fl - gk),\; bv = b\,(gh - el),\; cw = c\,(ek - fh),$$

and that of Q and R by

$$au' = a\,(kr - lq),\; bv' = (lp - hr),\; cw' = c\,(hq - kp).$$

If the faces P Q R are tautozonal, their sides must be to

each other in the ratio of a common constant, or, calling the latter C—

$$\text{C}\,(f\,l - g\,k) = k\,r - l\,q \qquad (1)$$
$$\text{C}\,(g\,h - e\,l) = l\,p - h\,r \qquad (2)$$
$$\text{C}\,(e\,k - f\,h) = h\,q - k\,p \qquad (3)$$

in which equations the axial lengths are eliminated by appearing on both sides as factors. Multiplying (1) by e, (2) by f, and (3) by g, and adding all three together, the resulting equation becomes

$$e\,k\,r - e\,l\,q + f\,l\,p - f\,h\,r + g\,h\,q - g\,k\,p = 0,$$ [1]

which is the condition of tautozonality for the three faces, P Q R. If, as in the previous case, we make

$$f\,l - g\,k = u,\ g\,h - e\,l = v,\ e\,k - f\,h = w,$$

the last equation becomes

$$u\,p + v\,q + w\,r = 0,$$

which shows that the condition required to bring three planes into the same zone depends only on their indices, and is completely independent of the lengths of their axes.

The quantities $u\ v\ w$ are called the indices of the zone, and to distinguish them from those of a face or form they are written in square brackets, thus $[u\ v\ w]$, forming the so-called zone symbol; the face, if any, to which this symbol belongs is called the zone plane, and is perpendicular to the

[1] This condition may be obtained directly from the consideration that three planes through the origin, parallel to the three faces, have for their equations, expressed as in ordinary geometry,

$$\frac{e\,x}{a} + \frac{f\,y}{b} + \frac{g\,z}{c} = 0, \text{ \&c.};$$

and the condition that the three planes should be parallel to a line is obtained by eliminating $\frac{x}{a}$, $\frac{y}{b}$, and $\frac{z}{c}$ from the three equations. This gives the determinant above written.

zone axis. It may be determined from the intersection of any two planes, P and Q, out of those forming the zone, and from it the whole number of possible tautozonal faces may be calculated by substituting for q and r successively all the simple rational numbers 0, 1, 2, . . . and calculating, in accordance with the above condition, the corresponding values of p.

Determination of a face by two zones. As a plane is determined when the positions of two straight lines parallel to it are given, that of the face of a crystal lying in two zones, and therefore parallel to the axes of both, is similarly determinable. If the symbols of the zones are

$$[u\,v\,w] \text{ and } [u'\,v'\,w'],$$

the indices $p\,q\,r$ of the face must satisfy the equation of condition in regard to both. Consequently

$$u\,p + v\,q + w\,r = 0,$$
$$u'\,p + v'\,q + w'\,r = 0,$$

whence we derive

$$p = r\,\frac{v\,w' - w\,v'}{u\,v' - v\,u'},$$

$$q = r\,\frac{w\,u' - u\,w'}{u\,v' - v\,u'}.$$

One of the three indices may, however, be made equal to any number at pleasure; say, for example,

$$r = u\,v' - v\,u',$$

which gives

$$p = v\,w' - w\,v',$$
$$q = w\,u' - u\,w',$$

as the three indices of the face common to both zones. These indices may be derived from those of the zones by the scheme of cross multiplication and subtraction, in the same way that those of a zone are found from those of two of its faces. Thus:

$$\begin{array}{c|ccccc|c} u & v & \times\ w & \times\ u & \times\ v & & w \\ u' & v' & w' & u' & v' & & w' \end{array}$$

$$vw' - v'w,\quad wu' - uw',\quad uv' - vu'$$
$$= p \qquad = q \qquad = r.$$

By this method the symbols of a face may be found when those of any two zones in which it lies, or of any two faces in each of those zones, are known. Suppose, for example, a face is observed to lie in one zone with the faces (1 2 3) and (1 1 3), and in another with (0 1 1) and (1 2 2), the corresponding zone symbols will be

$$\begin{array}{c|ccccc|c} 1 & 2 & \times\ 3 & \times\ 1 & \times\ 2 & & 3 \\ 1 & 1 & 3 & 1 & 1 & & 3 \end{array} \qquad \begin{array}{c|ccccc|c} 0 & 1 & \times\ 1 & \times\ 0 & \times\ 1 & & 1 \\ 1 & 2 & 2 & 1 & 2 & & 2 \end{array}$$

$$6-3,\ 3-3,\ 1-2=[3\ 0\ \bar{1}] \qquad 2-2,\ 1-0,\ 0-1=[0\ 1\ \bar{1}];$$

and that of the face common to both

$$\begin{array}{c|ccccc|c} 3 & 0 & \times\ \bar{1} & \times\ 3 & \times\ 0 & & \bar{1} \\ 0 & 1 & \bar{1} & 0 & 1 & & 1 \end{array}$$

$$0-\bar{1},\ 0-\bar{3},\ 3-0 = (1\ 3\ 3).$$

Any three faces intersecting at any angles may appear in the same zone, but a fourth (or any further number of faces), arbitrarily placed, is not possible according to the principle of rationality, or rather according to the particular development of it known as the principle of anharmonic ratios. This principle is of such importance that a general demonstration of it, obligingly furnished by the editor, Mr. Merrifield, is given in the following pages.

ANHARMONIC RATIOS.

Let A B C D (fig. 9) be any four points on a right line, and let O be an arbitrary point through which the right

FIG. 9.

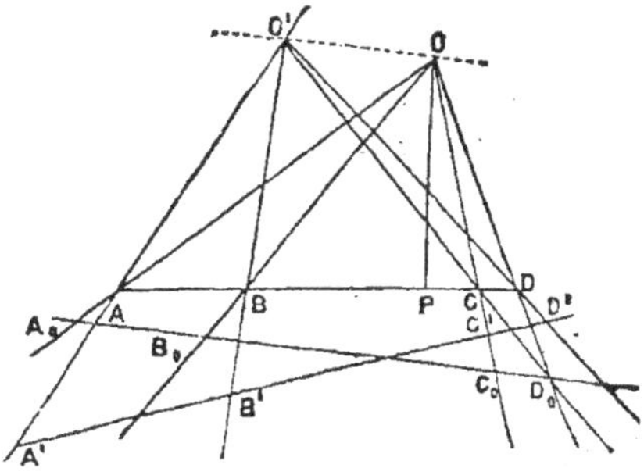

lines O A, O B, O C, O D are drawn. Also, let fall O P perpendicular on A D. Then we have

$$\left.\begin{aligned} OA.OB \sin AOB &= AB.OP \\ OB.OD \sin BOD &= BD.OP \end{aligned}\right\} \qquad \left.\begin{aligned} OA.OC \sin AOC &= AC.OP \\ OC.OD \sin COD &= CD.OP \end{aligned}\right\}$$

Since each side of the equation is a different expression for twice the area of a triangle.

Dividing out suitably—

$$\frac{\sin AOB}{\sin BOD} : \frac{\sin AOC}{\sin COD} = \frac{AB}{BD} : \frac{AC}{CD},$$

which is the anharmonic ratio of the four points or four lines: the second side of this equation shows that the ratio is independent of the position of the point O, which may therefore be at O′. The first side shows that, the pencil being given, the tr nsversal is immaterial. Thus the anharmonic ratios

$$\frac{AB}{BD} : \frac{AC}{CD}, \quad \frac{A_0B_0}{B_0D_0} : \frac{A_0C_0}{C_0D_0}, \quad \frac{A'B'}{B'D'} : \frac{A'C'}{C'D'},$$

are all equal, and this whether O A D, O′A D are in the same plane or not.

There is more than one anharmonic ratio of four points on a line. They are connected by means of the identity,

$$AB.CD + AC.DB + AD.BC = 0,$$

which is obtained by the cyclic permutation of the letters B C D in the expression A B . C D.

The ratio of any two of the terms in this formula may be taken as the anharmonic ratio of the pencil.

A further consequence of this property is, that if we consider O and O′ as in separate planes, O O′ A, O O′ B, O O′ C, O O′ D constitute a pencil of planes such that the anharmonic ratio of any transversal line is invariable, and equal to the corresponding sine-ratio of the dihedral angles. If any four points on a line are given, any pencil of planes through the four points will have the anharmonic ratio of the range, and every transversal of any such pencil will again have the same anharmonic ratio.

The sine-identity of a pencil is easily written down from the corresponding one of a range by simply inserting the letter at the vertex, and the word sine. Thus—

$$AB.CD + AC.DB + AD.BC = 0, \text{ gives}$$

$$\sin AOB.\sin COD + \sin AOC.\sin DOB + \sin AOD.\sin BOC = 0,$$

and the anharmonic sine-ratios are

$$\sin AOB.\sin COD : \sin AOC.\sin DOB, \text{ \&c.}$$

N.B.—Points on a right line are a *range*. Right lines meeting in a point are a *pencil*. Planes meeting in a right line are also a pencil.

ANHARMONIC PROPERTY OF ZONES.

The anharmonic prop rty of zones is this : that if we take in space four planes parallel to four of the faces of a zone, and meeting in one line, the anharmonic ratio of the pencil is rational.

It is to be remembered that all the faces of a zone are parallel to a line. Hence all their intersections, two and two, will be in parallel lines. The necessary construction is therefore at once obtained by drawing, through the intersection of any two faces, planes parallel to the other two.

The four faces of the zone, by the ordinary law of crystallography, cut off from each of the axes intercepts having rational numerical ratios; and what we have to prove is that, as a consequence of this rationality, the corresponding anharmonic ratio is also rational.

Confining our attention to one plane only, let us consider four transversal lines meeting the two axes: thus, in the annexed figure,

a_1b_1, a_2b_2, a_3b_3, a_4b_4, are four transversals to ox and oy, so taken that

$$oa_2 = l\,oa_1,\ oa_3 = m\,oa_1,\ oa_4 = n\,oa_1$$
$$ob_2 = p\,ob_1,\ ob_3 = q\,ob_1,\ \ ob_4 = r\,ob_1.$$

Through b_1 draw $b_1a'_2$ parallel to b_2a_2

" " $b_1a'_3$ " b_3a_3

" " $b_1a'_4$ " b_4a_4

FIG. 10

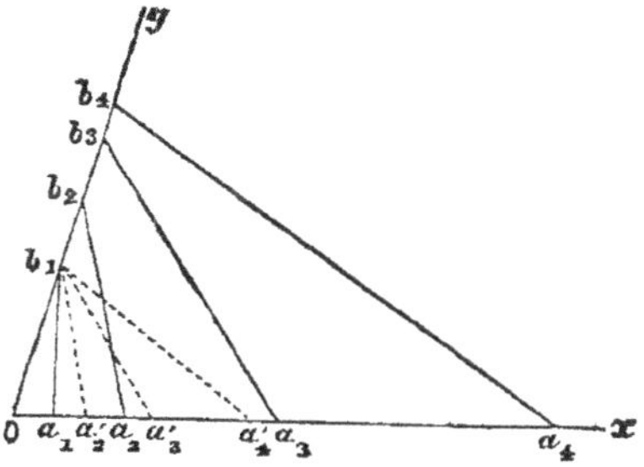

We want to show that if $l\,m\,n\,p\,q\,r$ are rational the anharmonic ratio of the pencil $b_1\ a_1\ a'_2\ a'_3\ a'_4$ is also rational.

This ratio is

$$\frac{a_1\,a'_2}{a'_2\,a'_4} : \frac{a_1\,a'_3}{a'_3\,a'_4} = \text{R}.$$

Now

$$a_1\,a'_2 = oa'_2 - oa_1 = \left(\frac{l}{p} - 1\right) oa_1$$

$$a_1\,a'_3 = oa'_3 - oa_1 = \left(\frac{m}{q} - 1\right) oa_1$$

$$a'_2 a'_4 = \mathrm{o}a'_4 - \mathrm{o}a'_2 = \left(\frac{n}{r} - \frac{l}{p}\right) \mathrm{o}a_1$$

$$a'_3 a'_4 = \mathrm{o}a'_4 - \mathrm{o}a'_3 = \left(\frac{n}{r} - \frac{m}{q}\right) \mathrm{o}a_1$$

$$\therefore \mathrm{R} = \frac{\frac{l}{p} - 1}{\frac{m}{q} - 1} : \frac{\frac{n}{r} - \frac{l}{p}}{\frac{n}{r} - \frac{m}{q}} = \frac{(l-p)(nq-mr)}{(m-q)(np-lr)},$$

which is rational, if $lmn\ pqr$ are so.

The whole of the anharmonic ratios of the zone are contained in the ratios two by two of the terms in the identity—

$$(l-p)(nq-mr) + (m-q)(lr-np) + (n-r)(mp-lq) = 0.$$

Since $lmn\ pqr$ are either integral, or else rational numerical fractions, each term of this identity must be so too, and the ratio of any two terms must be a rational fraction.

As a numerical example, let

$$l=2,\ m=3,\ n=4,\ p=3,\ q=5,\ r=8\ ;$$

then the above-written identity becomes

$$(-1)\times(-4)+(-2)\times(+4)+(-4)\times(-1)=4-8+4=0,$$

and the anharmonic ratios are any one of the following :

$$1,\ 2,\ \tfrac{1}{2},\ -1,\ -2,\ -\tfrac{1}{2}.$$

In practice, the dihedral angles will be measured, and the anharmonic ratio to be used will be the sine-ratio. The test of whether we have really got four faces of one crystal is the rationality of this anharmonic sine-ratio, when reduced to numbers. It will usually be a simple numerical fraction, in low terms.

CHAPTER III.

CUBIC SYSTEM.[1]

THE forms of this system, whose symmetry is completely exhibited in an actual cube, are referred to three principal axes at right angles to each other, whose unit lengths or

FIG. 11.

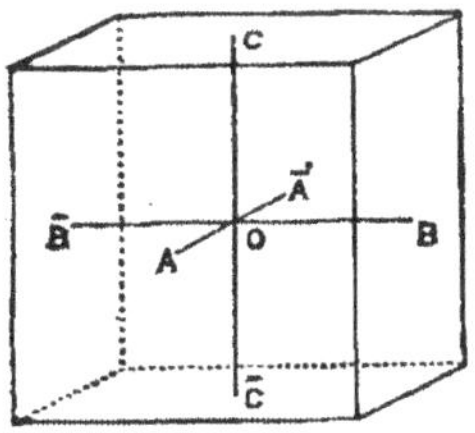

FIG. 12.

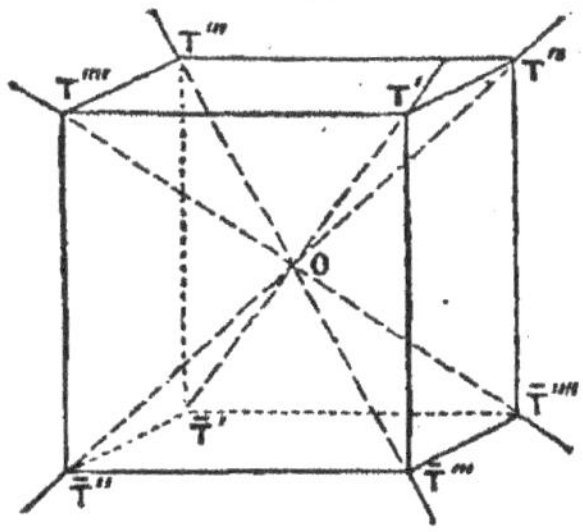

parameters are all equal. These are axes of quaternary symmetry, each one passing through the centre, O, and joining the centres of opposite faces perpendicularly : A O $\bar{\text{A}}$, B O $\bar{\text{B}}$, C O $\bar{\text{C}}$ (fig. 11). They lie two by two in planes parallel to the faces which are the principal planes of symmetry.

Next in order are the four axes of ternary symmetry—O T′, O T″, O T‴, O T⁗ (fig. 12), making with one another angles of 109° 28′ 16″. These are the four diagonals of the cube, and have no planes of symmetry corresponding to them.

Lastly, there are six axes of binary symmetry, the lines joining the middle points of opposite edges—B^1 to B^6 (fig. 13). They lie by pairs in the principal planes at right angles to each other, at 45° to the principal axes, and are

[1] Other names are Tesseral, Regular, Monometric, and Terquaternary.

normals to six corresponding planes of symmetry. These latter, which intersect the principal planes singly, at right angles in the binary axes, by pairs at 45° and 135° in the principal axes, and each other by threes in the ternary axes, are the rectangular sections of the cube whose sides are the edges and face diagonals, or the principal and binary axes, and whose diagonals are the lines joining opposite solid angles, or the ternary axes.

FIG. 13.

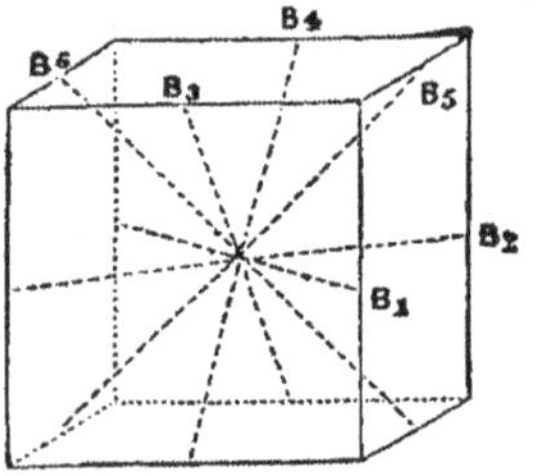

The symmetry of the cube is not altered by permuting the order of the principal axes, or by any inversion of sign, and the preservation of these symmetries involves the retention of all the others. Taking any face $(h\,k\,l)$, or $(a : n\,a : m\,a)$, the first condition requires that there should be six similar ones due to permutation of indices, or coefficients, namely—

$$hkl,\ klh,\ lhk,\ lkh,\ khl,\ hlk,$$

or

$$1\,n\,m,\ n\,m\,1,\ m\,1\,n,\ m\,n\,1,\ n\,1\,m,\ 1\,m\,n.$$

The second requires that for each of these combinations of letters there should be faces representing each of the following eight permutations of signs :

$$+++,\ +--,\ -+-,\ --+,$$
$$---,\ -++,\ +-+,\ ++-.$$

Combining all these together we find that to satisfy full cubic symmetry in the most general form, any face, $(h\,k\,l)$, requires to be associated with 47 others, constituting the so-called hexakisoctahedron. In its most regular development, with all the faces equally distant from the centre, it is contained by 48 plane scalene triangles whose edges all lie in planes of symmetry. These edges are all three kinds distinguished in fig. 14 as long, medium, and short (L, M, and S). Those of medium length all lie in the principal planes of symmetry, so that the sections of these planes are equilateral but not regular octagons, while the longer and shorter ones are arranged in alternate pairs in the other six planes of symmetry, the sections of the latter, therefore, are unequal eight-sided figures. The dihedral angle between adjacent faces, or the so-called interfacial angles, are always greater than 90° and less than 180°.[1] They are for the particular forms—

FIG. 14.

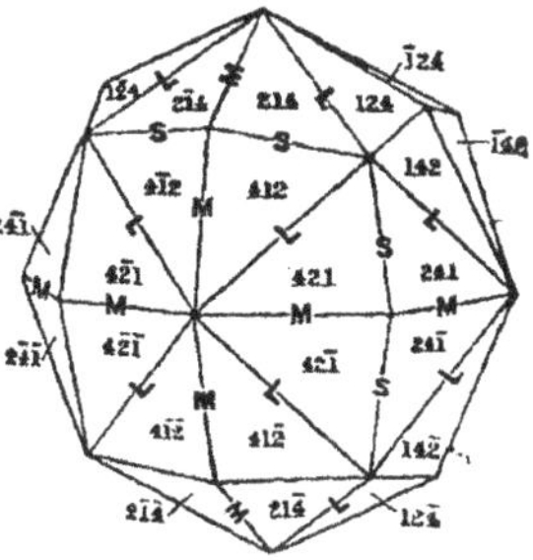

		L.	M.	S.
{3 2 1}	$3O\frac{3}{2}$,	158° 13′,	149° 00,	158° 13′;
{4 2 1}.	$4O2$,	162° 15′,	154° 47′,	144° 03′.

The points or solid angles are of three kinds—namely, six eight-faced, or formed by the meeting of four long and four medium edges in the poles of the principal axes; eight six-faced, formed by three long and three short edges in each of the axes of ternary symmetry, which are also known as trigonal interaxes; and twelve four-faced, formed by two short and two medium edges meeting in each of the axes of binary symmetry or rhombic interaxes.

Naumann's symbol for this form is *m On*, signifying that two parameters vary from the unit length, one being *m* and the other *n* times greater than that of the third axis.

The geometrical construction of the octant, including the

[1] That is, no face can be at right angles to a plane of symmetry, or the edges of symmetry are all effective edges of form.

six faces whose indices are all positive of the form {3 2 1}, is given in fig. 15.[1]

FIG. 15.

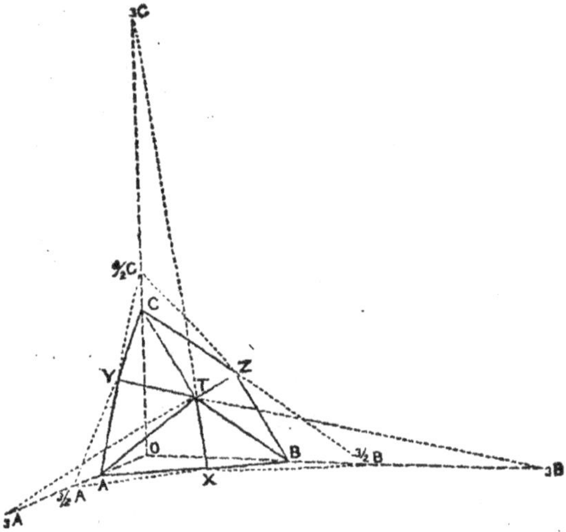

The arrangement of the symbols of the faces for all the positive values of the first parameter in the form {4 2 1} is

FIG. 16.

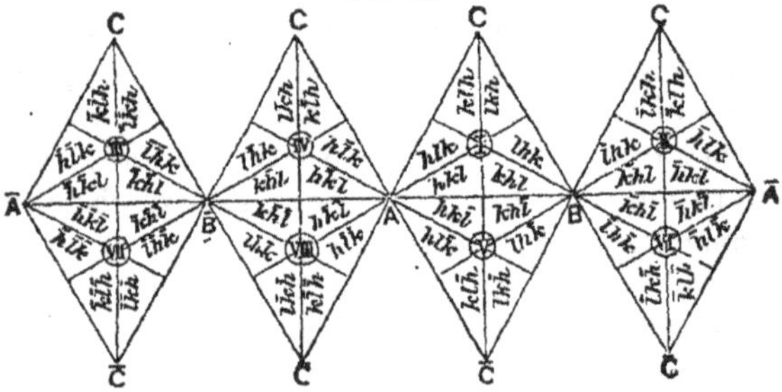

shown in fig. 19, and for those of the whole of the general form $\{h\,k\,l\}$ in the scheme fig. 16. This supposes the edges to be

[1] In this, and figs. 18, 21, 25, and 28, all the parts are similarly lettered to avoid detailed description.

projected on the surface of an octahedron, which is then flattened out, the octants being numbered from right to left, first above the equatorial or horizontal plane and then in the same order below, the positive pole of the first axis, A, being in the middle.

If, in the form $h\,k\,l$, two of the indices are equal, say $k = l$, one half of the letter permutations are lost, leaving

$$h\,k\,k,\quad k\,k\,h,\quad k\,h\,k,$$

which, with the eight permutations of sign, gives a solid of 24 faces, corresponding to two classes of forms, according as the two equal indices are smaller or larger than the third. In the former case, when $h > k$, the resulting form is of the kind shown in fig. 17, known as an icositetrahedron, or, more properly, trapezoidal icositetrahedron—from the shape of the faces—which is often contracted to trapezohedron. This may be regarded as a hexakisoctahedron, in which the interfacial angle over the long edges is 180°, or these edges are effective only as edges of symmetry. The actual edges are therefore 24 longer in the principal planes of symmetry, and 24 shorter in the other six planes of symmetry. The solid angles are, six four-faced, each formed by the meeting of four longer edges in the poles of the principal axes; eight three faced, formed by groups of three shorter edges in the axes of ternary symmetry; and eight four-faced (two and two edged) or formed by two longer and two shorter edges in the axes of binary symmetry.

FIG. 17.

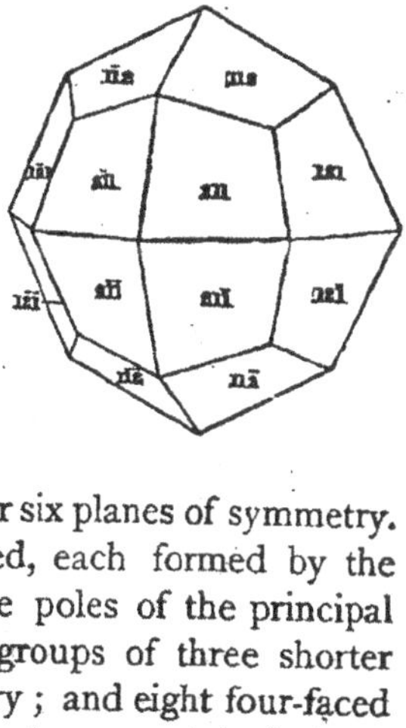

The values of the angles are, in the commonest kinds, $2\,O\,2$ and $3\,O\,3$, or {2 1 1} {3 1 1}

Over the longer edges . .	131° 49′	144° 54′
Over the shorter edges .	146° 27′	129° 31′

Weiss's symbol for this class of form is $a : ma : ma$, and Naumann's $m\,O\,m$.[1] The construction of the positive

FIG. 18.

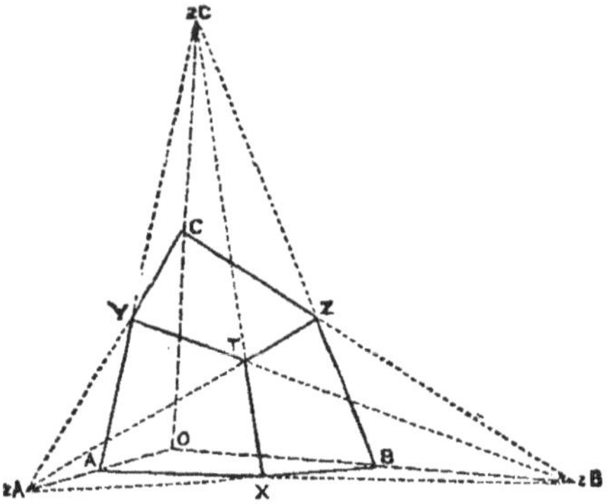

faces of $2\,O\,2$ is shown in fig. 18, the front half of the same form in fig. 17, and the general arrangement of the symbols of the whole of the faces in fig. 19.

FIG. 19.

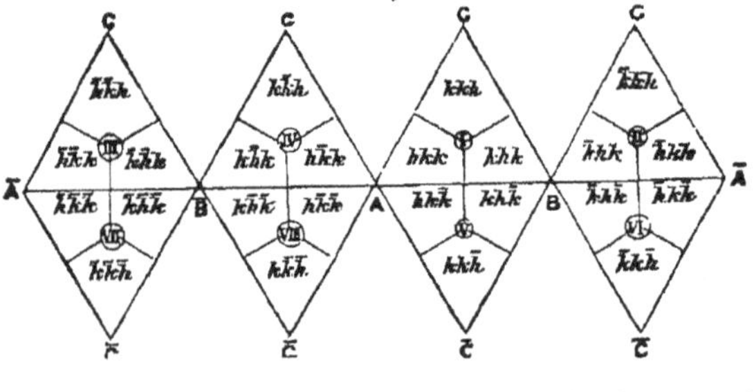

The second case, when the two equal indices are larger than the third, or as the symbol is more conveniently written $h\,h\,k$, is represented by the class of forms known

[1] The largest solid that can be inscribed in a sphere symmetrical to the nine planes of the cubic system is of this class; but it is not a possible crystallographic one, m having the irrational value 2·4142136 or tan 67° 30′, and these planes do not express its full symmetry.

as triakisoctahedra (fig. 20). These are contained by 24 isosceles triangles, whose shorter edges make eight three-faced solid angles in the axes of ternary symmetry, and with the longer ones six eight-faced (four are four-edged) solid angles in the poles of the principal axes. They are particular forms of hexakisoctahedra, having the dihedral angle of the shorter edges 180°; the plane angle between the two medium edges in any quadrant of a principal plane of symmetry is also 180°, or they lie in the same line. The effect of this is to make the principal sections[1] square or similar to those of the regular octahedron, so that the whole form may be compared to an octahedron having a low three-faced pyramid superposed upon each of its faces, a property which is indicated by the name. The edges formed by the meeting of these three planes represent by their position the longer edges of the hexakisoctahedron.

FIG. 20.

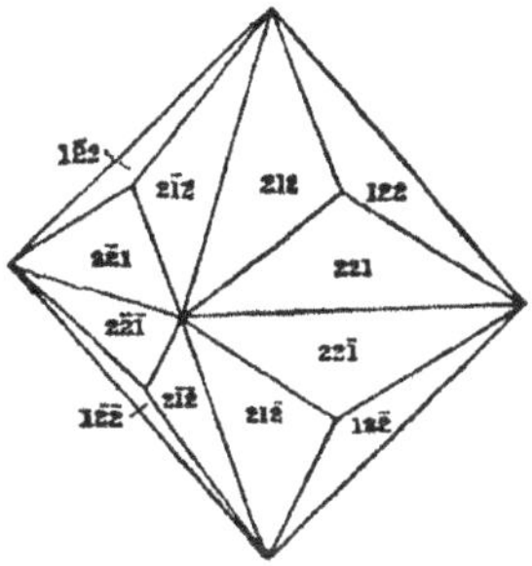

The interfacial angles of the more important of these forms are—

		Over the longer edges.	Over the shorter edges.
{2 2 1}	2 *O*	141° 03′	152° 44′
{3 3 1}	3 *O*	153 . 28′	142 . 08′.

Weiss's symbol is *a* : *a* : *m a*, and Naumann's *m O*, signifying that one of the parameters is *m* times the unit or octahedral length. The construction for *m* = 2 for the faces with positive indices is given in fig. 21, the front half of the same form with the faces noted in fig. 20, and the general scheme of notation for the whole form in fig. 22.

[1] The principal crystallographic sections are those upon planes containing principal axes of symmetry or form.

FIG. 21.

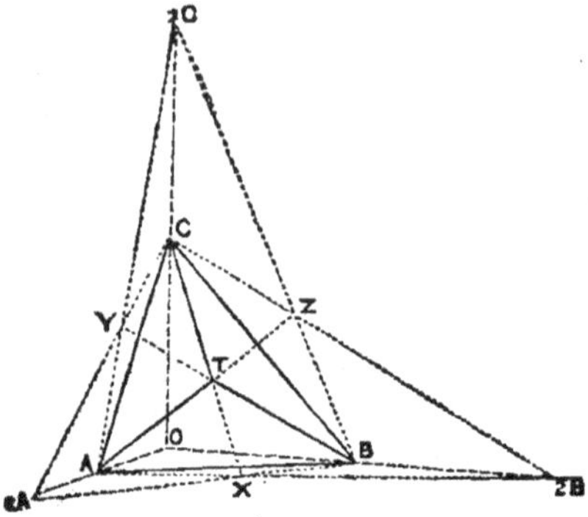

FIG. 22.

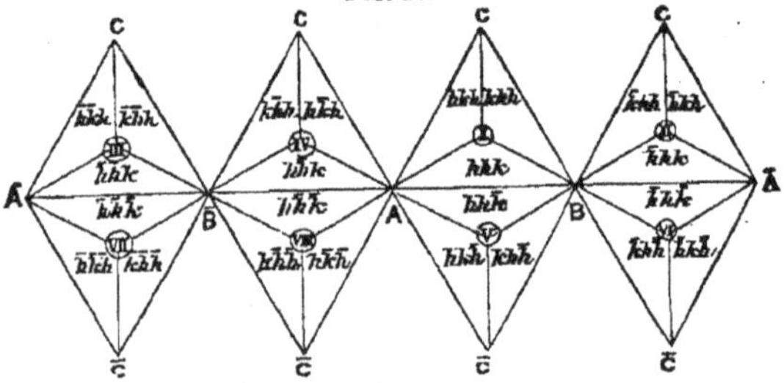

When the three indices are equal, or $h = k = l = 1$, there are no permutations of letters, and only the eight sign permutations remain, which give the regular octahedron (fig. 23).

FIG. 23.

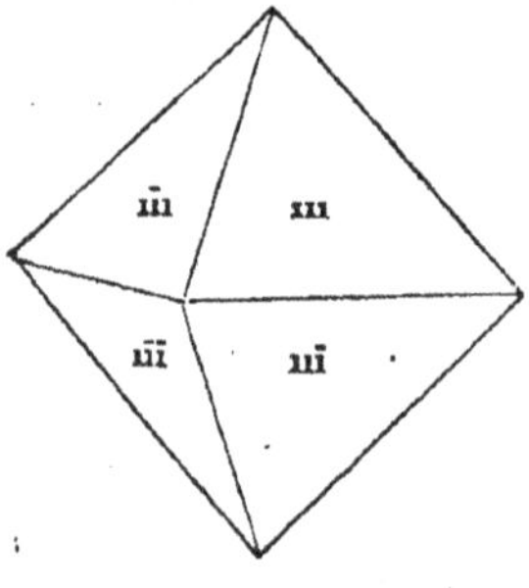

The eight faces of this form are equilateral triangles, their dihedral angle is 109° 28′ 16″. The twelve edges are all equal, and lie in the principal planes of symmetry, each one representing two of the medium edges of the hexakisoctahedron. The six solid angles, all four-faced and similar, lie in the poles of the axes of principal symmetry.

The axes of ternary symmetry are normal to the faces, and those of binary symmetry normal to the edges. The symbols are {1 1 1}. $a : a : a$ and O.

The sections upon the nine planes of symmetry of this solid, when put together, as shown in fig. 24, form a skeleton octahedron, made up of forty-eight trihedral cells, having a common apex in the central point. Models of this kind are useful as showing the relation of the special to the general forms of the system, and their common symmetry; the same cellular arrangement is characteristic of all, the difference being in the external contour of the constituent planes. For the octahedron the three principal sections are the squares described upon the edges, the other six are rhombs having the edges and crystallographic axes for shorter and longer diagonals respectively.

FIG. 24.

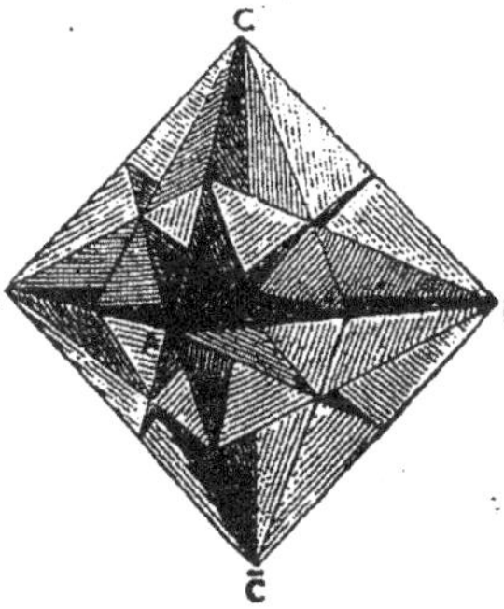

When one of the indices becomes zero, or the corresponding parameter co-efficient infinity, the change of sign corresponding to that index is lost, so that the number of sign permutations is only four, while the full number of letter permutations (six) are retained as follows:

$$hk\text{o},\ k\text{o}h,\ \text{o}hk,\ \text{o}kh,\ kh\text{o},\ h\text{o}k;$$

with the signs $++$, $+-$, $-+$, $--$ attached to each, giving twenty-four faces for the complete form, which is known as a tetrakishexahedron $\{hk\text{o}\}$. Weiss's symbol is $a : ma : \infty a$, and Naumann's ∞On. Fig. 25 shows the construction of one-eighth of the form $\infty O2$, or {2 o 1}; fig. 26 one-half of the same; and fig. 27 the general notation of the entire number of faces. The faces are isosceles triangles, whose shorter sides, representing the longer edges of the hexakisoctahedron, meet in four-faced solid angles in the poles of the principal axes, and by threes with three of the longer sides, forming six-faced solid angles in the ternary axes. Each of these longer edges, representing two of the shorter

FIG. 25.

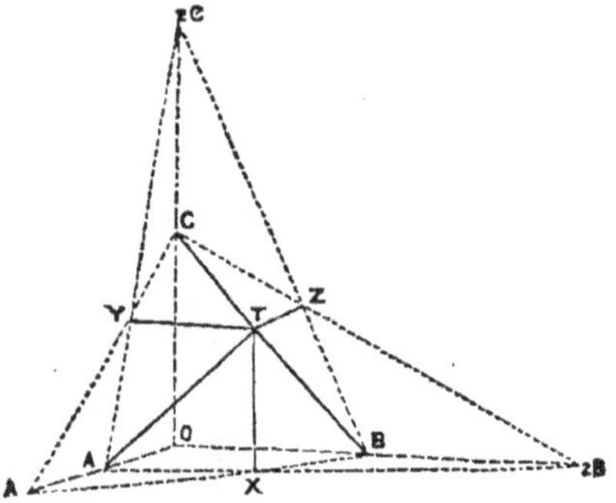

FIG. 26.

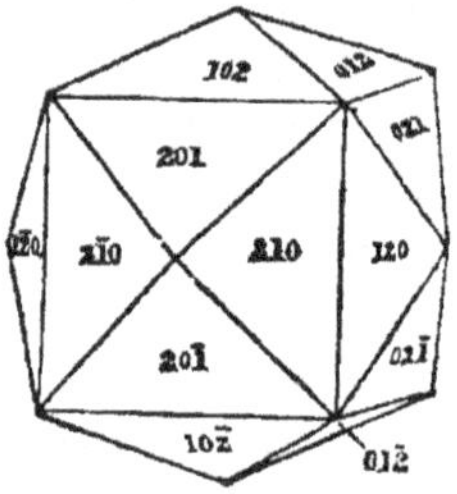

edges of a hexakisoctahedron, is parallel to one of the principal axes of symmetry, and together they enclose a cube, so that the character of the solid is that of a cube with a low

FIG. 27.

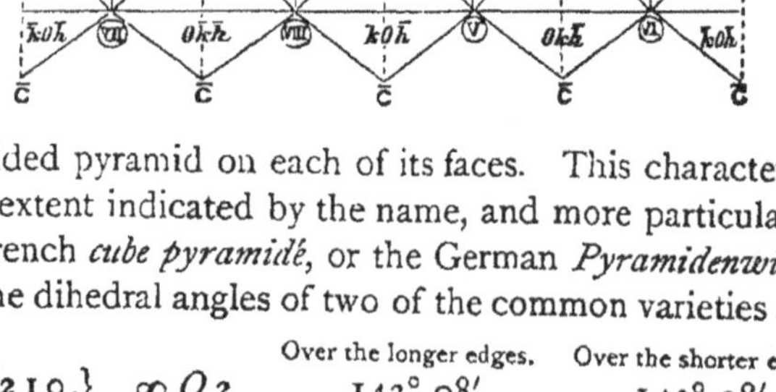

four-sided pyramid on each of its faces. This character is to some extent indicated by the name, and more particularly in the French *cube pyramidé*, or the German *Pyramidenwürfel.*

The dihedral angles of two of the common varieties are—

		Over the longer edges.	Over the shorter edges.
{2 1 0.}	$\infty O 2$	143° 08′	143° 08′
{3 1 0.}	$\infty O 3$	126° 52′	154° 09′.

When two of the indices are equal, and the third zero, which corresponds to a suppression either of one change of sign in the triakisoctahedron, or of one half of the permutations in the tetrakishexahedron, the resulting form has the symbols

$h\,h\,o$, $h\,o\,h$, $o\,h\,h$, with the signs $++$, $+-$, $-+$, $--$,

or twelve faces in all.

FIG. 28.

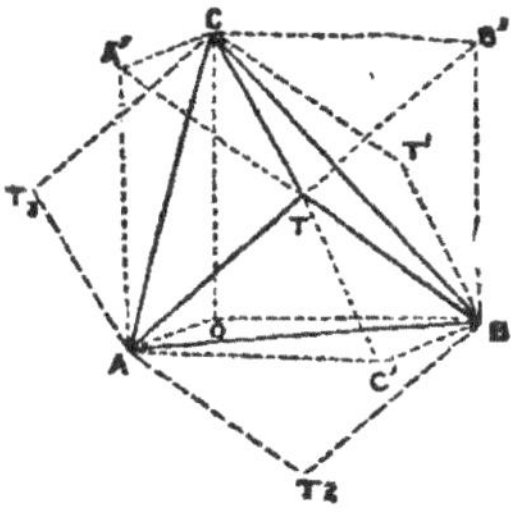

FIG. 29.

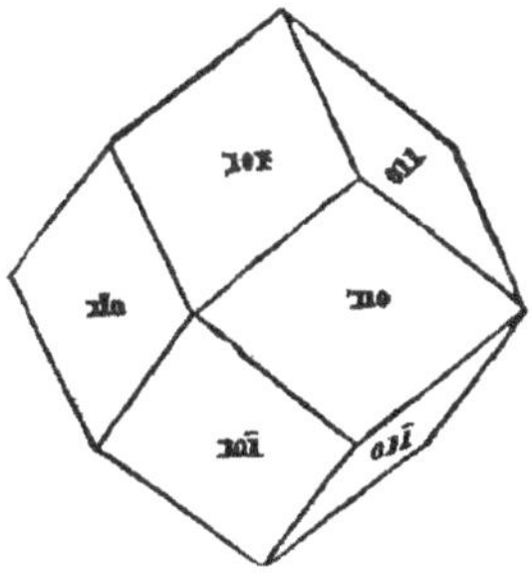

This is the right rhombic dodecahedron,[1] $\{1\,1\,0\}$, Weiss's $a : a : \infty\,a$, and Naumann's $\infty\,O$. Fig. 28 shows the con-

FIG. 30.

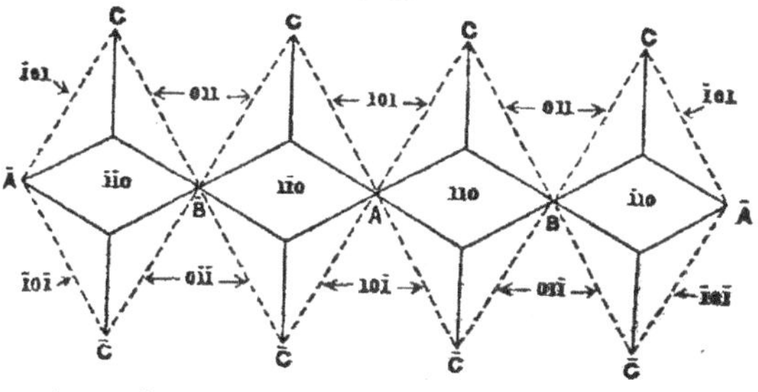

struction from the axes of the octahedron; fig. 29 the front half of the form; and fig. 30 the symbols of all the faces.

[1] As one index = o and the other two = h, any whole number may be substituted for the latter without changing its character; or, in other words, the symbol represents a single form, and not a series of forms. The same remark applies to the cube.

Each of the twelve faces is a rhombus whose plane angles are 70° 31′ 44″ and 109° 28′ 16″, and its longer and shorter diagonals are to each other as $\sqrt{2} : 1$. The dihedral angles of the faces are all 120°, the edges which represent the longer ones of the hexakisoctahedron form four-faced solid angles in the poles of the quaternary axes, and three-faced ones in the ternary axes. Each parallel pair of faces is normal to a binary axis, and consequently the whole form is parallel to the six corresponding planes of symmetry, the longer diagonals of the faces are parallel to the edges of an octahedron, and the shorter ones to those of a cube. This is one of the solids having the property of filling up space—that is, it will pack together with others of the same size without leaving any hollows.

Lastly, when two indices = 0, the third *h* may be 1, or any other whole number, leaving only three permutations, *h* 0 0, 0 *h* 0, 0 0 *h*, with + and − signs to the index *h* in each, or six in all. This is the cube (fig. 31), {100} $(a : \infty a : \infty a)$ or $\infty O \infty$. Considered as a special case of a hexakisoctahedron it is that having only the shorter edges effective as edges of form, their dihedral angle being 90°, the longer edges are represented by the diagonals of the faces and the medium ones by lines parallel to the edges passing through the centre of the faces. The principal planes of symmetry correspond to the axial planes, which are also parallel to the faces; this form, therefore, like the rhombic dodecahedron, is symmetrical to its own faces, or is contained by its principal planes of symmetry.

FIG. 31.

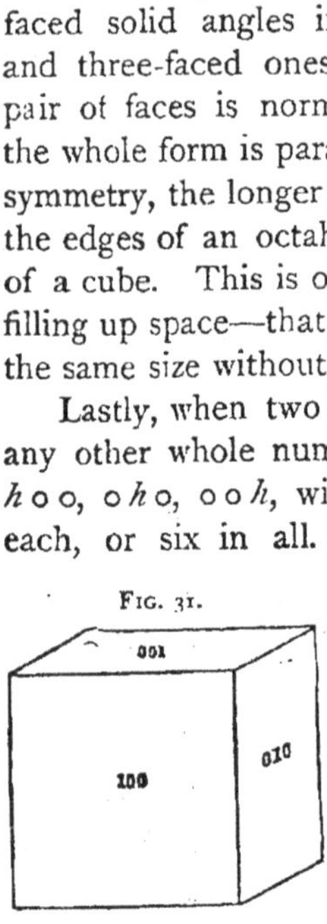

The above include all the possible cases of holohedral cubic forms, as will be seen by the following enumeration, in which a different order from that adopted in the description is followed:

1. *h k l.* Hexakisoctahedra. Indices all different, the least > o.

2. *h k* o. Tetrakishexahedra. Indices all different, the least = o.

3. *h k k.* Icositetrahedra. Two indices equal, less than the third, and > o.

4. *h* o o. Cube. Two indices equal, less than the third, and = o.

5. *h h k.* Triakisoctahedra. Two indices equal, and greater than the third, which is > o.

6. *h h* o. Rhombic dodecahedron. Two indices equal, and greater than the third, which is = o.

7. *h h h.* Octahedron. All three indices equal.

These relations may also be shown graphically by the diagram fig. 32, using Naumann's notation. The symbol of the unit form, *O*, is placed at the summit of a triangle, the left-hand side contains the forms obtained by varying one of the unit axes, or the triakisoctahedra, *m O*, which pass downwards into the rhombic dodecahedron ∞ *O*, when $m = \infty$; the right side contains the different icositetrahedra, *m O m* similarly passing downwards into the cube ∞ o ∞; when $m = \infty$, and the base contains forms intermediate between the cube and rhombic dodecahedron, or the tetrakishexahedra ∞ *O n*. The general symbol of the hexakisoctahedra, *m O n*, occupies the centre, and the guide lines connecting it with the sides indicate a passage to the triakisoctahedra when $n = 1$, to the icositetrahedra when $n = m$, and to the tetrakishexahedra when $m = \infty$.

FIG. 32.

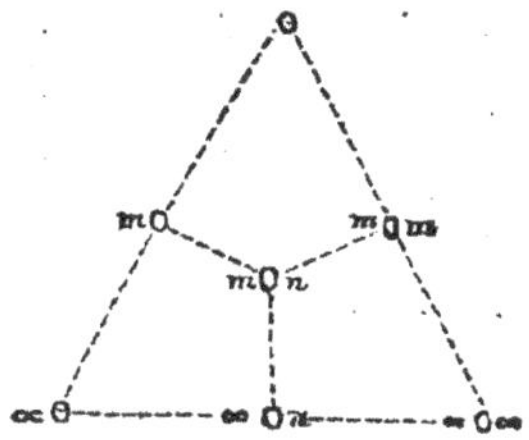

Hemihedral forms of the cubic system. These are obtained from the holohedral forms by the symmetrical suppression of one-half of their faces, which may be done in different ways. The relations of these to each other is best shown by considering first those derived from the general form $\{h k l\}$ of which all the others are particular cases. In this there are, as already shown, six permutations of letters and eight of signs, giving the forty-eight faces. But these divide symmetrically into four groups—namely, the letter permutations into the two following cyclical groups :

Direct . . . . . $h k l$, $k l h$, $l h k$;
Inverse . . . . . $l k h$, $k h l$, $h l k$;

and the signs into two groups of four :

Direct . . $+ + +$, $+ - -$, $- + -$, $- - +$;
Inverse . . $- - -$, $- + +$, $+ - +$, $+ + -$;

the full arrangement being represented in the following table given by Miller :

	A				
$h k l$	$k l h$	$l h k$	$l k h$	$k h l$	$h l k$
$h \bar{k} \bar{l}$	$k \bar{l} \bar{h}$	$l \bar{h} \bar{k}$	$l \bar{k} \bar{h}$	$k \bar{h} l$	$h l \bar{k}$
$\bar{h} k \bar{l}$	$\bar{k} l \bar{h}$	$\bar{l} h \bar{k}$	$\bar{l} k \bar{h}$	$\bar{k} h \bar{l}$	$h l k$
$\bar{h} \bar{k} l$	$\bar{k} \bar{l} h$	$\bar{l} \bar{h} k$	$\bar{l} \bar{k} h$	$k \bar{h} l$	$\bar{h} \bar{l} k$
	C			D	
$\bar{h} \bar{k} \bar{l}$	$k \bar{l} \bar{h}$	$\bar{l} \bar{h} k$	$\bar{l} k \bar{h}$	$k \bar{h} \bar{l}$	$\bar{h} \bar{l} \bar{k}$
$\bar{h} k l$	$\bar{k} l h$	$l h k$	$\bar{l} k h$	$\bar{k} h l$	$\bar{h} l k$
$h k l$	$k \bar{l} h$	$l h k$	$l \bar{k} h$	$k \bar{h} l$	$h \bar{l} k$
$h k \bar{l}$	$k l h$	$l h \bar{k}$	$l k \bar{h}$	$k h l$	$h l \bar{k}$

There are three ways in which the four parts of the above table, distinguished by the letters A B C D, can be grouped so as to divide the whole into two symmetrical parts, namely :

Chequerwise, or A with D and B with C ;
Into right and left halves, or A with C and B with D ;
Into upper and lower halves, or A with B and C with D.

The first of these, corresponding to the two orders—

1. Direct letters and signs + inverse letters and signs (A + D),
2. Direct letters and inverse signs + inverse letters with direct signs (B + C)

—gives the forms represented in figs. 33 and 35, the former being derived from the extension of the white faces, and the latter of the shaded ones, in the hexakisoctahedron {321} (fig. 34). It corresponds to an extension of alternate faces

FIG. 33.

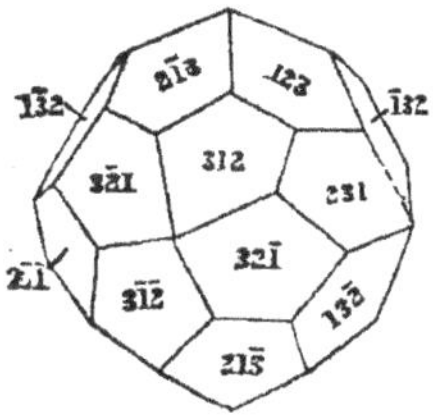

FIG. 34.

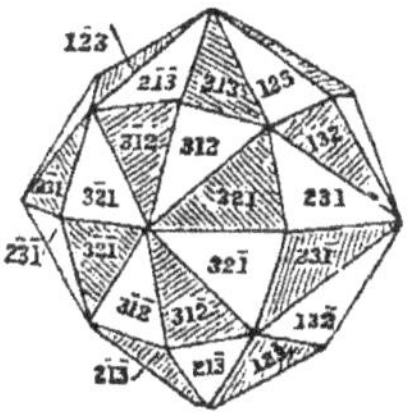

FIG. 35.

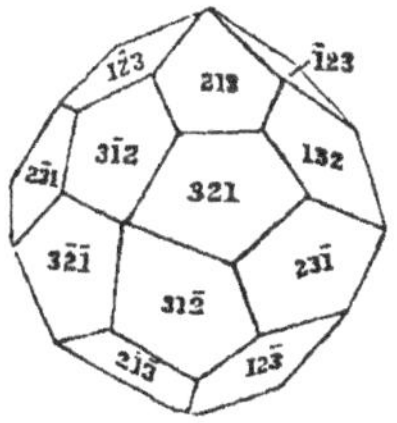

of the holohedral form, so that all the edges of the latter are obliterated, the only points common to the original and derived forms being the ends of the principal and ternary axes; and as in the form {*h k l*} any face is inclined to each of the three planes of symmetry in which its edges lie, at less than a right angle, it will by extension lose its symmetry to all of them, and the resulting forms will be plagihedral (skew-faced) or asymmetric, which names are used to indicate this particular kind of hemihedrism. There is, however, no change in the axes of symmetry, which are of the same number and kind as in the holohedral forms. Geometrically, their faces are irregular pentagons, which are so arranged that the two correlated (direct and inverse or positive and negative[1]) hemihedra, derived from the same holohedral form, cannot be superposed or made to corre-

[1] Either may be considered as the positive or negative one; but the choice, when made, applies to all similar forms in the same substance. Generally that containing the face *h k l* is taken as positive.

spond with each other by rotation. It will be seen that the edges forming the four-faced solid angles at the extremities of the principal axes have a right handed inclination to the vertical and horizontal lines in one form, and a left-handed one in the other, which cannot be altered by change of position. Such solids are said to be non-superposable, or enantiomorphous, *i.e.* permanently right- and left-handed, like gloves. This hemihedrism is not applicable to any of the specialised forms of the hexakisoctahedron, or the other six holohedral forms of the system, for it requires the four groups of signs and letters to be unbroken, which they cease to be if two letters are interchangeable, or if a sign becomes ambiguous. This may be seen in another way by considering each of these six forms as a hexakisoctahedron, in which the dihedral angle over one or more of the three kinds of edges is 180°, when, supposing one of the faces meeting in such an edge to be removed, the extension of the other will only fill up its place, and the particular shape will be restored. This would be expressed, in ordinary mineralogical language, as follows:—The plagihedral hemihedral forms of the cubic system are, with the exception of those derivable from hexakisoctahedra, undistinguishable from the holohedral forms. These particular hemihedral forms are not known to exist either in natural or artificial crystals, and therefore no special class of symbols are required for them. They may be indicated by $\alpha\{h k l\}$. $\alpha\{l k h\}$, the prefix α representing asymmetric.

Parallel hemihedrism. The second kind of hemihedrism represented by the division of the table on page 50 into right- and left-hand halves, which corresponds to the application of both kinds of sign permutations to each group of letter permutations taken separately, gives rise to hemihedral forms with parallel faces. For the hexakisoctahedron these are shown in figs. 36 and 38, the former being derived from the white, and the latter from the shaded faces in fig. 37. These are twenty-four faced solids, known as dyakisdodecahedra; the faces are trapezoidal with two equal and two unequal sides. In each of the principal sections half the edges represent those of the holohedral form extended, the other half being replaced by shorter

FIG. 36.

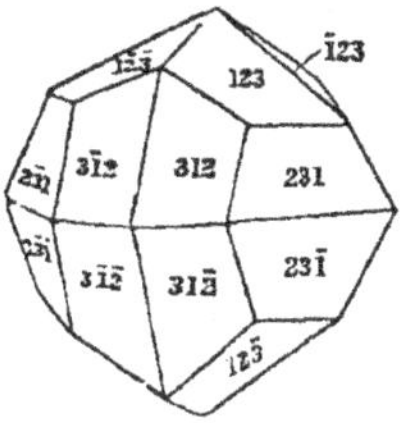

FIG. 37.

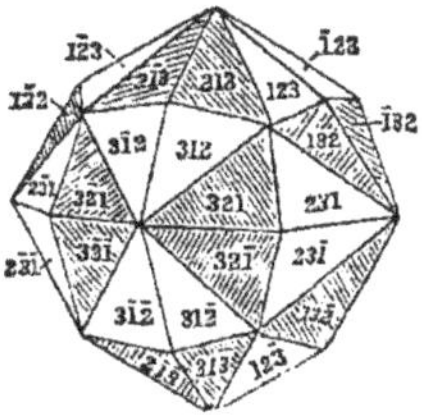

FIG. 38.

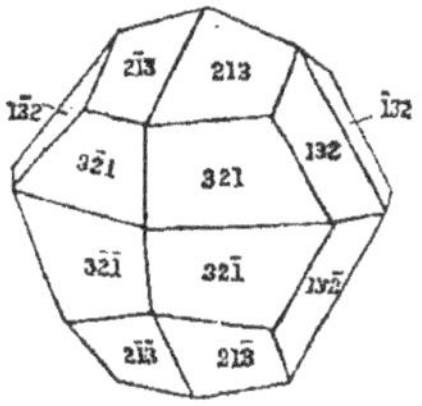

ones, which are less steeply inclined to the principal axis in which they meet than the former. The remaining edges make three-faced solid angles in each of the axes of ternary symmetry, having no symmetrical relation to those of the holohedral form. The general result of this is, that the symmetry to the binary axes is lost and that to the principal axes reduced from quaternary to binary, while that to the ternary axes is unchanged, and only the three principal sections remain planes of symmetry. The general symbols for these forms are, for fig. 36 $\pi\ \{l\,k\,h\}$, and for fig. 37 $\pi\ \{h\,k\,l\}$. Naumann's are $\left[\frac{m\,O\,n}{2}\right]$ and $-\left[\frac{m\,O\,n}{2}\right]$; the prefix π and the square brackets respectively indicating parallel hemihedrism. The positive and negative forms derived from the same hexakisoctahedron are superposable, or either one may be brought in the position of the other by a quarter revolution about a principal axis.

The dihedral angles in the particular case given $\pi, \{3\ 2\ 1\}$ are:

Over the longer edges in the principal sections .	150° 0′
,, shorter edges in the principal sections .	115° 23′
,, unsymmetrical edges in the principal sections	141° 47′

The tetrakishexahedron, when similarly developed, produces fig. 39 from the white faces, and fig. 41 from the shaded ones, in {2 1 0} fig. 40. These are pentagonal dodecahedra with irregular faces, one of the sides being prominently longer [1] than the other four, which are equal to each

FIG. 39.

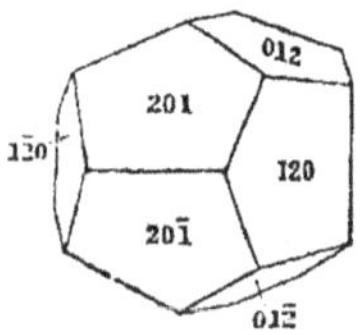

FIG. 40.

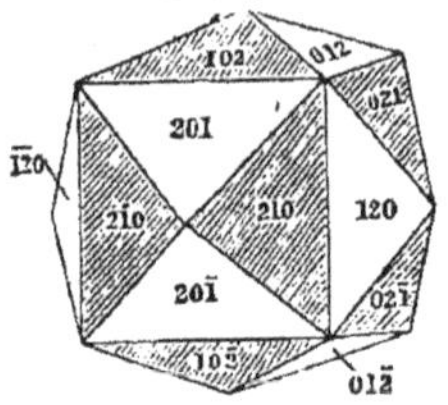

FIG. 41.

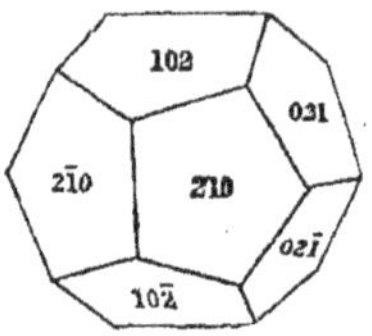

other. These longer sides are the only effective edges in the principal sections, and as they are parallel to the principal axes the partial cubic symmetry is at once apparent. The three-faced solid angles mark the symmetry to the ternary axes, and that to the binary axes is wanting, as in the preceding instance. The symbols are:

$$\pi\{0\,k\,h\} \text{ and } \pi\{h\,k\,0\} \text{ or } \left[\frac{\infty O n}{2}\right] \text{ and } -\left[\frac{\infty O n}{2}\right].$$

For the particular case given, $\pi\{2\ 1\ 0\}$, the dihedral angles are:

Over the longer edges	126° 52′
Over the shorter edges . . .	113° 35′

The dihedral angle over the edges in the principal sections of the hexakisoctahedron increases as the value of m in its symbol is increased, becoming 180° when $m = \infty$, or when it becomes a hexakistetrahedron; the same relation holds good with its hemihedral form, the dyakisdodecahedron, which under similar conditions passes into a

[1] Or shorter when n approximates to 1. The regular or Platonic dodecahedron is an impossible form, as the indices of the hexakistetrahedron producing it are $\sqrt{5} + 1$, 2, 0, which are not admissible on account of irrationality. π {850} is very near to it.

pentagonal dodecahedron. In the same way, when *n* is the symbol of the hexakistetrahedron, and its hemihedral form becomes ∞, they pass into the cube.

This mode of hemihedrism is not applicable to the other five holohedral forms, or rather it does not produce geometrically different forms from them.

Inclined hemihedrism. When the faces are so selected that the whole of the letter permutations with only one-half of the sign permutations appear in a hemihedral form, the latter is said to be hemihedral with inclined faces. This corresponds to the division of the table on page 50 into upper and lower halves, the former containing the symbols with an odd number—one or three—of positive indices, and the latter those with one or three negative indices, or, in other words, a face and its counterpart can never appear on the same form. Geometrically, this signifies the extension of all the faces in alternate octants, or the suppression of those in opposite and adjacent octants of the holohedral form, as shown in fig. 42 for the white, and in fig. 44 for the

FIG. 42.

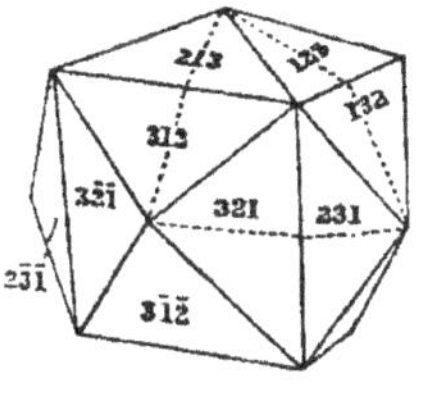

FIG. 43.

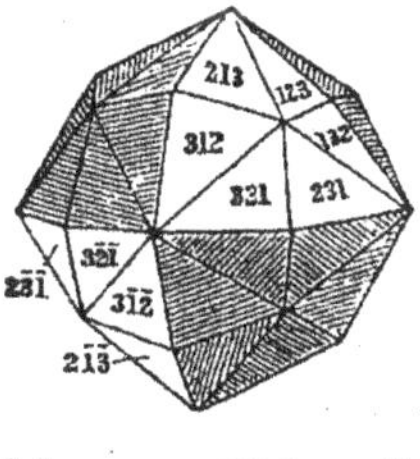

FIG. 44.

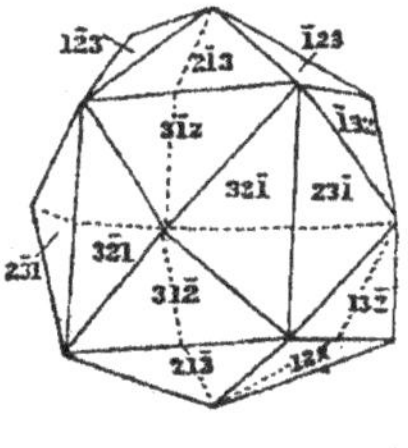

shaded faces of {3 2 1} fig. 43. This solid is called a hexakistetrahedron, from its resemblance to a tetrahedron enclosed by four groups of six-faced pyramids. From the construction it will be easily seen that the edges forming the six-faced solid angles are the same as the longer and shorter edges of the holohedral form, and preserve their characteristic inclination, but the new ones formed by the meeting of the extended faces have no symmetrical relations to the original form. The highest symmetry is therefore ternary about the normals of the octahedron, that about the principal axes is reduced from quaternary to binary, and the normals to the rhombic dodecahedron are not axes of symmetry. As a consequence, the symmetry to the faces of the cube is lost, while that to the faces of the rhombic dodecahedron remains.[1]

The symbols are :

$$\kappa\,\{h\,k\,l\} \text{ and } \kappa\,\{\bar{h}\,\bar{k}\,\bar{l}\} \text{ or } \frac{m\,O\,n}{2} \text{ and } -\frac{m\,O\,n}{2}.$$

The dihedral angles of the particular form given, $\kappa\,\{3\,2\,1\}$, will be the same as those over the longer and shorter edges in $\{3 2 1\}$; those over the special hemihedral edges are 120° 55′.

The inclined hemihedral forms of the icositetrahedron are called triakistetrahedra, fig. 45 being that derived from the white, and fig. 47 from the shaded faces of fig. 46.

FIG. 45.

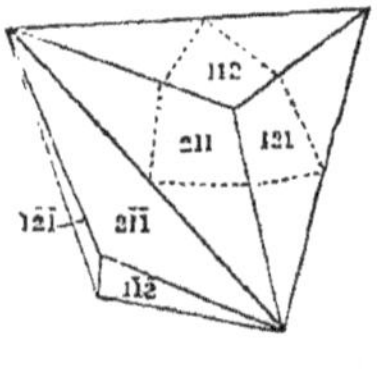

FIG. 46.

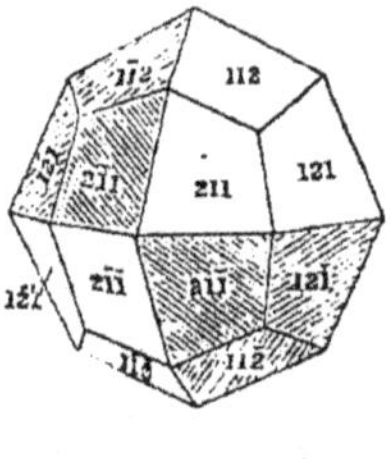

FIG. 47.

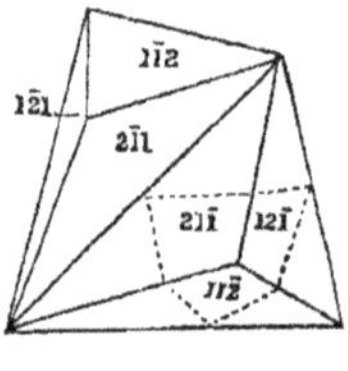

In this the new edges formed by the extended faces enclose a regular tetrahedron, upon each of whose faces a triangular pyramid formed by the original faces is superposed. The symbols are :

$$\kappa\,\{h\,k\,k\} \text{ and } \kappa\,\{\bar{h}\,k\,k\}, \text{ or } \frac{m\,O}{2} \text{ and } -\frac{m\,O}{2}.$$

For the particular case, $\kappa\,\{2\,1\,1\}$, the dihedral angles over the

[1] As these relations, which hold good for all inclined hemihedral forms, are not readily seen in perspective figures, it will be well for the student to study them upon a model. The regular tetrahedron is the most convenient form for this purpose.

tetrahedral edges are 109° 28′, or the supplements of those of the regular tetrahedron; those over the shorter edges have the same value as in {2 1 1}.

The inclined hemihedral forms of the triakisoctahedron are called deltoid dodecahedra (figs. 48 and 50), of which the first is derived from the white, and the second from the shaded faces in {2 2 1} fig. 49. In these the extended faces

FIG. 48.

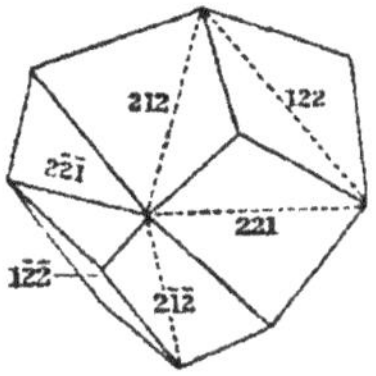

FIG. 49.

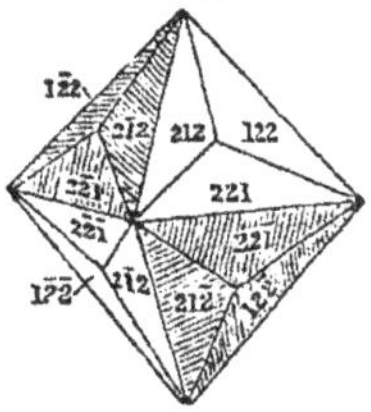

FIG. 50.

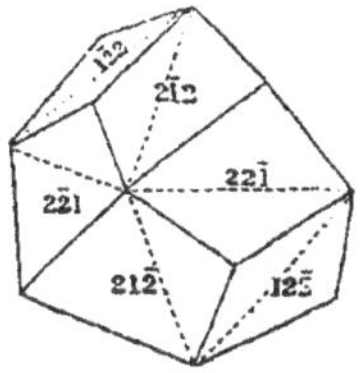

form a new three-edged pyramid above the enclosed octahedron, but of different proportions to those of the original form, the faces, when regularly developed, being deltoids or four-sided figures, one of whose diagonals joins similar and the other dissimilar angles. The symbols are :

$$\kappa\{h\,h\,k\} \text{ and } \kappa\,\{h\,\bar{h}\,k\}, \text{ or } \frac{mO}{2} \text{ and } -\frac{mO}{2}.$$

The dihedral angles over the hemihedral edges in κ {2 2 1} are right angles.

The octahedron (fig. 52) by this method produces the

FIG. 51.

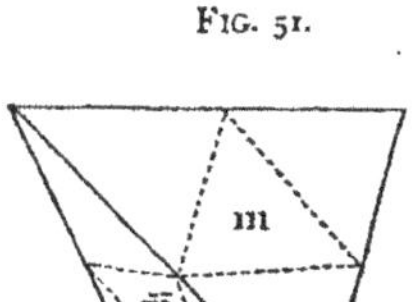

FIG. 52.

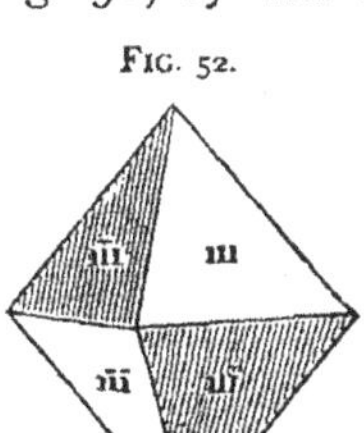

FIG. 53.

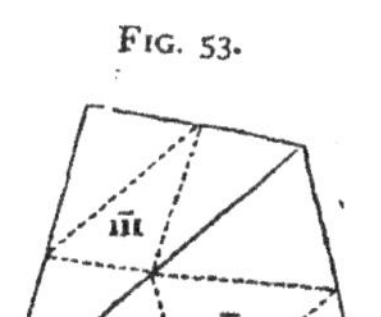

two regular tetrahedra (figs. 51 and 53) as its hemihedral forms. Their faces are all equilateral triangles of four times the area and twice the length of edge of those of the octahedron from which they are derived. The dihedral angles are 70° 32′, or supplements to those of the octahedron; the edges correspond to diagonals of the faces of a cube described about the axes of the enclosed octahedron. The principal axes join the middle points of opposite edges; the ternary axes are, as in the holohedral form, normals to faces, but on one side only, the opposite sides meeting the extended faces in the solid angles.

Neither the cube nor the rhombic dodecahedron can produce forms dissimilar from themselves by any of the three methods of hemihedrism. The former, when considered as a particular form of hexakisoctahedron, has

FIG. 54.

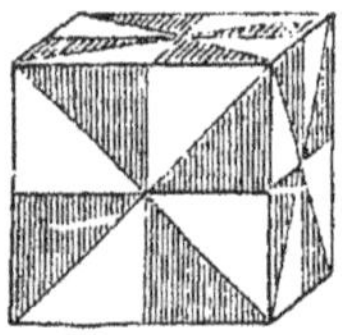

FIG. 55.

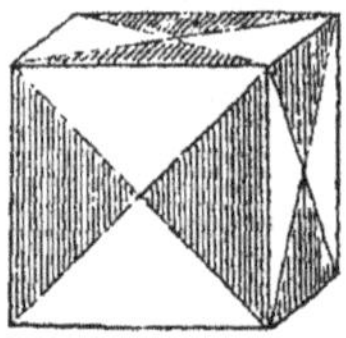

FIG. 56.

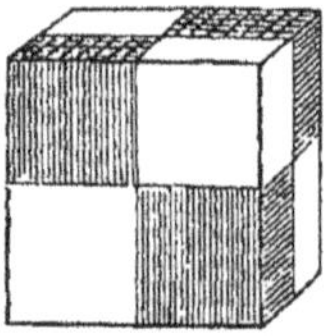

eight of its faces in the same plane, and therefore the removal of half of these, whether by alternating eighths, as in fig. 54, or by quarters, as in figs. 55, 56, which correspond to the three methods of hemihedral selection, can

FIG. 57.

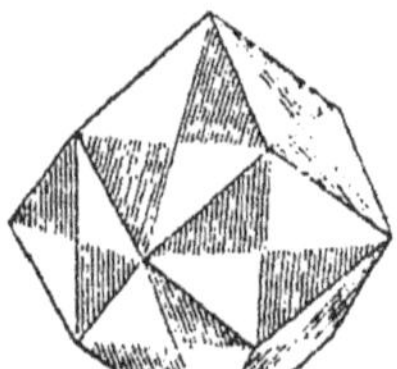

FIG. 58.

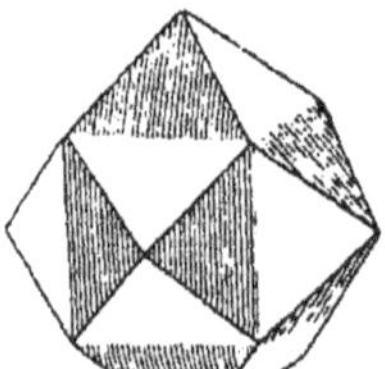

FIG. 59.

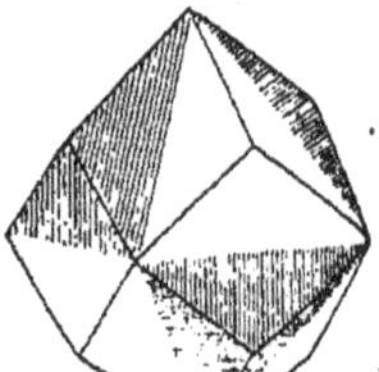

have no effect, as the extension of the remaining parts, whether white or shaded, will restore the original form. In the same way the rhombic dodecahedron in each of its faces contains four of those of $\{h\,k\,l\}$; and therefore their removal by quarters or halves, as in figs. 57, 58, and 59, and the extension of the remainder, does not change the form.

It may therefore be said that there are no hemihedral forms of these solids; but this is only true as a geometrical proposition, and is at variance with a general principle, deduced from observation, that the forms making up the crystals of any particular substance are all of the same kind —*i.e.* either all holohedral or of the same class of hemihedrism. For example, in iron pyrites a very large number of forms are known, the principal ones being pentagonal- and dyakis-dodecahedra, which often appear alone as well as in combination with the cube and octahedron, but never with a hexakisoctahedron or tetrakishexahedron, or any inclined hemihedral form. It is therefore necessary to consider the cube and rhombic dodecahedron as common to the holohedral and hemihedral classes of forms alike, their true character being only determinable by the nature of their combinations. The relation and derivation of the different hemihedral forms may be expressed by their symbols, according to Naumann, in diagrams similar to that already given for the holohedral form. Fig. 60 gives the scheme for the parallel, and fig. 61 for the inclined, class.

FIG. 60.

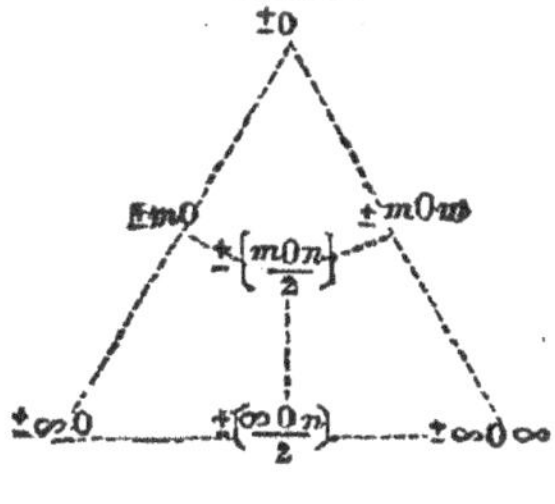

FIG. 61.

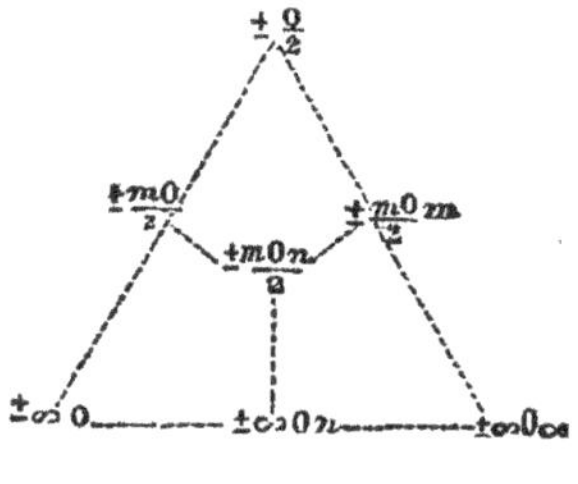

In Weiss's notation, parallel hemihedral forms are distinguished by the symbol $\frac{1}{2}$ preceding that of the type face; and the inclined ones in the same way, with the addition of a positive or negative sign. Thus the two dyakisdodecahedra, $\pi\{h k l\}$ and $\pi\{l k h\}$, are respectively $\frac{1}{2}(a : m a : n a)$ and $\frac{1}{2}(n a : m a : a)$, and the hexakis-tetrahedra, $\kappa\{h k l\}$ and $\kappa\{h \bar{k} l\}$ $+\frac{1}{2}(a : n a : m a)$ and $-\frac{1}{2}(a : n a : m a)$.

Tetartohedral forms. If any hemihedral form of the hexakisoctahedron be subjected to one of the other modes of hemihedry, a new form containing only one-fourth of the full number of faces, but otherwise satisfying the general conditions of systematic symmetry, is obtained. This is said to be tetartohedral. By reference to the table on page 50 it will be easily seen that the result is the same whichever methods of hemihedry be employed, the whole series of symbols being resolved into the four groups, each containing one group of letter and sign permutations, and representing a separate form. Of these, only two are geometrically distinct otherwise than by position; that containing direct letters with inverse signs is superposable to that containing inverse letters with direct signs, or B with C, as are also the groups A and D, where letters and signs are both direct or both inverse, but not otherwise—that is, the tetartohedra resulting from the same hemihedron are right- and left-handed to each other.

Figs. 62 and 63 represent the forms obtained by the

FIG. 62.

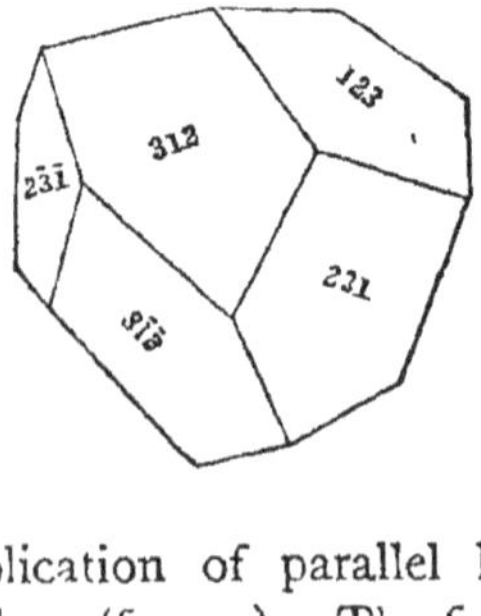

FIG. 63.

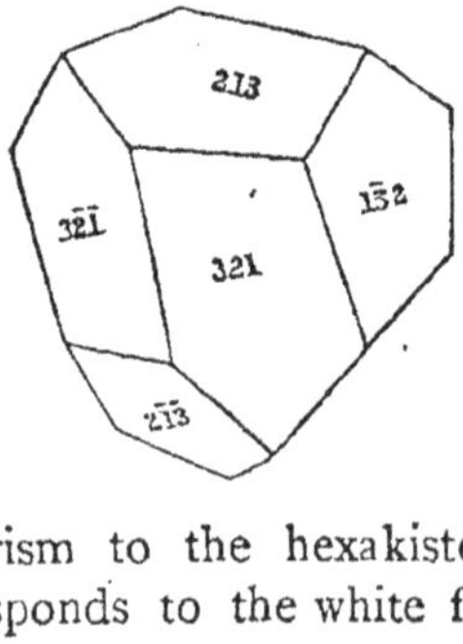

application of parallel hemihedrism to the hexakistetrahedron (fig. 42). The first corresponds to the white faces and the second to the shaded ones, in alternate (the 1st, 3rd, 6th, and 8th) octants of the hexakisoctahedron (fig. 34), or to the divisions A and B of the table on page 50. These are called right and left tetartohedral pentagonal dodecahedra, their faces, when most regularly developed, being irregular pentagons. The only points in common with the holohedral forms are alternate extremities of the ternary axes, which retain their original positions: those of the principal axes meet in six edges which are oblique to the principal sections. As both kinds of hemihedrism are involved in their production, the symmetry to both series of planes is lost, or the forms are plagihedral, but that to the axes is of the same kind as in the hemihedral forms, or to three binary and four ternary axes. This, therefore, may be regarded as the fundamental axial symmetry of the cubic system, it being the minimum common to all classes of forms, the higher quaternary kind being added in the case of the holohedral and plagihedral hemihedral classes.

The inverse tetrakishexahedron gives rise to two similar forms differing only in position, and if the former be considered as positive they will be the negative tetartohedra. The general symbols will be:

$$\pi\kappa\{h\,k\,l\} \quad \pi\kappa\{l\,k\,h\}$$
$$\pi\kappa\{\bar{h}\,\bar{k}\,\bar{l}\} \quad \pi\kappa\{\bar{l}\,\bar{k}\,\bar{h}\}$$

and

$$+\frac{m\,O\,n}{4}r, \quad +\frac{m\,O\,n}{4}l$$
$$-\frac{m\,O\,n}{4}r, \quad -\frac{m\,O\,n}{4}l$$

These are the only geometrically distinct tetartohedral forms, and, although crystallographically possible, they are not known to occur independently in either natural or artificial crystals. The actual existence of the condition of tetartohedrism is, however, known from the fact of a few substances appearing in crystals showing both kinds of hemihedrism—a point that will be considered in treating of the combinations of the system.

Combinations of the cubic system. Any number of simple forms may appear with their full number of faces in a single crystal, which will then have a sort of composite character, the faces of the constituent forms being, as a rule, recognisable not by their shapes, but by their positions. Such crystals are termed combinations. That the shapes of the faces must be altered, even with the most exact regularity of position, will be apparent when it is considered that the simple forms described about any definite lengths of axes are all exterior to the octahedron, or enclosed by the cube, having only points or lines in common, and therefore no combination is possible between them in this condition. If, however, the size of any one of them be altered relatively to another by shifting its faces parallel to the original positions, either nearer to the centre in the case of an exterior, or further from it in an interior form, which does not alter its crystallographic significance, a new solid will be produced, with edges and solid angles either wholly or

FIG. 64.

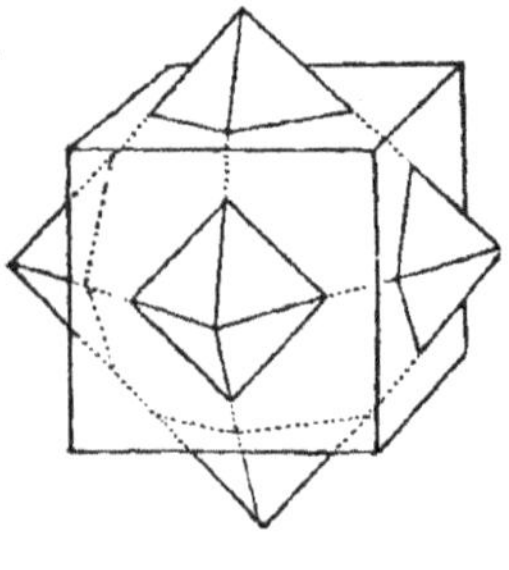

FIG. 65.

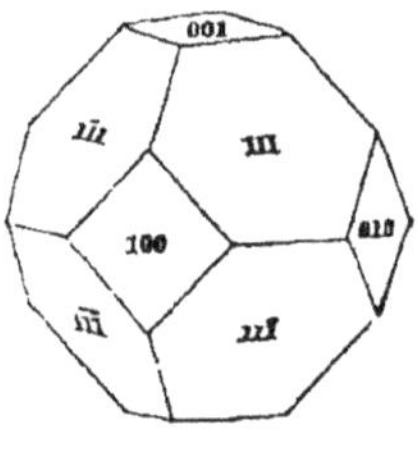

partly different from those of the component forms. For instance, fig. 64 represents the complete interpenetration of a cube by an octahedron, their principal axes being common as regards position, but differing in length; the points of the latter form project as four-faced pyramids above the faces of the former, and similarly the solid angles of the cube form three-faced pyramids above the faces of the octahedron. If these projecting portions be removed, which is necessary to produce convex angles, the form is reduced to fig. 65, which may be considered as an octahedron with its solid angles cut off or truncated[1] by the faces of a cube. Fig. 66 is another example in which the faces of the two forms are evenly balanced, so that it may be equally well

FIG. 66.

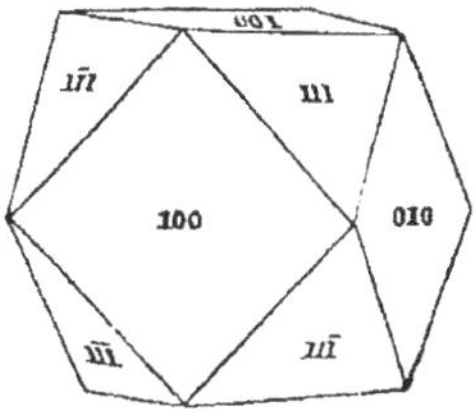

FIG. 67.

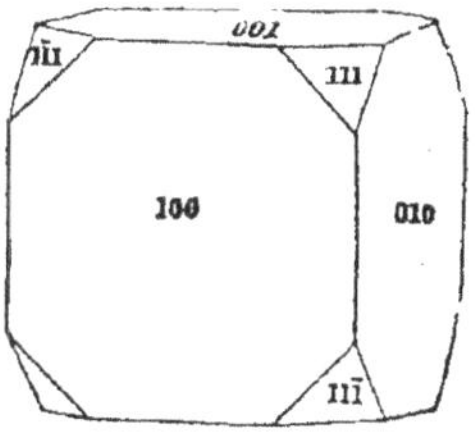

regarded as a cube modified by an octahedron, or the reverse; while in fig. 67 the cube is the principal or dominant form, the octahedron being only represented by a small triangular plane in each of the corners. In neither of these are the characteristic shapes of the faces seen in the simple forms apparent, but they are nevertheless easily recognisable from their constant position, and the parallelism of the new edges of combination.

As a rule, special names are not applied to combinations, but they are described by their symbols, that of the most prominent form being placed first. Thus, fig. 65 is indicated by the symbols $O \,.\, \infty O \infty$, and fig. 67 as $\infty O \infty \,.\, O$, while either order applies equally well to fig. 66.

The faces of the octahedron when in combination with the rhombic dodecahedron appear as triangular planes

[1] An edge replaced by a single plane making equal angles with the adjacent faces is truncated; when the replacement is by two planes it is bevelled. Similarly, a solid angle is truncated by a single plane, bevelled by two, and acuminated or blunted by three or more.

truncating the three-faced solid angles of the latter, as in fig. 68. When the octahedron is the dominant form, the faces of the rhombic dodecahedron truncate its edges as in

FIG. 68.

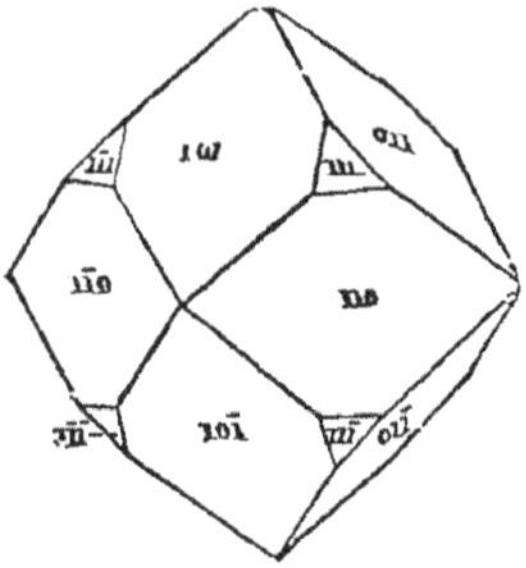

FIG. 69.

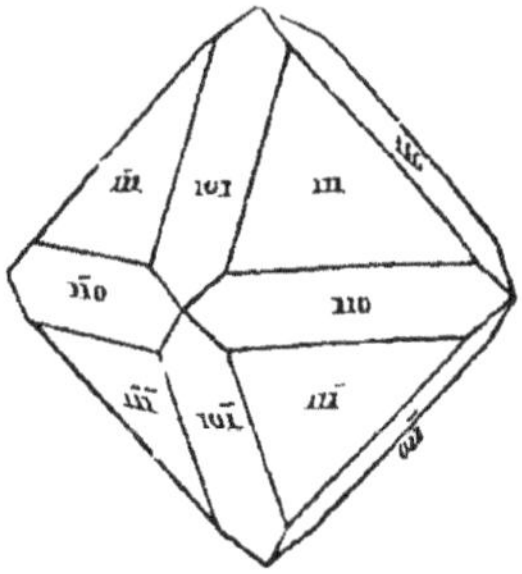

fig. 69. The cube truncates the four-faced solid angles of the rhombic dodecahedron, as in fig. 70, and conversely the latter truncates the edges of the former, as in fig. 71.

FIG. 70.

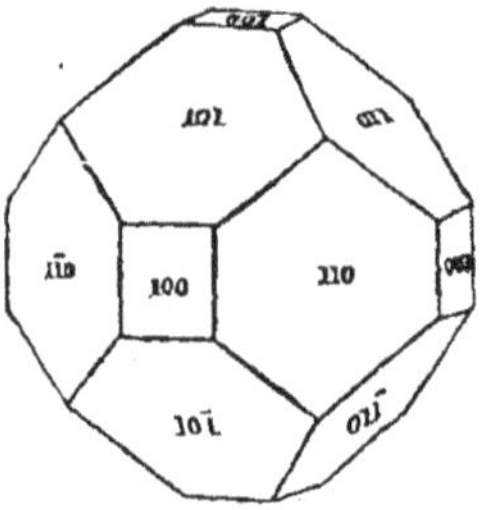

FIG. 71.

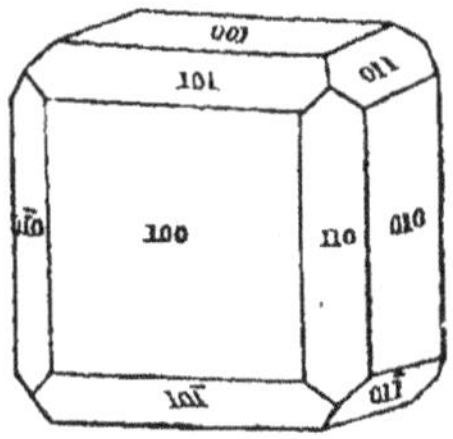

Examples of these combinations are of very common occurrence in nature, especially in magnetite, galena, and fluor-spar.

The faces of the icositetrahedron, 2 *O* 2 or {2 1 1}, lie in zones whose axes are parallel to the edges of the octahedron and the plane diagonals of the cube, and therefore form blunt four-faced pyramids upon the solid angles of the former (fig. 72), and three-faced ones upon the latter (fig. 73).

In a combination of these three forms, when the first predominates, as in fig. 74, the cube truncates its four-faced, and the octahedron its three-faced, solid angles.

The combinations of icositetrahedra with the rhombic dodecahedron vary very considerably in appearance, ac-

FIG. 72.

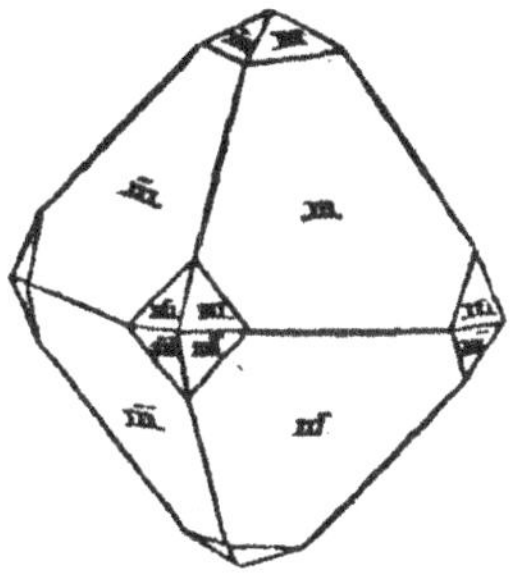

FIG. 73.

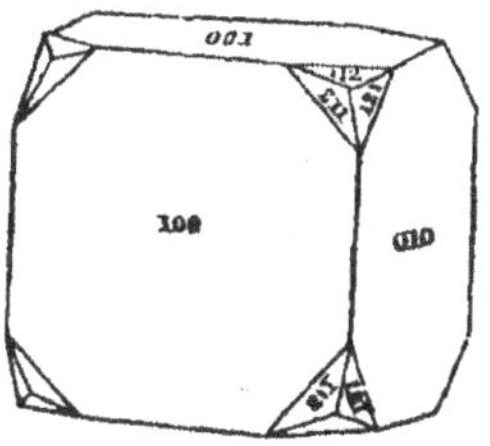

cording to the value of *m* or the relation of *h* to *k* in their symbols. The form most commonly observed, which is $2 O 2$ or $\{211\}$, truncates the whole of the edges of $\{110\}$; as the two forms are tautozonal (fig. 75). This combination

FIG. 74.

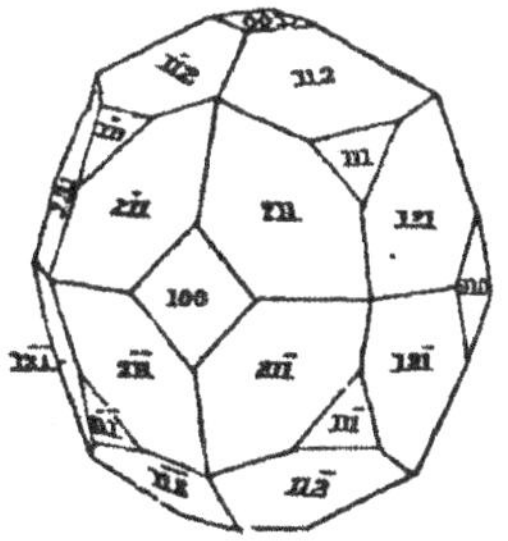

FIG. 75.

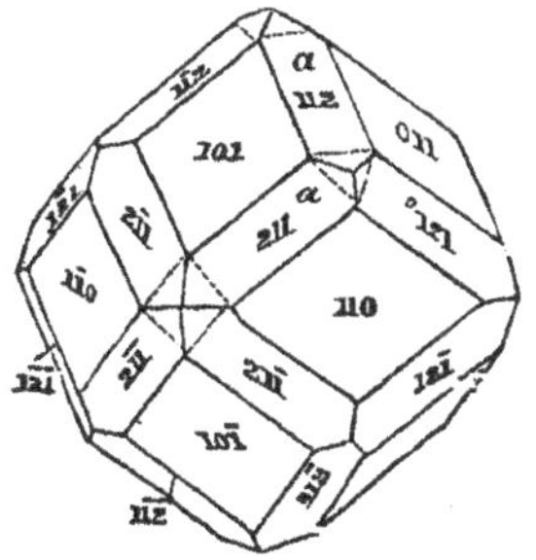

gives a good example of the determination of the symbols of a form from those of its zones. If the edges of the cube and octahedron be added, as shown by the dotted lines, it will be apparent that any face such as the upper one, marked *a*, lies in two zones, the first containing 1 0 1 and 0 1 1 or [$\bar{1}$ $\bar{1}$ 1] and the second 0 0 1 and 1 1 1 or [$\bar{1}$ 1 0]. From these symbols we get, by the method given on page 32, a symbol ($\bar{1}$ $\bar{1}$ $\bar{2}$) or the counterpart of that of the particular face, and similarly ($\bar{2}$ $\bar{1}$ $\bar{1}$) for the second face, marked *a*. In such a case, therefore, no special calculation is required for the determination of the form, when the parallelism of the edges is exhibited, by measurement.

When $m > 2$ or $h : k > 2 : 1$, the faces of the icositetrahedron appear as deltoids in groups of four upon the four-

FIG. 76.

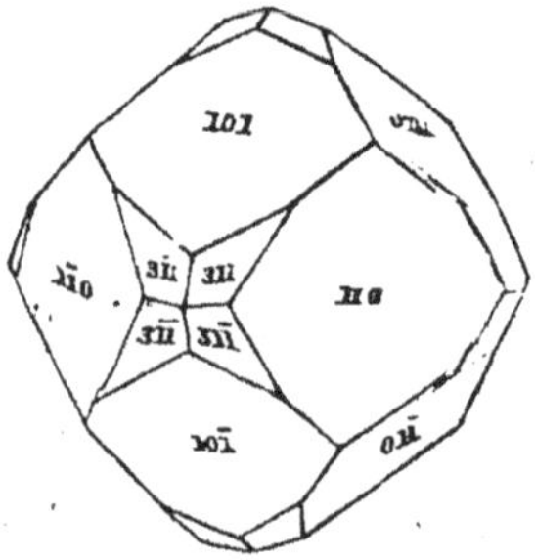

FIG. 77.

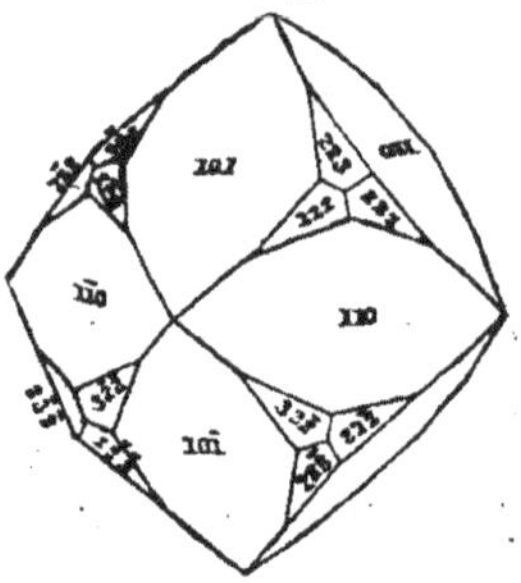

faced solid angles of the rhombic dodecahedron, as shown for the combination $\infty O . 3 O 3$, or {1 1 0}, {3 1 1} in fig. 76.

When $m < 2$ or $h : k < 2 : 1$, similar planes appear in groups of three upon the three-faced solid angles, as in the combination $\infty O . \frac{3}{2} O \frac{3}{2}$ or {1 1 0}, {3 2 2} (fig. 77).

FIG. 78.

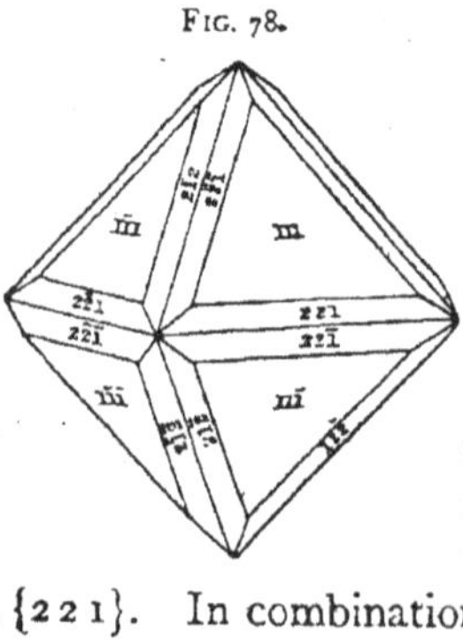

The triakisoctahedron, having its longer edges in common with the octahedron, bevels the edges of the latter and conversely the octahedron truncates the three-faced solid angles of the former, both of which relations are apparent in fig. 78, containing the forms $O . 2 O$ or {1 1 1}, {2 2 1}. In combination with the rhombic dodecahedron a triakisoctahedron forms obtuse pyramids blunting the three-faced solid angles (fig. 79), and with the cube groups of three deltoid planes upon the solid angles (fig. 80).

FIG. 79.

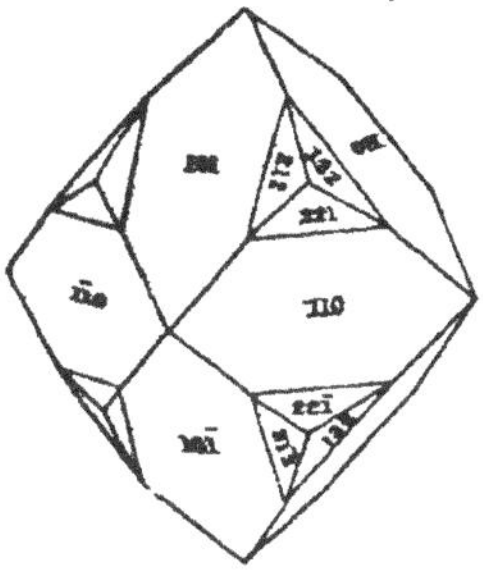

FIG. 80.

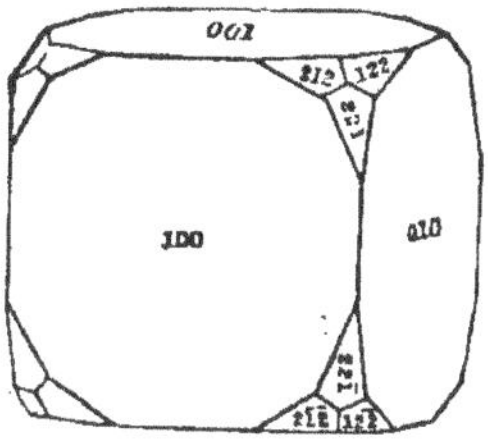

The triakisoctahedron, in combination with the icositetrahedron, modifies the edges lying in the dodecahedral planes of symmetry (the short edges of $h\,k\,l$). In the particular case represented in fig. 81, $2\,O\,2$, $\frac{3}{2}\,O$, or $\{2\,1\,1\}$, $\{3\,3\,2\}$ the two forms are tautozonal, and therefore the new edges are parallel to the original ones. When m is greater than $\frac{3}{2}$, or the form is nearer to a rhombic dodecahedron, the replacing faces appear as very acute triangles whose summits meet in the ternary axes.

FIG. 81.

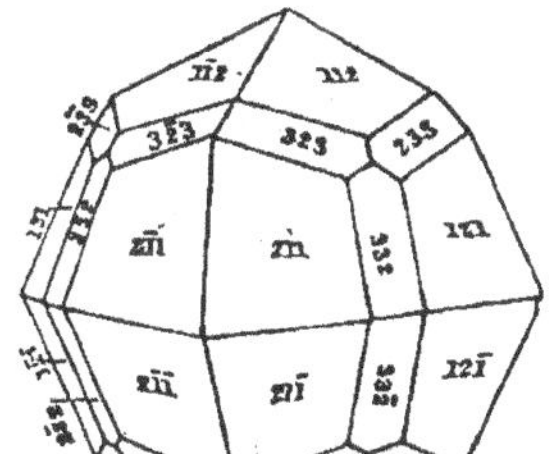

The tetrakishexahedron having its longer edges in common with the cube, will in combination bevel these edges, and conversely the cube will truncate the four-faced solid angles of the first form, as shown in fig. 82, representing the combination $\infty O \infty$. $\infty O 2$ or $\{1\,0\,0\}$ $\{2\,1\,0\}$, which is commonly observed in fluorspar. The same forms being also tautozonal to the rhombic dodecahedron, the latter will have its four-faced solid angles blunted by the faces of $\infty O 2$ (fig. 83). If the cube were to be added its faces would truncate these new solid angles, and any more obtuse tetrakishexahedron would bevel the edges between {2 1 0} and {1 0 0}, and any more acute one those between {2 1 0} and {1 1 0}.

FIG. 82.

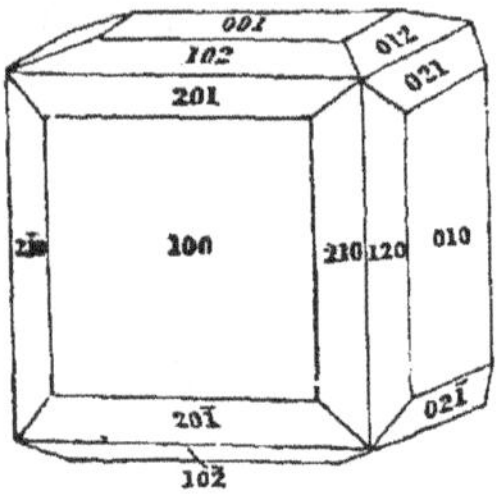

FIG. 83.

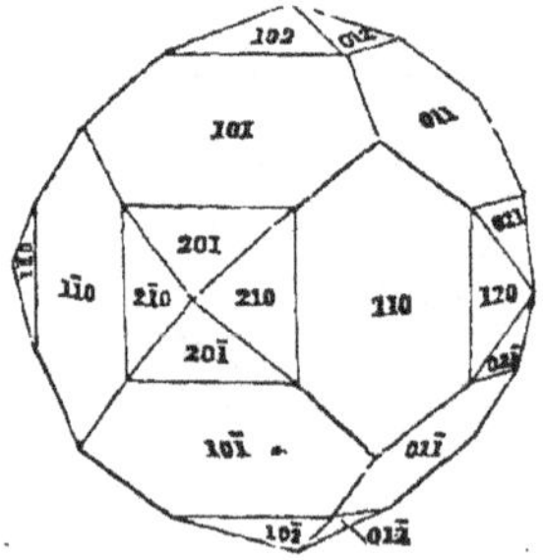

The tetrakishexahedron and octahedron are not tautozonal forms, and therefore in combination their faces assume irregular shapes, as seen in fig. 84. By comparing this with figs. 65 and 69 it will be seen that these shapes will vary with the value of m or the inequality of h and k in the symbols of the tetrakishexahedron. As these values diminish, the octahedial faces become more nearly triangular, until when $m = 1$ or $h = k$ they are equilateral triangles, as in fig. 69; and in the reverse direction, when $m = \infty$ or $h : k = 1 : 0$, they are equiangular hexagons of 120°, as in fig. 65. The first of the cases, however, is the condition of a rhombic dodecahedron, and the second that of a cube, which are the respective limiting forms of the tetrakishexahedron in these directions. The acute angles of the deltoid planes in fig. 84 will alter in a corresponding manner to a minimum of 0° in fig. 69, and to a

FIG. 84.

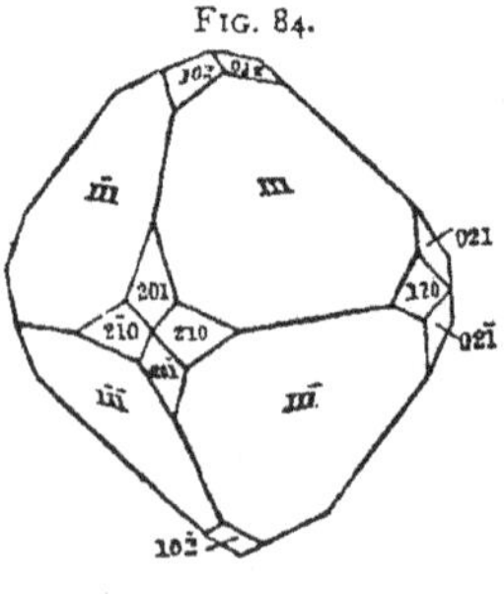

maximum of 90° in fig. 65. By comparing analogous combinations between limiting forms, their relations may often be more readily appreciated than by the most elaborate verbal explanations; and therefore the construction of forms with different parameters to those given in the figures may be recommended to the learner as a useful exercise.

The combinations of the tetrakishexahedron with the other twenty-four-faced forms vary very considerably in appearance, accordingly as different values are assigned to the variable parameters in their symbols, and to illustrate them properly a larger number of figures would be required than can be given here. One special case deserves mention—namely, that by regular truncation of the edges lying in the principal sections, the icositetrahedron {2 1 1} may be changed into the tetrakishexahedron {2 1 0}.

The hexakisoctahedron appears in combination with the octahedron in groups of eight triangular faces, blunting the

FIG. 85.

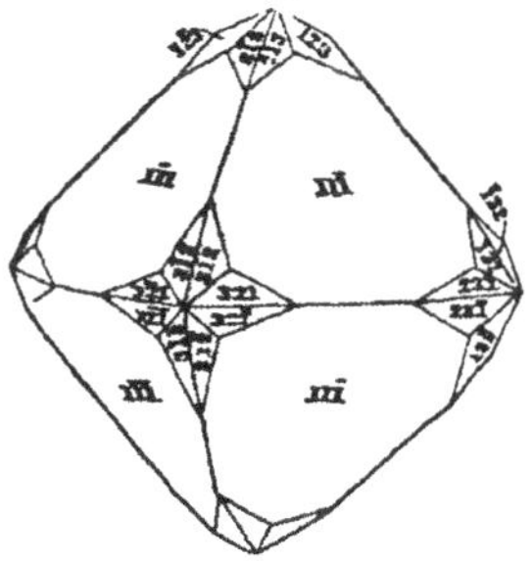

FIG. 86.

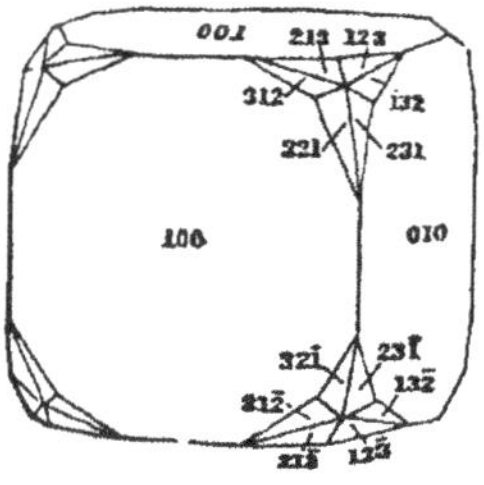

solid angles (fig. 85), and in similar groups of six faces upon the solid angles of the cube (fig. 86). From these it will also be apparent that the octahedron truncates the six-faced, and the cube the eight-faced, solid angles of the hexakisoctahedron.

In combination with the rhombic dodecahedron, the hexakisoctahedron may appear either as bevelling the edges or as modifying the three- or four-faced solid angles. The first case is that in which the forms are tautozonal, represented in fig. 87, for {1 1 0}, {3 2 1} or $\infty O, 3 O \frac{3}{2}$ corresponds to the condition $h = k + l$ in $\{h k l\}$ or $m n = m + n$, or $n = \frac{m}{m-1} n, = \frac{m}{m+1}$ in $m O n$. The known forms of this character are {3 2 1} . {4 3 1} and {64 . 63 . 1}, the first being that most frequently observed in nature. When

$$h < k + l, \ n < \frac{m}{m-1} \text{ or } m n < m + n,$$

as in {543} . {432} or {15 . 11 . 7}, the replacement is by six faces upon the three-faced solid angles of {1 1 0} ; and lastly, when

$$h > k + l, \ n > \frac{m}{m-1} \text{ or } m n > m + n,$$

as in {4 2 1} . {7 3 1}, &c., the replacement is by eight faces upon the four-faced solid angles of {1 1 0}. By far the larger

FIG. 87.

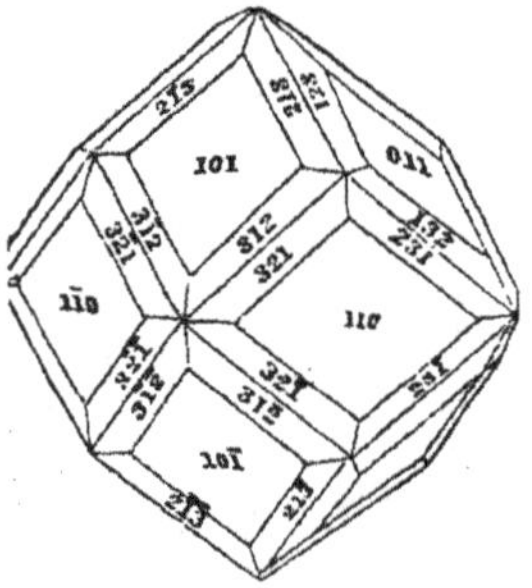

FIG. 88.

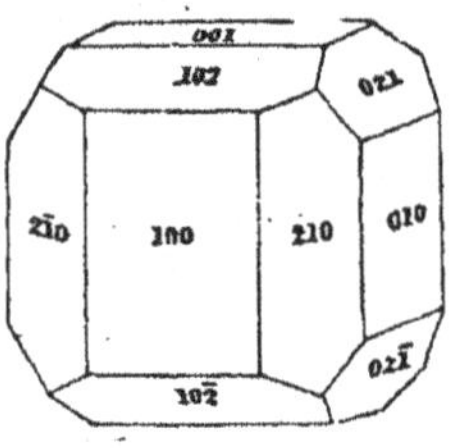

number of known hexakisoctahedra are of the latter kind, but they are not easily recognised, as they only occur in very subordinate combinations in crystals containing numerous other forms. For such complex combinations the reader is referred to Schrauf's Atlas, and the larger treatises on Mineralogy.

Hemihedral combinations. Among the simpler and more important cases, that of the cube and pentagonal dodecahedron is seen in fig. 88. This differs from fig. 82, by the omission of the alternate bevelling planes in each zone, or the faces of $\pi\ \{2\ 1\ 0\}$ truncate the edges of $\{1\ 0\ 0\}$ unsymmetrically, being unequally inclined to adjacent faces; but this inequality diminishes with that between h and k or k and l, or as we approach the cube and rhombic dodecahedron respectively. This class of combination is very characteristic of the pyrites group of minerals. In fig. 89, the pentagonal dodecahedron inverse to that in fig. 88 modifies the solid angles of the octahedron symmetrically. If these planes be extended to the complete obliteration of the octahedral edges, a solid is obtained with twenty faces which are very nearly equilateral triangles, and approximating in appearance to the regular icosahedron, which, however, we have seen, is not possible in crystallography.[1]

The dyakisdodecahedron appears as an obtuse three-faced pyramid upon the ternary solid angles of the pentagonal

FIG. 89.

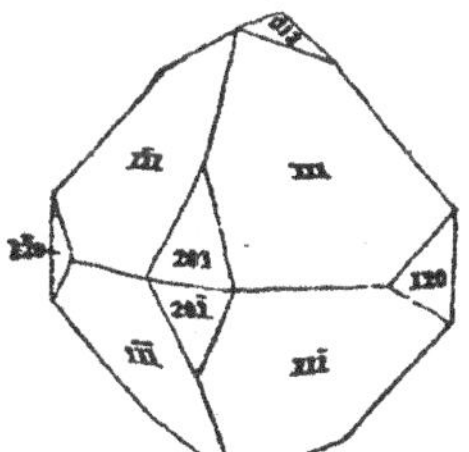

FIG. 90.

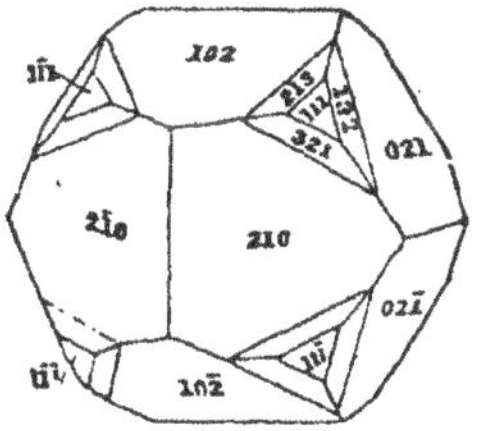

dodecahedron, and has its own solid angles of the same kind truncated by the octahedron, as in fig. 90. It also forms groups of three irregular four-sided planes upon the solid angles of the cube, as in fig. 91.

In the inclined hemihedral forms, the cube truncates the edges of a tetrahedron (fig. 92), and has half its own solid angles, or one in each pair joined by a ternary axis truncated by the faces of the latter (fig. 93). Two opposite tetrahedra in combination appear as in fig. 94, the faces of one being

[1] $\{1\ 1\ 1\}\ \pi\{8\ 5\ 0\}$ or $O.\ \frac{1}{2}\ [\infty\ O\frac{8}{5}]$ is a very close approximation, possible but not actually observed.

prominently larger than the other. It will be easily seen that when they are equally developed the combination will

FIG. 91.

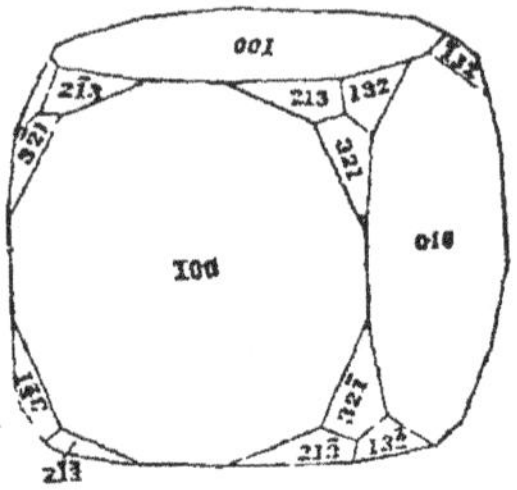

FIG. 92.

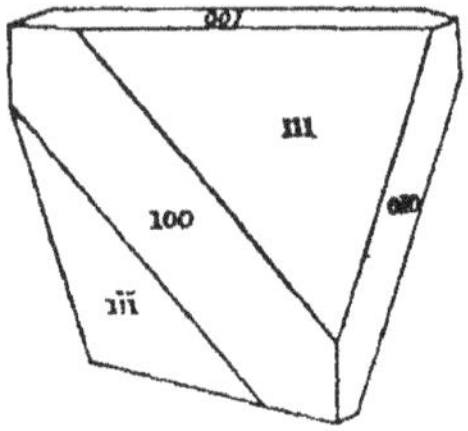

be an octahedron, and therefore indistinguishable from a holohedral form. When this does occur, the hemihedral nature of a substance is often apparent from some physical dissimilarity in the two classes of faces. such as one set being more brilliant than the other, or striated when the others are smooth, and therefore indicating that the crystal is not a true octahedron.

FIG. 93.

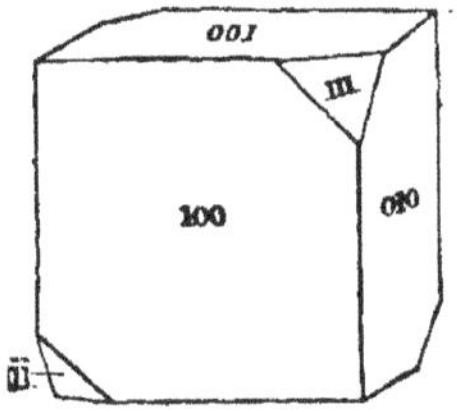

The rhombic dodecahedron forms a low three-faced pyramid upon each of the solid angles of a tetrahedron, as in fig. 95.

FIG. 94.

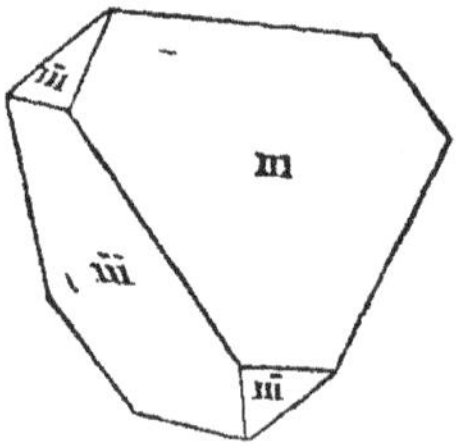

FIG. 95.

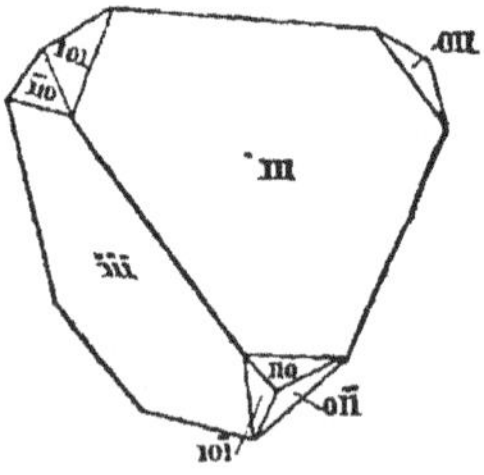

The triakistetrahedron bevels the edges of the tetrahedron of the same direction as in the combination $\kappa\ \{2\ 1\ \bar{1}\}$. $\kappa\ \{\bar{1}\ 1\ 1\}$ (fig. 96), which is very characteristic of the antimonial copper ore known as fahlerz.

Fig. 97 represents one of the few known cases of cubic tetartohedrism, as evidenced by the occurrence of inclined

FIG. 96.

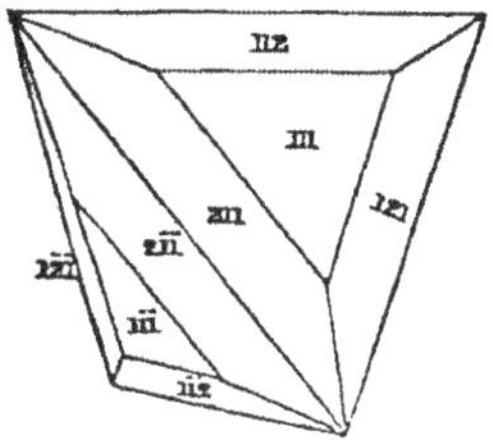

FIG. 97.

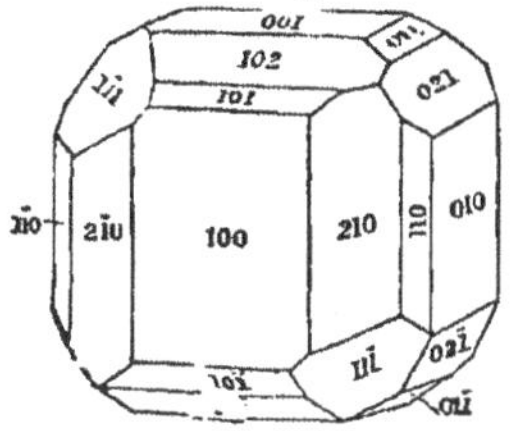

and parallel hemihedral forms in the same combination. It is not a natural substance, being a crystal of an artificial salt—chlorate of sodium. The same class of development characterises the crystals of the nitrates of lead, strontium, and barium, an actual tetartohedron $+\frac{1}{4}(5\ O\ \frac{5}{3})\ r$, or $\kappa\pi\{3\ 5\ 1\}$, having been determined by Lewis in the latter salt.

CHAPTER IV.

HEXAGONAL SYSTEM.

THE symmetry characteristic of the most completely developed forms in this system is seen in the regular hexagonal prism with parallel end faces (fig 98). This has binary symmetry about six axes, making angles of 30° and 150° to each other, the strong lines a_1, a_2, a_3, and the dotted ones alternating with the same plane, and senary or hexagonal about a seventh axis, which is normal to the other six, and indicated by the vertical line *c*. These axes correspond to seven planes of symmetry: a principal one, or that containing the six binary

FIG. 98.

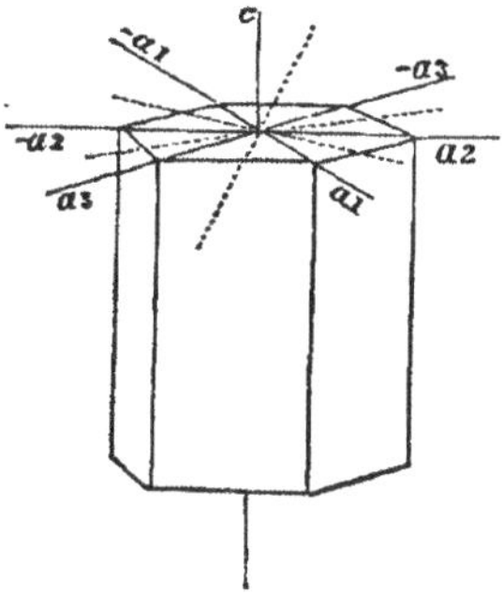

or lateral axes, normal to the principal or vertical axis, and six others corresponding to the diametral sections of the prism upon each of the lateral and the vertical axes. Three of these, or those parallel to the faces of the prism, are distinguished as lateral axial planes, and the alternate ones containing the dotted lateral axes as lateral interaxial planes.

The geometrical relations of faces of the above kind may be expressed in four different ways, each of which has been adopted as the basis of a system of notation. These are :

1. *Weiss's system.*[1] With four reference axes, three lateral at 60° and 120° to each other, and perpendicular to a fourth or principal axis, taken in the order of fig. 98, or a_1, a_2, a_3, c.

2. *Schrauf's system.* If in fig. 98 the faces meeting the right and left lateral axis a_2 be produced until they meet both in front and behind, the result is a four-faced prism upon a rhombic base, whose diagonals are at right angles, and in the ratio of $1 : \sqrt{3}$, or that of the side of an equilateral triangle to its altitude. This therefore gives a method whereby hexagonal forms may be referred to three independent axes all at right angles to each other, as in the rhombic system ; if the constant relation of $1 : \sqrt{3}$ be assumed for the lateral axes. This method is adopted by Schrauf, who calls it the orthohexagonal system ; by it the prism in fig. 98 is not a simple form, but a combination of the unit rhombic prism {1 1 0}, and a form containing two faces parallel to the right and left diametral plane, or {1 0 0}.

3. *Miller's system.* If a cube be placed with a ternary axis upright, the three edges meeting in either of the poles of that axis will be parallel and opposite to the three at the other pole, so that if lines parallel to them be drawn through the centre a system of three axes will be obtained, inclined to the vertical, but making equal angles with each other, and having equal parameters. In this particular case these angles will be right angles, but the same relation holds good for any analogous form contained by six equal rhombic faces,

[1] Weiss calls the system *sechsgliedrig*, or six-membered.

(*rhombohedron*) whose solid angles are formed by edges meeting at a greater or less angle than 90°, which inclination will also be characteristic of their axes. The relation of such a system of axes to the unit faces is that of the legs of a table formed by three sticks, tied together in the middle, to the table top and the ground respectively—the first being the face 1 1 1, and the second $\bar{1}\,\bar{1}\,\bar{1}$. This principle is adopted in Miller's rhombohedral notation.

4. *Bravais-Miller system.* The three semi-axes, having like signs in Miller's system, when represented by their orthogonal projections, are resolved into a common vertical and three horizontal lines, the latter making angles of 120° with each other. This brings us back to the system of four axes with the difference that the positive and negative semi-axes alternate with, instead of succeeding, each other at 60° in the horizontal plane; and in this way we obtain the hexagonal notation of Bravais as adapted to Miller's system. It has the advantage of maintaining the relation between the notation by indices and that by parameter-coefficients subsisting in the other systems, as well as of expressing the physical symmetry of the forms more readily than the otherwise preferable rhombohedral notation of Miller. The properties of any face are determined by three indices, one referring to the independent vertical axis and the others to two of the lateral axes; but to determine its position in the form, a fourth index, referring to the third lateral axis, is required, giving a general symbol of the form $\{h\,k\,l\,i\}$, in which the position of the last letter, referring to the vertical axis, is invariable, while the other three are interchangeable with positive and negative signs, subject to the condition that their algebraical sum is always equal to zero, or $h + k + l = 0$. This property will be apparent from fig. 99, where Ox, Oy, and Oz represent the distances

FIG. 99.

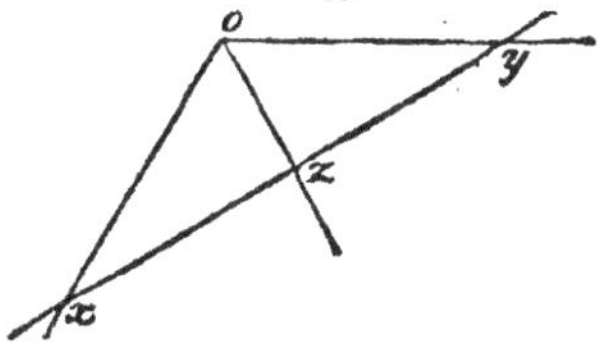

at which a right line cuts three axes, making angles of 60° to each other conjointly at O.

For,

$$\Delta\, O x z + \Delta\, O z y + \Delta\, O y x = 0,$$

or,

$$O x \,.\, O z \sin . x O z + O z \,.\, O y \,.\, \sin . z O y + O y\, O x \sin . y O x = 0;$$

but since

$$x O z = 60^\circ,\ z O y = 60^\circ,\ y O x = 120^\circ,$$

the sines are all equal. Dividing them out, and also dividing by the product $O x.\ O y.\ O z$, we obtain—

$$\frac{1}{O x} + \frac{1}{O y} + \frac{1}{O z} = 0,$$

which quantities are represented by the three indices in the order given above. Further, if we consider the axes x and y as positive, z will be negative, and *vice versâ*; and as the intercept of the latter is the shortest, it will have the largest

FIG. 100.

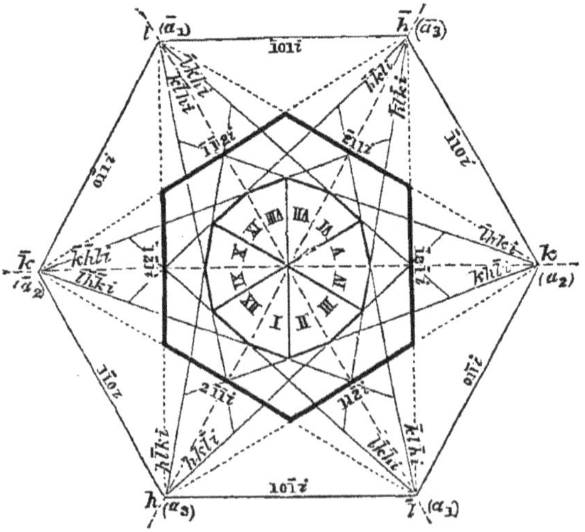

of the three indices, but with the opposite sign to that of the other two; so that the general symbol becomes $\{h k \bar{l} i\}$, in which the first two indices are independent, the third being equal to their sum, but with the opposite sign. The number of permutations of letters and signs satisfying these conditions is twelve, where order is shown by the twelve divisions numbered like a clock-face, making up the interior polygon in the diagram, fig. 100. They represent the horizontal projection of the faces corresponding to any value of i, and when this is greater than 0 there will be twelve similar faces below the horizontal plane having $\bar{i}$ in their symbols, the whole making up the twenty-four-faced solid known as a dihexagonal pyramid, which is the general representative form of the system. These symbols are given in full on page 88.

In Weiss's system, the lateral axes are noted as $a_1\ a_2\ a_3$, their positions being also shown in fig. 100. According to this, the symbols determining a face, those of the two independent lateral and the vertical axes, are $a : na : mc$, while that of the third lateral axis, which determines its position, is $\frac{n}{n-1}a$; the full symbol therefore will be $a : na : \frac{n}{n-1}a : mc$,[1] the last term referring to the vertical axis; but in order to express the relation between these symbols and the Bravais notation it is necessary to invert the order of the first three and change the sign of the parameter corresponding to l, which gives the form $\frac{n}{n-1}a : na : a : mc$, subject to which alteration the indices in $\{hk\bar{l}i\}$ will be the reciprocals of Weiss's symbols. Naumann's contracted symbol is mPn, which is of analogous signification to that of the hexakisoctahedron in the cubic system, with the dif-

[1] A notation by indices corresponding to this order, of the form ξhkl, in which h and k refer to the two independent lateral axes ξ, which is always $= h - k$, to the third, and l to the vertical axis, is adopted in Groth's treatise. This has the advantage of using the index letters in the same general order as in the other systems, but it has not been adopted in any subsequent work.

ference that *P*, the initial of pyramid, is substituted for *O* as the type-unit form ; *m* is the parameter coefficient of the vertical, and *n* that of the longer lateral axis.

The lateral axes are related to the vertical in some constant ratio expressed by $a : c$, where if $a = 1$, c is always some irrational number greater or less than, but never equal to, unity. This is known as the *fundamental, or unit-axial ratio*, and the form from which it is derived as the *unit-form* of the species, as in all other forms of the same substance the lengths of lateral axes are expressed as rational multiples or submultiples of unity, and that of the vertical as similar rational modifications of *c*. The particular ratio is, however, only constant for the crystals of the same substance ; and therefore, unlike the cubic system, in which the forms constitute only a single series, there are as many series of hexagonal forms as there are different minerals crystallising in that system, which is also true for all the remaining systems.

The dihexagonal pyramid is represented in perspective

FIG. 101.

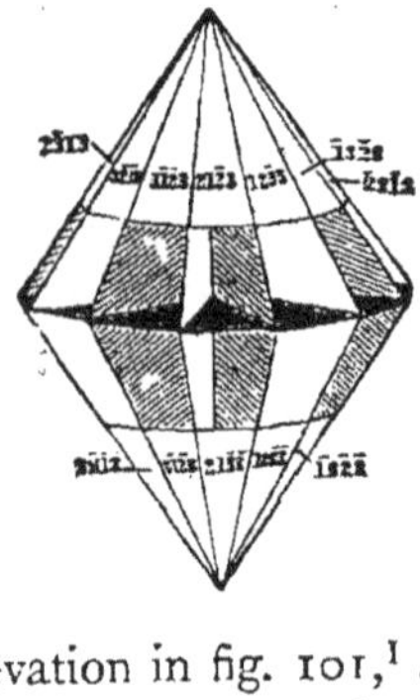

FIG. 102.

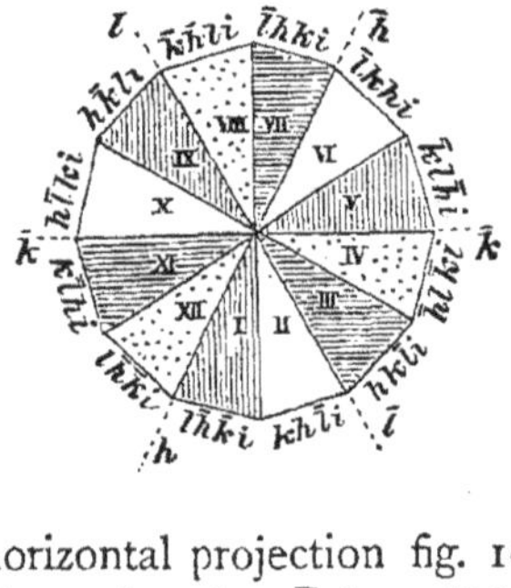

elevation in fig. 101,[1] and in horizontal projection fig. 102, the first being noted as an imaginary form $\{1\ 2\ \bar{3}\ 3\}$ or $P\frac{3}{2}$, in

[1] This is intended to represent the effect of a model whose faces are a mere skin covering the edges of the planes of symmetry which are exposed by the removal of the middle portion of the covering. The principal plane is indicated by strong, and the lateral axial planes by light, shading ; the interaxial planes are white.

which $a : c = 1 : 1 \cdot 1$, while the second has the general notation, the faces being further shaded in accordance with the scheme of triads in the table on page 88. Its twenty-four faces meet in thirty-six edges of three different kinds—namely, twelve middle or basal edges in the principal plane of symmetry, and twenty-four terminal or polar edges, which lie alternately in the lateral, axial, and interaxial planes, and meet the vertical axis in a twelve-faced solid angle at an equal distance on either side of the basal plane; their other extremities form with the basal edges twelve four-faced middle or basal solid angles. The dihedral angles over the basal edges are all similar; those over the polar edges are alternately larger and smaller, but never equal, as that requires the basal section to be a regular dodecagon,[1] which is excluded as giving for n the irrational value $\sqrt{2}$ sin. 75° $= 1 \cdot 3666 \ldots$ For any value of n lower than this, the more obtuse polar edges lie in the interaxial planes, and $n = 1$, their angle $= 180°$, or the twelve faces are reduced to six, giving the normal hexagonal pyramid, or that of *the first order*, shown in elevation in fig. 103[2] and in plan in fig. 104, and in the outer hexagonal figure of fig. 100. This is contained by twelve isosceles triangles, forming a six-faced pyramid on either side of the regular hexagonal base. The dihedral angle over a basal edge is equal to twice the oblique angle at the base of a right-angled plane triangle whose perpendicular and base are respectively the vertical

[1] By constructing this solid, or, what is easier, projecting it on its base, it will be seen to be symmetrical about twelve instead of six lateral axes, and to have twelve- instead of six-part symmetry about its principal axis, or to have higher symmetry than that defined above as characteristic of the system. This illustrates the statement on page 9, that solids of higher than senary symmetry are excluded by the principle of rationality.

[2] The edges of the inte axial planes are shown as dotted lines, to signify that these are no longer effective edges of form, but only of symmetry.

and a lateral interaxis, and if the former be called c, the latter a', and the observed basal angle β, we have :

$$\frac{c}{a'} = \tan. \frac{\beta}{2}.$$

But a' is inclined at 30° to an adjacent lateral axis a; therefore when a is made $=1$, $a' = \frac{1}{2}\sqrt{3}$ and $a : c = 1 : \frac{1}{2}\sqrt{3} \tan. \frac{\beta}{2}$, which corresponds to the fundamental ratio of the species, if the particular pyramid measured be assumed

FIG. 103.

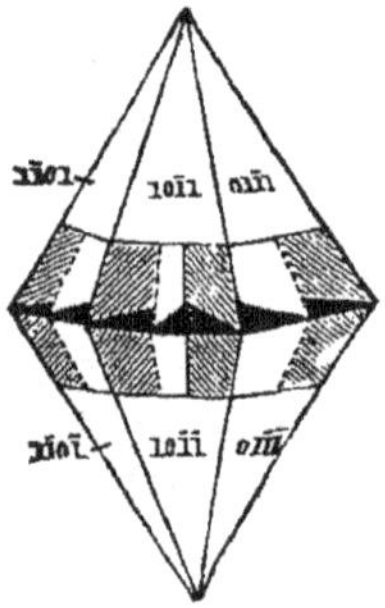

FIG. 104.

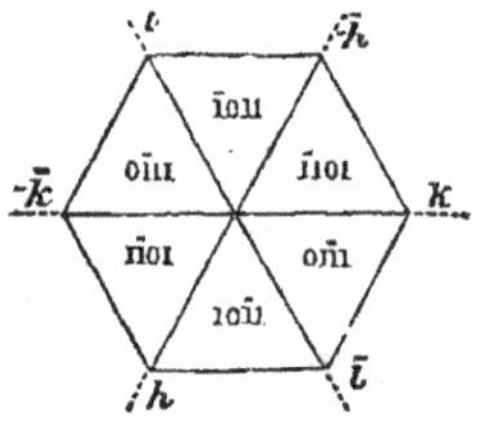

as the unit form of the series. The same quantity may also be found from the dihedral angle over a polar edge by the expressions :

Sin. $\xi = \text{cotan.} \frac{\gamma}{2} \sqrt{3}$ and tan. $\xi = m\,c$, or $= c$ when $m = 1$, where $\gamma =$ the measured angle and ξ the inclination of that edge to the lateral axis lying in the same plane with it.

The hexagonal pyramid is the particular form of $\{h\,k\,\bar{l}\,i\}$, in which $h = 1$ and $k = 0$, whence $\bar{l} = \bar{1}$. If i also $= 1$, the general symbol becomes $\{1\,0\,\bar{1}\,1\}$, or that of a face parallel to the right and left axis k. To obtain uniformity with other systems it is customary, however, to take not this, but the face meeting the positive pole of that axis, or

$\{0\bar{1}11\}$, as representative of the form. The notation of the six upper faces, in which i is positive, is given in figs. 104 and 100. Weiss's symbol is $\infty a : a : a : c$, and Naumann's nP or P when $n = 1$ signifying the unit form.

When n in the dihexagonal pyramid is greater than 1.366. . . ., the more obtuse polar edges are those in the lateral axial planes, their dihedral angles becoming 180° when $n = 2$. This gives the hexagonal pyramid of the *diagonal position* or *second order*, seen in elevation in fig. 105 and plan in fig. 106, and the strong-lined hexagon in fig. 100. It has the same general geometrical properties as

FIG. 105.

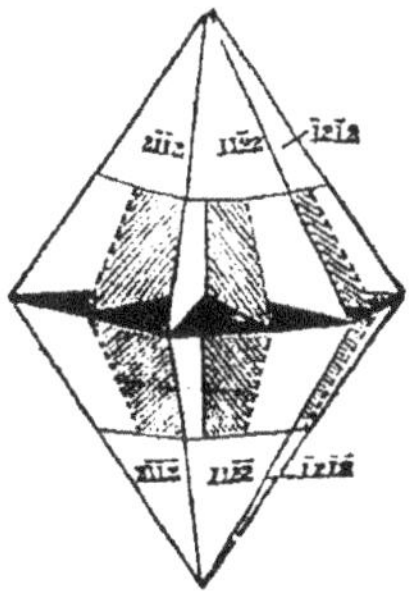

FIG. 106.

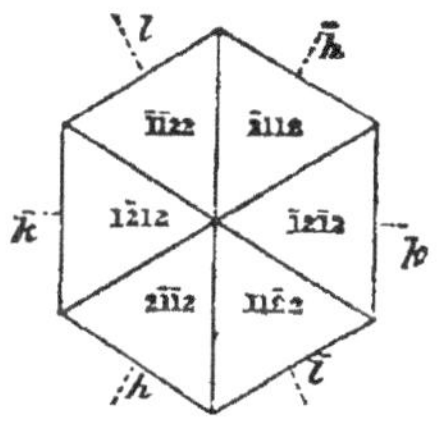

the pyramid of the first order from which it differs in the position and relative lengths of the axes; half the measured angle across a basal edge gives the oblique angle at the base of a right-angled triangle whose base and perpendicular are in the ratio of $a : c$.

Any face cuts two adjacent positive semi-axes at the same, and the intermediate negative one at half that distance or $h = k = 1$, whence $\bar{l} = \bar{2}$. The general symbol is therefore $\{11\bar{2}i\}$, or for the form with the unit vertical axis $\{11\bar{2}2\}$, corresponding to Weiss's $2a : 2a : a : c$. Naumann's general symbol is $mP2$ or $P2$ for the unit form. The full notation of the upper half is given in fig. 106.

If the length of the vertical axis in either class of

pyramid be varied by multiplying its unit-value by any rational coefficient, *m*, which may be either greater or less than unity, other pyramids upon the same base will be obtained which will be steeper, or their basal angle will increase proportionately with that of *m*, and *vice versâ*, as in fig. 107, representing three normal hexagonal pyramids of different altitudes. If the middle one of these be regarded

FIG. 107.

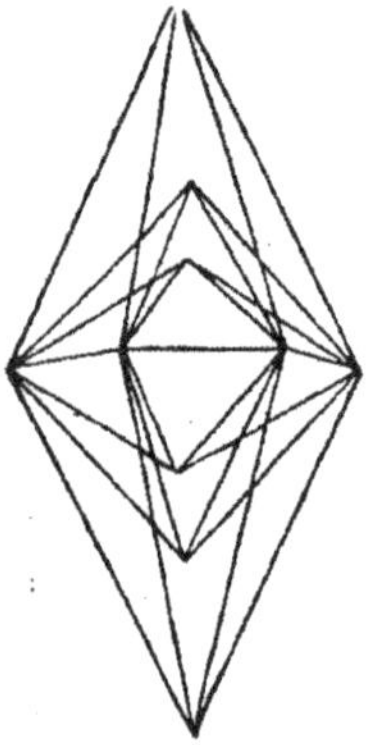

FIG. 108.

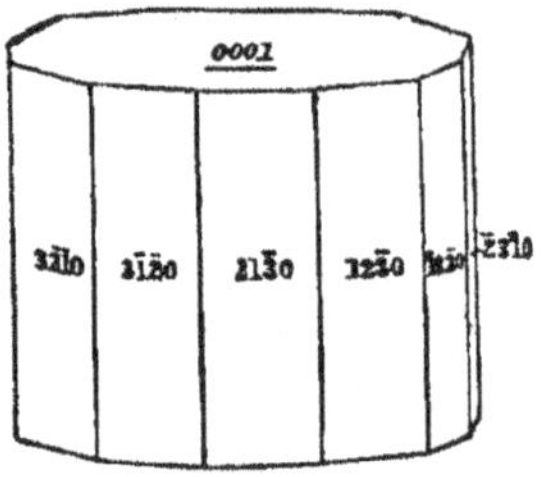

as P or $\{0\,1\,\bar{1}\,1\}$, the outermost will be $2\,P$ or $\{0\,2\,\bar{2}\,1\}$, and the innermost $\frac{1}{2}\,P$ or $\{0\,\bar{1}\,1\,2\}$; if the latter is considered as P the others will be $2\,P$ and $4\,P$ or $\{0\,4\,\bar{4}\,1\}$; and lastly, if the outermost is the unit, the others will be $\frac{1}{2}\,P$ and $\frac{1}{4}\,P$ or $\{0\,1\,\bar{1}\,4\}$ respectively. When $m = \infty$ or $i = 0$ the basal angles become 180°, and as those of the polar edges change in the reverse order they diminish until they correspond to the plane angle of the base; or, in other words, the pyramid becomes a prism. The forms of this class corresponding to the three kinds of pyramids are represented in figs. 108, 109, and 110. The first of these, derived from fig. 101, is a dihexagonal prism contained by twelve similar rectangular faces meeting in edges parallel to the vertical axes at angles which are alternately greater and less than 150° (that being the angle of a regular dodecagon, which is excluded by the irrationality of its parameters). The symbols, as will be apparent from the derivation, are

$$\{h\,k\,\bar{l}\,0\},\ \frac{n}{n-1}a : n\,a : -a : \infty\,c, \text{ and } \infty P n.$$

Fig. 109 is the hexagonal prism of the first order, contained by six faces meeting at 120° in the lateral axial

FIG. 109.

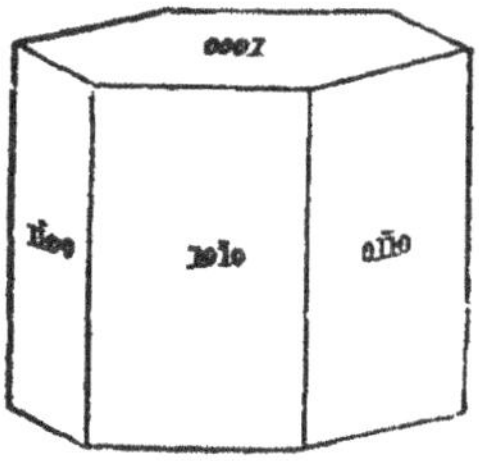

FIG. 110.

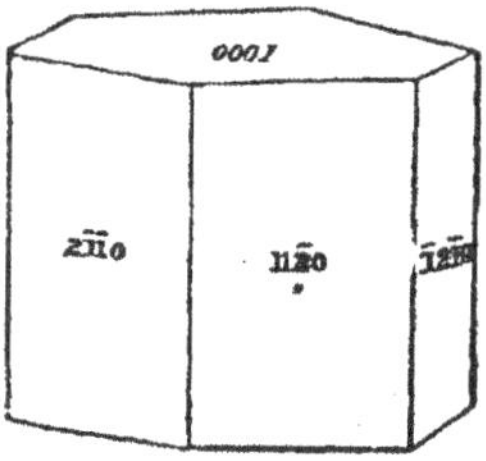

planes, or its base is a regular hexagon similar in position to that of the unit-pyramid. It is represented by

$$\{0\,1\,\bar{1}\,0\},\ \infty a : a : -a : \infty c, \text{ and } \infty P.$$

Fig. 110 is the hexagonal prism of the second order, derived from the corresponding pyramid, fig. 105. It is also on a regular hexagonal base, its edges lying in the lateral interaxial planes, and is represented by

$$\{1\,1\,\bar{2}\,0\},\ 2a : 2a : -a : \infty c, \text{ and } \infty P2.$$

As all the faces of prisms lie in single zones which are unlimited in the direction of their axes, they cannot of themselves form complete crystals, but can only appear in combination. These are said to be *open forms*, and are common in all except the cubic systems, where the simple forms are necessarily closed.

When m is less than 1, the basal edges of the pyramid become sharper and the polar ones blunter than those of the unit form ; and when $m = 0$ the angles of the former are 0° and of the latter 180°, or the whole of the faces meeting the vertical axes at either extremity fall into the same surface, and the form is reduced to the single unlimited plane containing the lateral axes. When, however, this is shifted from the central position, it will intercept some length upon the vertical axis, and require a corresponding parallel face on the opposite side of the origin ; the complete form is therefore represented by two faces parallel to the basal section. This, known as the *basal* or *terminal pinakoid*, is another form only possible in combination, having no proper geometrical form, its shape being conditioned by the edges formed in combination. For instance, in fig. 108, it is dihexagonal, in fig. 109 hexagonal of the first order, and in fig. 110 of the second, these differences being obviously due to the different prisms with which it is combined. The symbols are :

$$\{0001\}, \infty a : \infty a : \infty a : c, \text{ and } 0P.$$

The seven classes of forms described above—namely, the three pyramids, with their corresponding prisms and the terminal pinakoid, are the only kinds possible with full hexagonal symmetry to seven axes, as will easily be seen by giving special values to the indices in the general symbol $h\,k\,\bar{l}\,i$, and working them out subject to the condition $h + k + l = 0$. Their relation to each other is best seen in Naumann's diagram, formed by arranging their symbols as in the margin. The first vertical line contains the hexagonal pyramids of the first order ; the second, the dihexagonal pyramids ; and the third, the pyramids of the second order. The unit forms of these three classes are in the third horizontal line ; their obtuse modifications

$$\begin{array}{ccccc}
0P & & 0P & & 0P \\
\vdots & & \vdots & & \vdots \\
\frac{1}{m}P & \ldots & \frac{1}{m}Pn & \ldots & \frac{1}{m}P2 \\
\vdots & & \vdots & & \vdots \\
P & \ldots\ldots & Pn & \ldots\ldots & P2 \\
\vdots & & \vdots & & \vdots \\
mP & \ldots & mPn & \ldots & mP2 \\
\vdots & & \vdots & & \vdots \\
\infty P & & \infty Pn & & \infty P2
\end{array}$$

having $m < 1$, or generally a proper fraction expressed as $\frac{1}{m}$ in the second; the fourth line contains the more acute pyramids having $m > 1$; the fifth, the corresponding infinitely acute form or prisms, and the top line the common limit of obtuse form, or the terminal pinakoid. Any one of these forms lies in the same zone with any other in the same line, whether horizontal or vertical. In the use of this scheme it must always be remembered that the rational coefficient m applies not to the natural unit number 1, but to the fundamental arbitrary ratio $a : c$. From what has been said concerning the properties of pyramids of varying altitudes, it will be easily seen that the basal angles vary more rapidly than those over the polar edges, as they may range from 0° to 180°, while the latter can only vary between 120° and 180°. In the description of hexagonal minerals, therefore, the angle of the basal edge is usually given as that characteristic of the species, and determining the ratios $a : c$ most accurately.

FIG. 111.

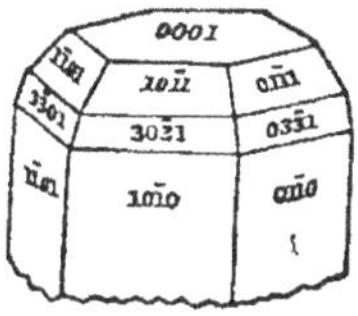

FIG. 112.

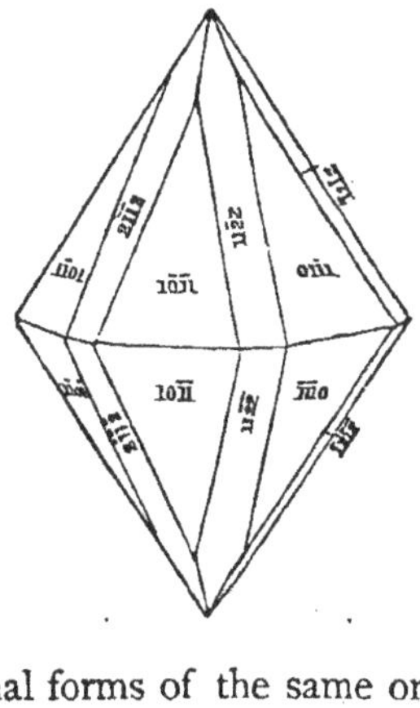

In combinations of hexagonal forms of the same order, the more acute, or those with the highest m or lowest i, bevel the basal edges of the more obtuse, and, conversely, the latter acuminate or blunt the polar summits of the former. The steepest form of any series, the prism, and the flattest, the terminal pinakoid, respectively truncate the basal edges and polar summits of any pyramid, as in fig. 111, containing ∞P, $3P$, P, $0P$, or $\{01\bar{1}0\}$, $\{03\bar{3}1\}$, $\{01\bar{1}1\}$, $\{0001\}$.

FIG. 113.

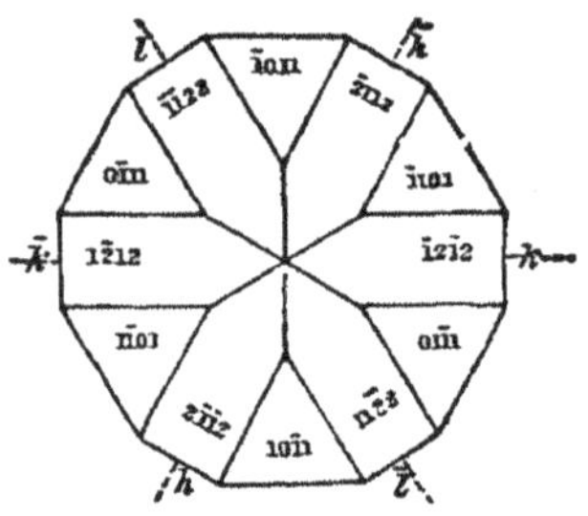

FIG. 114.

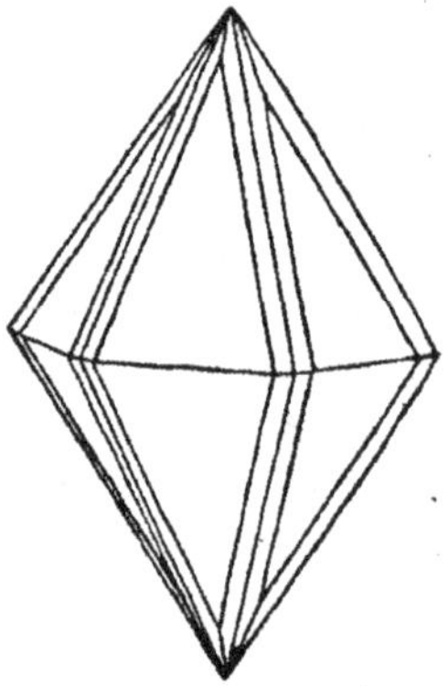

In combinations of forms of different orders, having m in common, or of the same altitude upon different bases,

FIG. 115.

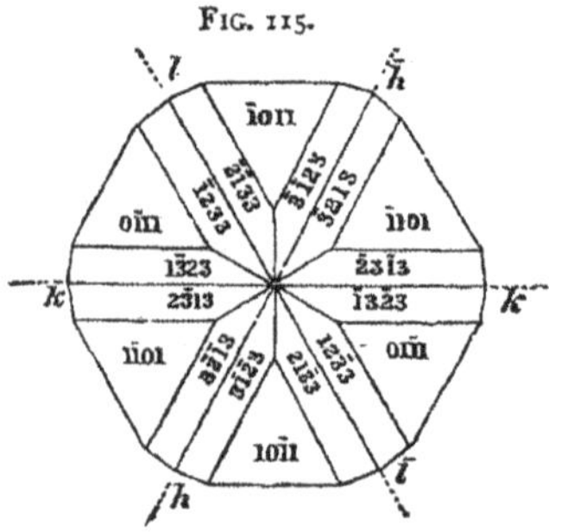

FIG. 116.

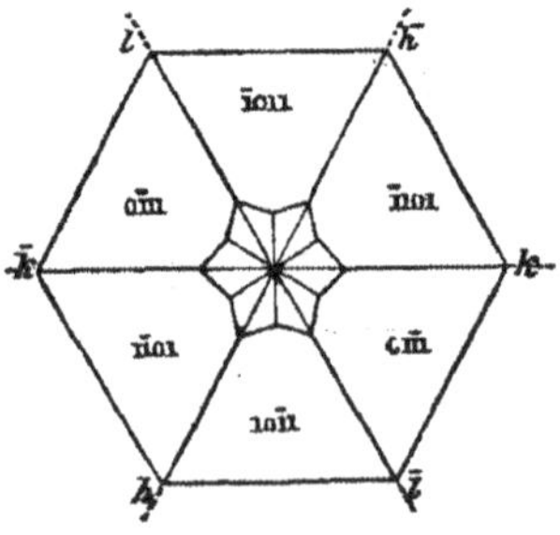

$m P 2$ truncates the polar edges of $m P$, and the latter bevels the basal solid angles of the former, as in figs. 112, 113, while $m P$ truncates the edges of $m P n$ lying in the lateral interaxial planes (figs. 114,[1] 115), and $m P 2$, those in the lateral axial planes, the relations of the three classes of prisms being similar to those of the corresponding pyramids.

In combinations of forms in which both m and n, or base and altitude, are different, $m P n$, when combined with a more acute hexagonal pyramid of either order, appears as an obtuse twelve-faced point upon the polar summit of the latter, as seen in plan (fig. 116), for $P \frac{1}{m} Pn$. A pyramid of the second order upon a more acute form of the first forms an analogous six-faced point (fig. 117), the plan of $n P$, $P 2$. In the reverse condition, when the pyramid of the second

FIG. 117.

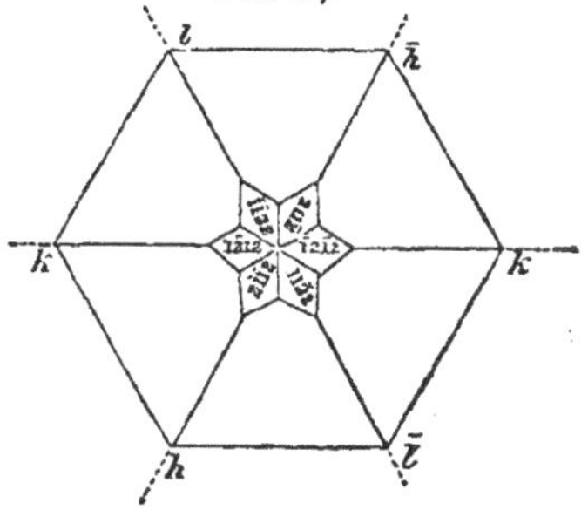

FIG. 118.

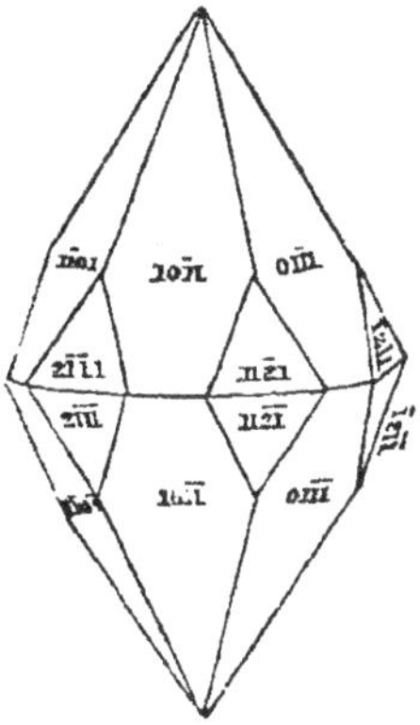

FIG. 119.

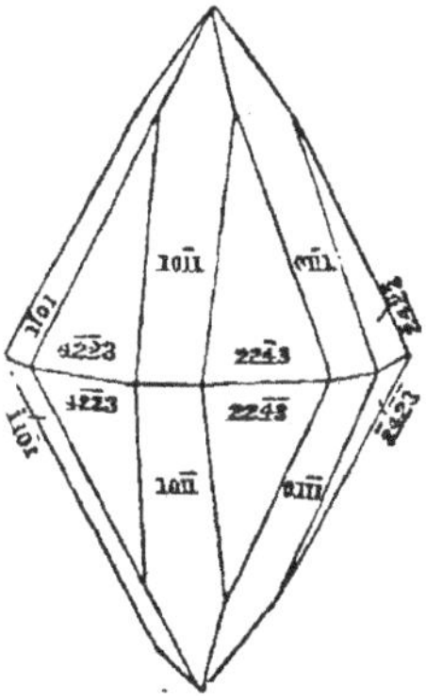

order is the more acute form, its faces modify the solid angles in the basal section of that of the first order, the

[1] This is noted as $P P \frac{3}{2}$, or $\{0 1 \bar{1} 1\}$ $\{1 2 \bar{3} 3\}$.

aspect of such combinations varying with the differences in the parameters of the two forms. For instance, in fig. 118 the faces of $2\,P\,2$ or $\{1\,1\,\bar{2}\,1\}$ make new edges, which are parallel to the polar edges of P or $\{0\,1\,\bar{1}\,1\}$; while in fig. 119 the combination edges are parallel to the polar ones of the replacing form[1] $\frac{4}{3}\,P\,2$, or $\{2\,2\,\bar{4}\,3\}$. Lastly, when $m = \infty$, the basal solid angles of the less acute form are truncated, or, in other words, the prism of either order truncates the basal solid angles of the pyramid of the other, as in fig. 120, which is noted as $P, \infty P\,2, 0\,P$, but would do equally well for $P\,2, \infty\,P\,0\,P$, if the order of the first two constituents be supposed to be reversed. The above include all the simpler cases of holohedral hexagonal combinations, but they are not very commonly met with in nature, there being but few species in which full hexagonal symmetry prevails, and these are generally remarkable for the large number of forms present, the most of which are, however, as a rule, very subordinate to the dominant form, generally a prism. A few examples of such combinations, which are usually best represented in horizontal projections, will be found in the volume on 'Descriptive Mineralogy.'

FIG. 120.

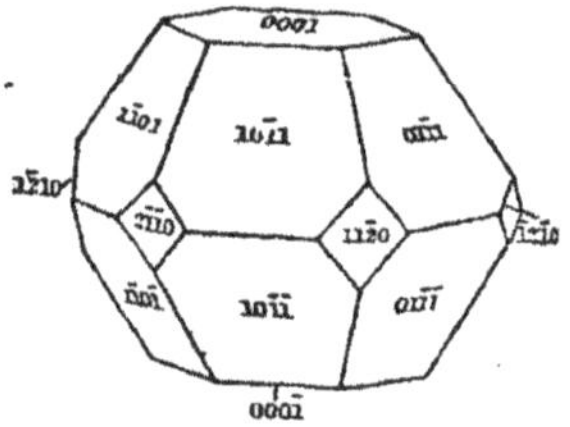

Hemihedral hexagonal forms. The faces of the dihexagonal pyramid divide symmetrically into four groups of six, which, as the corresponding faces above and below the basal section differ only in the sign of their fourth index, may be represented by the four triads indicated by different shadings in the horizontal projection (fig. 102), corresponding to the following table, in which the faces are

[1] In other words P truncates the polar edges of $\frac{4}{3}\,P\,2$, a relation which is very commonly observed in natural crystals.

numbered in order from left to right, commencing on the right side of the positive semi-axis h :—

A	B	C	D
I. $l \bar{h} \bar{k} i$	II. $k h \bar{l} i$	III. $h k \bar{l} i$	IV. $\bar{h} l \bar{k} i$
V. $\bar{k} l \bar{h} i$	VI. $\bar{l} k h i$	VII. $\bar{l} h k i$	VIII. $\bar{k} \bar{h} l i$
IX. $\bar{h} k l i$	X. $h \bar{l} k i$	XI. $k \bar{l} h i$	XII. $l \bar{k} \bar{h} i$
E	**F**	**G**	**H**
XIII. $l \bar{h} \bar{k} \bar{i}$	XIV. $k h \bar{l} \bar{i}$	XV. $h k \bar{l} \bar{i}$	XVI. $\bar{h} l \bar{k} \bar{i}$
XVII. $\bar{k} l \bar{h} \bar{i}$	XVIII. $\bar{l} k h \bar{i}$	XIX. $\bar{l} h \bar{k} \bar{i}$	XX. $\bar{k} \bar{h} l \bar{i}$
XXI. $\bar{h} k l \bar{i}$	XXII. $h \bar{l} k \bar{i}$	XXIII. $k \bar{l} h \bar{i}$	XXIV. $l \bar{k} \bar{h} \bar{i}$

This may be halved symmetrically in the three following ways, each corresponding to a possible case of hemihedrism :

1	A C F H	and	B D E G
2	A B G H	and	C D E F
3	A C E G	and	B D F H

In the first case, one form contains uneven numbered faces above alternating with even numbered ones below, and the other even ones above and uneven below. This is known as *plagihedral* or *trapezohedral hemihedrism* ; in the second, or *rhombohedral hemihedrism*, the grouping is by alternate pairs of faces above and below the basal section, or the 1st,

3rd, and 5th, with the 8th, 10th, and 12th pairs; and the 2nd, 4th, and 6th with the 7th, 9th, and 11th pairs. In the third, or *pyramidal hemihedrism*, the grouping is symmetrical to the base, one form containing all even and the other all odd numbered faces.

Of the two most obvious methods of dividing the table into upper and lower, and right and left halves, the first is excluded by not satisfying the conditions of symmetry, giving forms having the indices of the vertical axis either all positive or all negative. The second, a one-sided distribution of faces, has been called *trigonotype* hemihedrism by some writers, while others say that it does not possess true hemihedrism. This is possibly rather a question of terms than of fact. It need not be discussed, as the form does not occur in minerals.

Trapezohedral hemihedrism. The dihexagonal pyramid, when divided hemihedrally in the manner shown in fig. 122, corresponding to the first case, as defined above, gives rise to the two hemihedral forms (figs. 121 and 123), the first

FIG. 121.

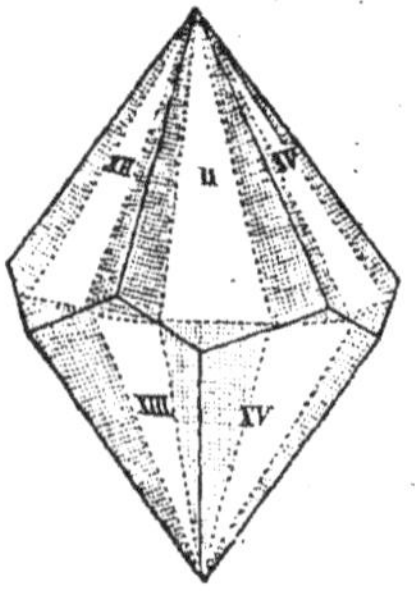

FIG. 122.

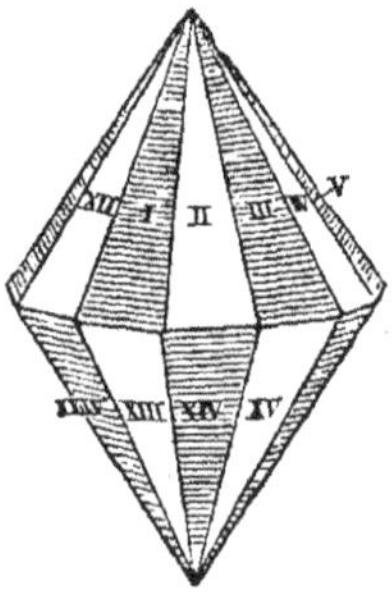

FIG. 123.

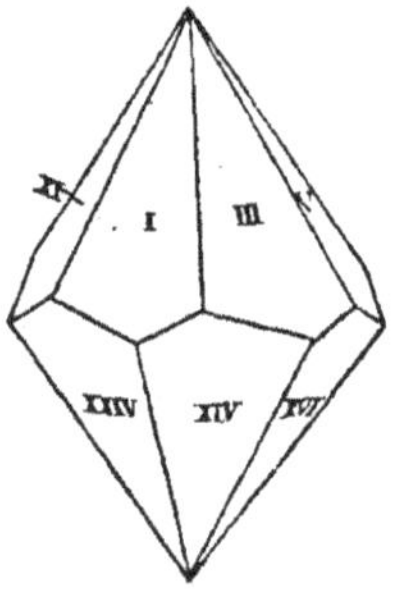

from the white, and the second from the shaded faces of fig. 122. Fig. 124 is a horizontal of projection of fig. 123; the faces are numbered according to the table on page 89, the edges below the horizontal plane being shown in dotted lines. These known as hexagonal trapezohedra are contained by twelve trapeziform faces meeting in twelve similar polar and twelve dissimilar middle edges, the latter being alternately longer and shorter. Like the plagihedral hemihedra of the cubic system, they have the same number and kind of axes of symmetry as the holohedral forms, but no planes of symmetry, and are therefore non-superposable.

FIG. 124.

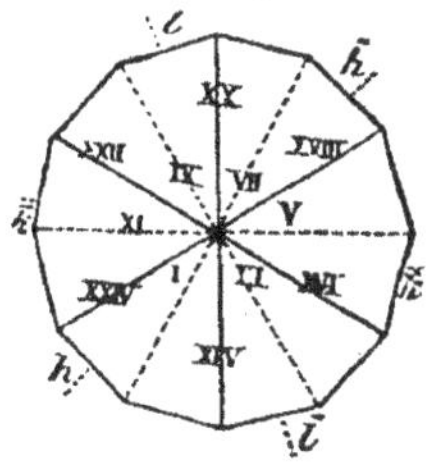

As a face of a hexagonal pyramid of either order or a dihexagonal prism contains two, that of a hexagonal prism four, and the terminal pinakoid six, faces of the general form $\{h\,k\,l\,i\}$, it is clear that none of these holohedral forms will be geometrically changed by this kind of hemihedry, which is therefore only effective in producing new forms in the dihexagonal pyramid. No examples of crystals of this kind of hemihedrism, whether of natural or artificial origin, are known, so that as yet they only represent a geometrical possibility. It may be represented by the symbols

$$\kappa''\ \{h\,k\,l\,i\},\ \tfrac{1}{2}\left(\frac{n}{n-1}a^{1} : na : -a : mc\right)^{1}$$

and

$$\frac{mPn}{2}r,\quad \frac{mPn}{2}l.$$

Rhombohedral hemihedrism. The second method of hemihedral division of the dihexagonal pyramid, that by alternate pairs of faces above and below the base, as in fig. 126, produces from the white faces fig. 125, and from the shaded ones fig. 127. The corresponding horizontal projections are seen in figs. 128 and 129. A form of this class, known as a

[1] The hemihedrism is here shown by the prefix ½ to the holohedral symbol without special indication of the particular case meant, and is therefore general for all kinds.

scalenohedron, a contraction of scalene dodecahedron, is contained by twelve faces, all similar scalene triangles meeting

FIG. 125.

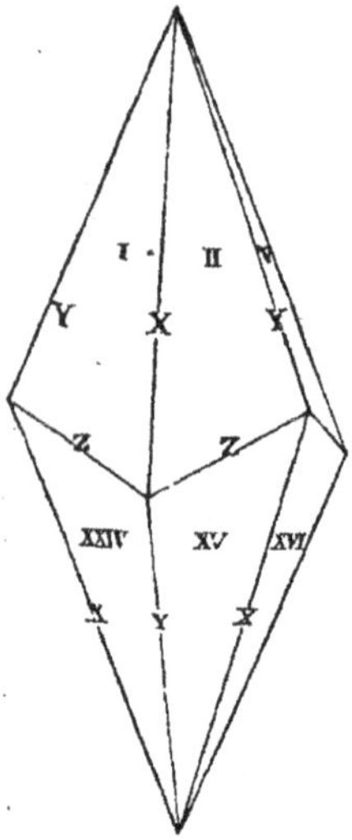

FIG. 126.

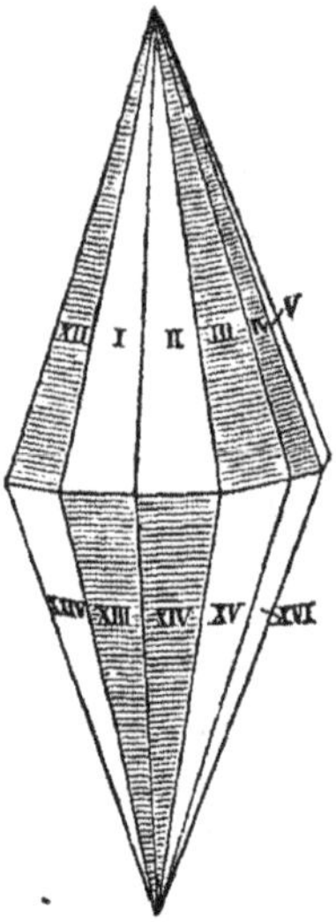

FIG. 127.

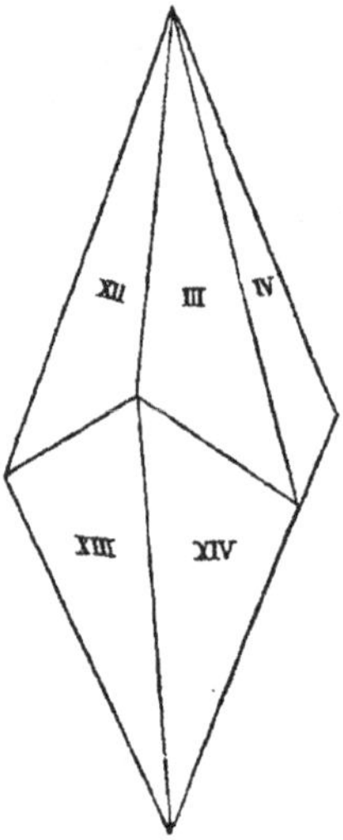

in two kinds of polar edges, six longer or more obtuse (X), six shorter or more acute (Y), and six middle edges (Z), lying in zigzag order about the basal section.

From the derivation of this form it will be apparent that

FIG. 128.

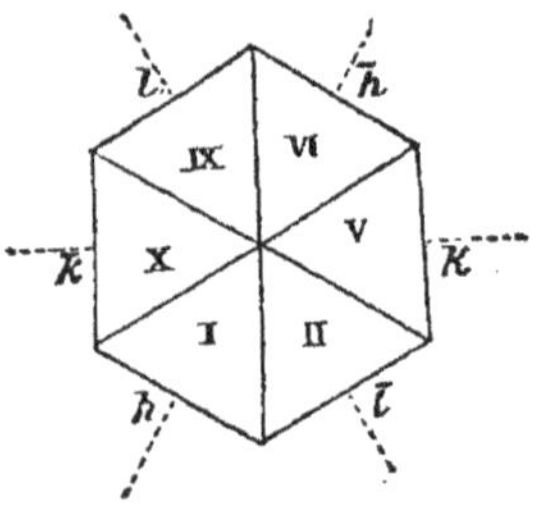

FIG. 129.

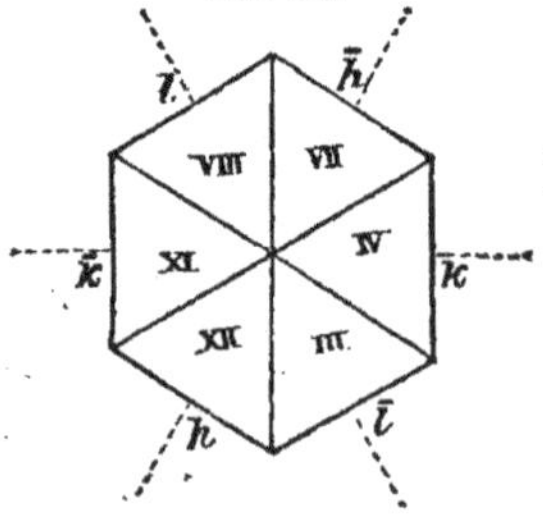

the longer polar edges have the same character as in the holohedral form, and that new shorter ones (Y) lie in the same transverse sections with them—that is, in the lateral interaxial planes, which are therefore planes of symmetry, while the lateral axial planes are not; and from the obliquity of the middle edges (z) to the original basal section, the latter cannot be a plane of symmetry. The lateral crystallographic axes being normals to the interaxial planes, they will be axes of binary symmetry, and as these planes also intersect in the vertical axes, the latter will also be an axis of symmetry, but reduced from hexagonal to ternary. We therefore have ternary symmetry about the principal axis, binary about the lateral crystallographic axes, and three planes of symmetry inclined at 60° to each other as characteristics of this class of hemihedrism.

The two scalenohedra derivable from the same dihexagonal pyramid are superposable—that is, either may be brought into coincidence with the other by rotation through 60° or 180° about the vertical axis. They are distinguished as positive and negative, or direct and inverse forms, according to position. The choice of position is, however, arbitrary, and usually depends upon structural peculiarities.

The general symbols of the scalenohedron are:

$$\kappa\ \{h\,k\,l\,i\} \text{ and } +\frac{m\,P\,n}{2},\ -\frac{m\,P\,n}{2}.$$

The same selection, applied to the hexagonal pyramid of the first order, as in fig. 131, produces the two hemihedral forms, figs. 130 and 132—the former from the white and the latter from the shaded faces. Figs. 133 and 134 are the corresponding horizontal projections. A form of this class, known as a rhombohedron, a contraction for rhombic hexahedron, is contained by six faces, all equal rhombs, meeting in three polar edges at either end of the vertical axis, and six middle edges in *zy* order about the basal section. The polar diagonals of the faces meeting the vertical axis, as shown in the dotted line in fig. 130, represent the obtuse edges (x) of the scalenohedron, as will be readily seen when it is remembered that the hexagonal pyramid is that particular dihexagonal pyramid whose faces meet at 180° in the obtuse polar edges. The symmetrical relations

FIG. 130.

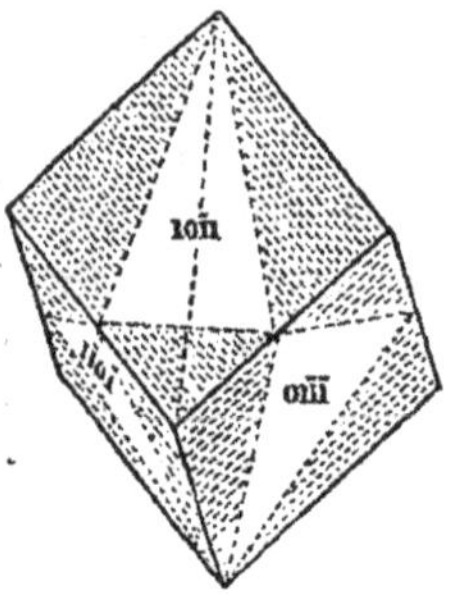

FIG. 131.

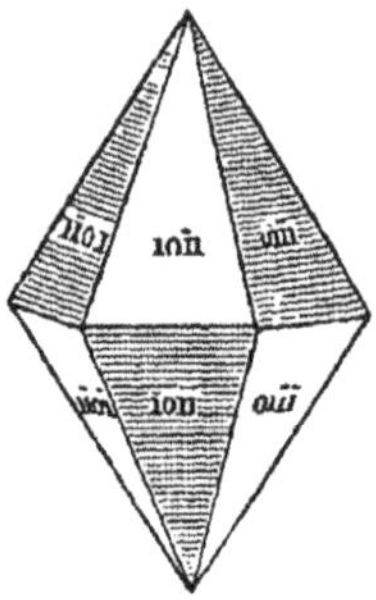

FIG. 132.

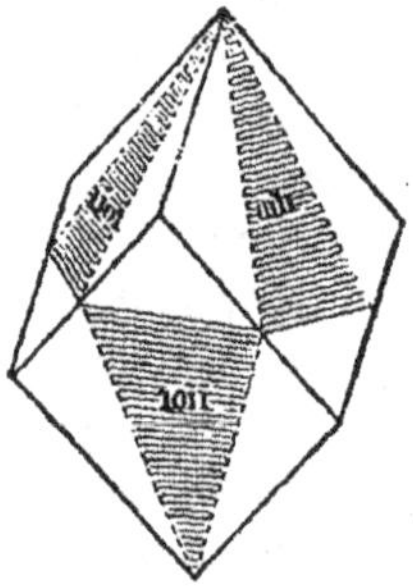

are therefore in every respect similar to those of the scalenohedron. The rhombohedron, being the most important form

FIG. 133.

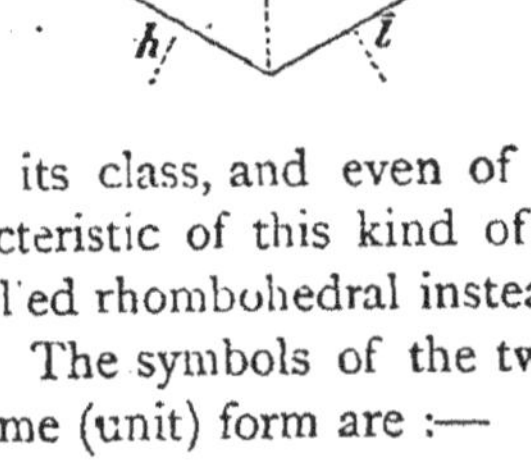

FIG. 134.

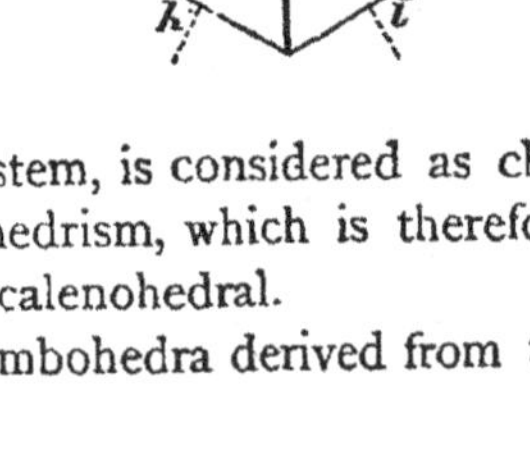

of its class, and even of its system, is considered as characteristic of this kind of hemihedrism, which is therefore called rhombohedral instead of scalenohedral.

The symbols of the two rhombohedra derived from the same (unit) form are :—

$$\kappa\{0 1 \bar{1} 1\}, \kappa\{1 0 \bar{1} 1\}, \text{and} + \frac{P}{2} - \frac{P}{2}.$$

The dihedral angles over the two kinds of edges in the rhombohedron are mutually supplemental, or one kind is as much above as the other is below 90°. When the larger angle is in the polar edges, the rhombohedron is *obtuse*, but when it is in the middle edges it is *acute*. The form occupying the middle position, or having the angles of both polar and middle edges right angles is the cube, which is a possible rhombohedron, deriving from the hexagonal pyramid, having $a : c = 1 : 1 \cdot 2247$; and although it is not known to exist in nature, there are several forms very nearly approaching it. This is an example of what are called limiting forms, where the same geometrical solid may arise in two systems. In this case it is obviously possible, from the circumstance that the ter-quaternary symmetry of the cube includes the lower ter-binary kind of the rhombohedron, and the real test of the nature of the form is to be looked for not in the presence of the lower, but the absence of the higher symmetry, which is usually apparent in the character of its combinations; but assuming it to appear as a simple form, its true nature could only be determined by optical or other investigation of the structural peculiarities of the substance.

The symbols of the two positive and negative, or direct and inverse, rhombohedra, originating from the same hexagonal pyramid, where the latter is a unit form, are

$$\kappa\{0\,1\,\bar{1}\,1\},\ \kappa\{1\,0\,\bar{1}\,1\},\ \text{and}\ +\frac{P}{2}.\ -\frac{P}{2};$$

or, generally,

$$\kappa\{0\,0\,\bar{1}\,i\},\ \kappa\{1\,0\,\bar{1}\,i\},\ \text{and}\ +\frac{mP}{2}.\ -\frac{mP}{2}.$$

Like the scalenohedra they are superposable, either one being brought into coincidence with the other by rotation through 60° or 180° about the vertical axis. From their derivation it will be apparent that in combination they truncate each other's solid angles obliquely, as in fig. 135; and when the two are exactly balanced, the hexagonal pyramid (fig. 131) is reproduced.

A rhombohedron of either position and any length of vertical axis has its polar edges truncated by the faces of another of the same series of the opposite position, whose height is one half or breadth of base twice that of the first,

FIG. 135.

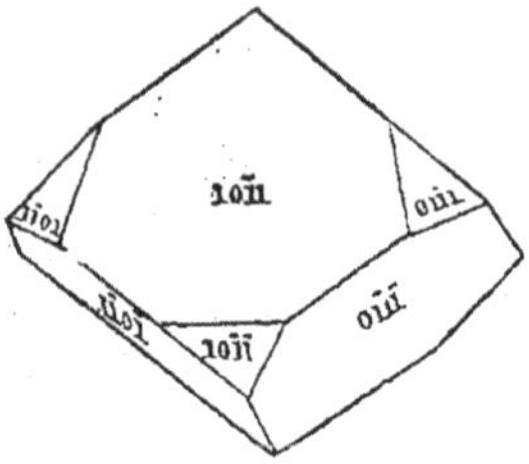

FIG. 136.

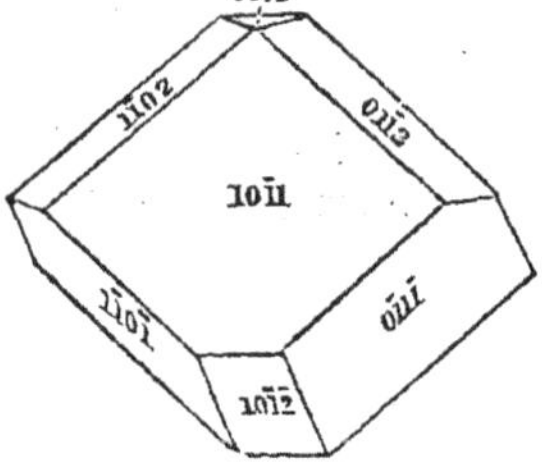

as in fig. 136. This case is commonly observed in calcite, whose principal rhombohedron $+\frac{P}{2}$ has its polar edges replaced by faces of $-\frac{\frac{1}{2}P}{2}$, and similarly modifies those of $-\frac{2P}{2}$, and in the same way $-\frac{\frac{1}{2}P}{2}$ is modified by $+\frac{\frac{1}{4}P}{2}$, $-\frac{4P}{2}$ by $+\frac{2P}{2}$, and so forth.

By varying the value of m in the symbol of a rhombohedron, a series of other forms of greater or less altitude upon the same base are obtained in the same way as with the pyramid. As m decreases, the angles of the middle edges, as well as their obliquity to the basal section, diminish as those of the polar edges increase, the latter becoming 180° when $m = 0$, producing the basal pinakoid. In the opposite direction the middle edges become more obtuse and increase their inclination to the basal plane; and when $m = \infty$ they fall into the same vertical lines with the polar edges, or the hexagonal prism ∞P is produced. The limits of the series of the rhombohedron are therefore the same as those of the hexagonal pyramid of the first order upon the same base.

The same kind of relation subsists in one direction

between the scalenohedron and its holohedral form, and therefore, by increasing the value of m in $\frac{mPn}{2}$, more acute forms upon the same base may be obtained up to a dihexagonal prism ∞Pn; but when m is diminished, the obtuse dihedral angle of the edges (X) increases more rapidly than that of the acute ones (Y) (fig. 125), and becomes 180° when the latter has still a measurable inclination, while that of the middle ones is unchanged. The twelve faces at either end are therefore changed into six, producing a rhombohedron, when the value of m is still far in excess of 0; or the inferior limit of the series of the scalenohedron is not the basal pinakoid, but a rhombohedron whose symbol is $\frac{\frac{m(2-n)}{n}P}{2}$, when that of the scalenohedron is $\frac{mPn}{2}$. As these forms have their middle edges in common, the rhombohedron is completely enclosed in the scalenohedron, as

FIG. 137.

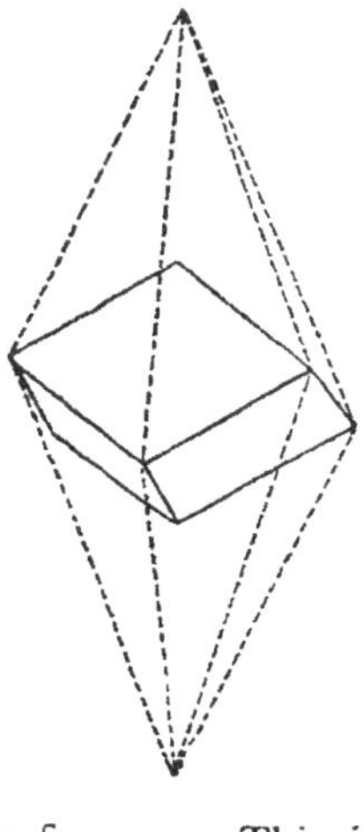

FIG. 138.

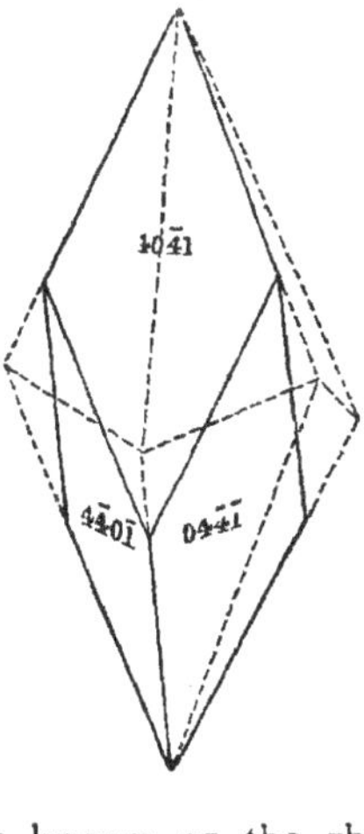

FIG. 139.

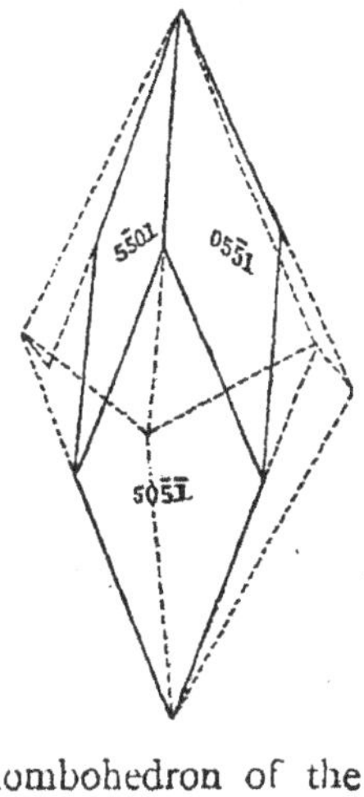

in fig. 137. This is known as the rhombohedron of the middle edges.

There are two other rhombohedra included in any scalenohedron, each having its polar edges parallel to one or other kind of the same edges in the latter form. The first of these, or rhombohedron of the shorter polar edges (fig. 138), is similar in direction to the scalenohedron, and is represented by the symbol $\frac{\frac{m(2n-1)}{n}P}{2}$, while the second, or rhombohedron of the longer polar edges (fig. 139), which is the most acute one that can be included, is inverse in direction, having the symbol $-\frac{\frac{m(n+1)}{n}P}{2}$, when $\frac{mPn}{2}$ is positive, and *vice versâ*. In this the length of the vertical axis is the sum of those of the other two, as will be seen by comparing the factors measuring this dimension in the three symbols, as—

$$n + 1 = (2n - 1) + (2 - n).$$

The special values of these three kinds of rhombohedra for the most commonly observed scalenohedra are as follows:

Scalenohedra	$\frac{3F\frac{3}{2}}{2}$,	$\frac{4P\frac{4}{3}}{2}$,	$\frac{5P\frac{5}{4}}{2}$,	$\frac{6P\frac{6}{5}}{2}$
Rhombohedra of middle edges .	$\frac{P}{2}$,	$\frac{2P}{2}$,	$\frac{3P}{2}$,	$\frac{4P}{2}$
Rhombohedra of shorter polar edges	$\frac{4P}{2}$,	$\frac{5P}{2}$,	$\frac{6P}{2}$,	$\frac{7P}{2}$
Rhombohedra of longer polar edges	$-\frac{5P}{2}$,	$-\frac{7P}{2}$,	$-\frac{9P}{2}$,	$-\frac{11P}{2}$

In the combinations of these four correlated forms, the most obtuse one, the rhombohedron of the middle edges, modifies the polar summits of the scalenohedron, producing blunt three-faced points, the new edges being parallel to the original middle edges, as in fig. 140, which represents the common scalenohedron of calcite, reduced by cleaving away its points, the faces produced by cleavage being those of the unit rhombohedron of the species. The rhombohedron of the shorter polar edges truncates the longer ones of the scalenohedron of the same direction; and that of the longer

FIG. 140.

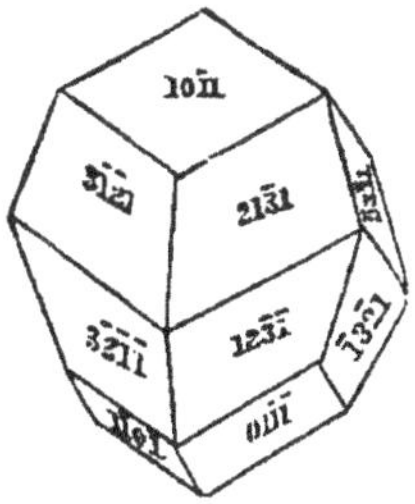

FIG. 141.

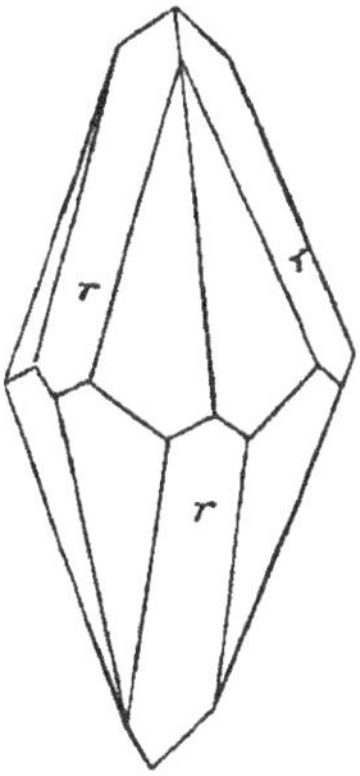

FIG. 142.

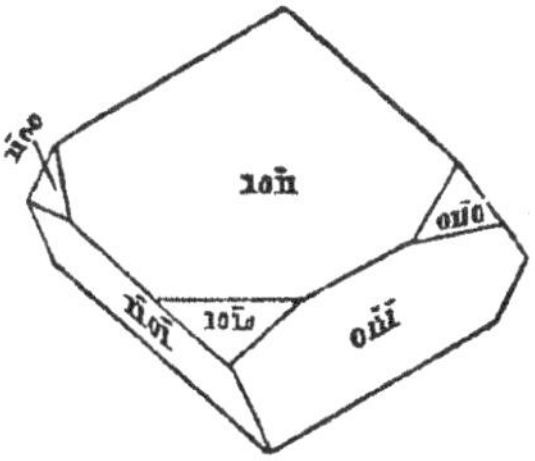

polar edges the shorter ones of the inverse scalenohedron, as in fig. 141, where the faces of *r r r* will be those of $-\frac{5P}{2}$, if the scalenohedron is considered as $+\frac{3P\frac{3}{2}}{2}$.

The pyramid of the second order, the three classes of prisms, and the basal pinakoid are not changed in appearance by rhombohedral hemihedrism.

In combination, the prism of the first order truncates the middle solid angles of rhombohedra and scalenohedra alike; in the first case the new faces are triangular planes, as in $\frac{P}{2}.\infty P$ (fig. 142), and in the second deltoidal, as in $\frac{3P\frac{3}{2}}{2}.\infty P$ (fig. 143). When, however, the prismatic edges proper, those parallel to the vertical axis, are apparent, both prism and rhombohedron appear as five-sided figures (fig. 144), representing one of the commonest kinds of Calcite crystals, $\infty P . - \frac{\frac{1}{2}P}{2}$. The prism of the second order being the common limit of both rhombohedra and scalenohedra,

FIG. 143.

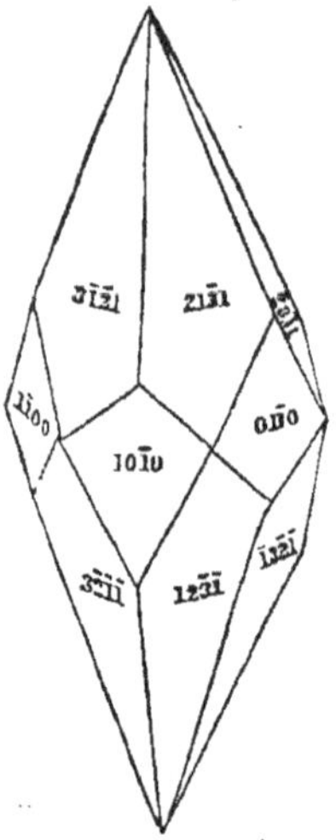

FIG. 144.

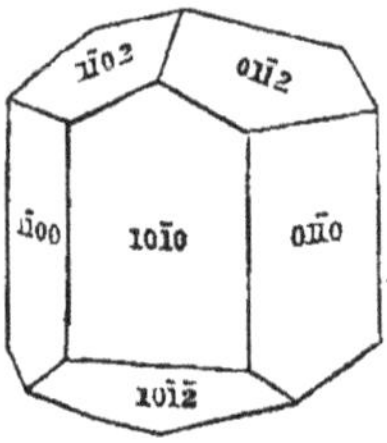

FIG. 145.

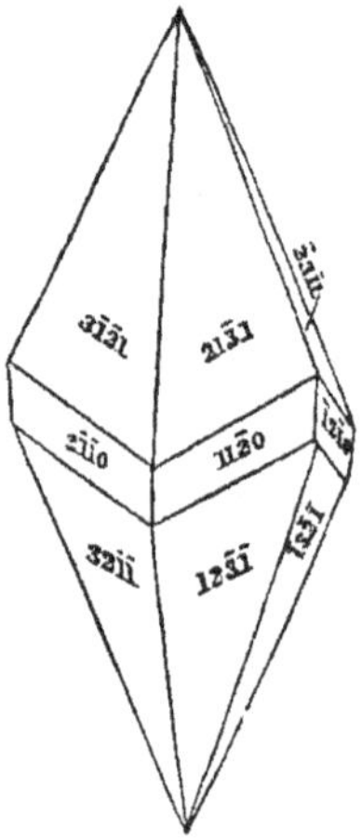

it will, in combination, truncate their middle edges, as in $\frac{3P\frac{3}{2}}{2} . \infty P2$ (fig. 145), and $\frac{P}{2} . \infty P2$ (fig. 146), both being observed cases in Calcite.

The basal pinakoid truncates the polar summits both of scalenohedra and rhombohedra, producing either six- or three-sided faces; the latter case is illustrated in fig. 136.

FIG. 146.

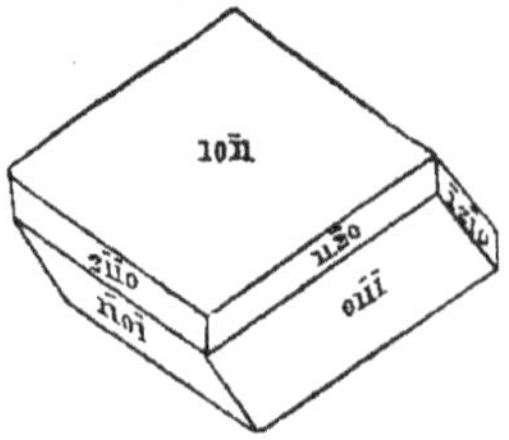

When many rhombohedra and scalenohedra of diffe ent values are combined together, the resulting solid is often of great complexity, and its proper symmetry is not always easily seen. In such cases projections of the faces upon the plane of the base are very useful, and are generally to be preferred to perspective figures.

Naumann's rhombohedral notation. As by far the larger number of hexagonal minerals are rhombohedrally hemihedral, it is generally convenient to adopt the rhombohedron as the unit form of the series rather than the pyramid, which is less common, and as a rule has its two kinds of faces so unequally developed as to be more properly regarded as a combination of two rhombohedra. It is therefore customary to write the symbol $\frac{P}{2}$ as $\pm R$, which represents the two rhombohedra deducible from any particular value of the ratio $a : c$. This may also be derived from the angle over a polar edge in a rhombohedron, or from the supplement of that over a middle edge, by the expression

$$\cos.\ \alpha = \frac{\cos.\ \frac{1}{2} r}{\sin.\ 60^\circ},$$

where $r =$ the measured angle, α the side of a spherical triangle, corresponding to the angle at the vertex of the right-angled plane triangle whose perpendicular and base are the vertical axis, and a lateral interaxis, or c and a' respectively, whence

$$\frac{c}{a'} = \text{cotan. } \alpha.$$

But as the lateral axes and interaxes $a : a'$ are in the proportion of $1 : \frac{1}{2}\sqrt{3}$, when $a = 1$, the required length of the vertical axis will be

$$c = \tfrac{1}{2}\sqrt{3}\ \text{cotan. } \alpha.$$

From any rhombohedron $\pm m R$ by altering the value m, making the vertical axis longer or shorter, a series of forms of the same kind are obtained, ranging from the basal pinakoid $0 R$ to the prism ∞R, which, when arranged in order, give a series analogous to that of the pyramid on page 83, or

$$0R \ldots \pm \frac{1}{m} R \ldots \pm R \ldots mR \ldots \infty R.$$

The most obtuse form of scalenohedron being the rhombohedron having the same middle edges, the symbol of the latter will serve to indicate any scalenohedron upon the same base, if another sign be added, marking the number of times that its unit vertical axis is lengthened. This class of symbol has the following forms:—

$$\pm mRn, \pm m'Rn', \text{ or } \pm mR^n;$$

in which m refers to the vertical axis of the rhombohedron of the middle edges, and n to the same axis in the corresponding scalenohedron. These lengths are, however, related to each other in the constant ratio of $1 : \frac{n}{2-n}$, and therefore the rhombohedral symbol of any scalenohedron whose dihexagonal notation is $\frac{mPn}{2}$ will be $\frac{m(2-n)}{n} R \frac{n}{2n}$, and conversely, $m'Rn'$ will be $m'n'P\frac{2n'}{n'+1}$.

In this notation the rhombohedra enclosed in any scalenohedron, mRn, are—1. That of the middle edges, mR; 2. That of the shorter polar edges, $\frac{m}{2}(3n-1)R$; and 3. That of the longer polar edges, $-\frac{n}{2}(3n+1)R$; and the table on page 98 becomes—

Scalenohedra	$R3$	$2R\frac{3}{2}$	$3R2$	$4R\frac{5}{2}$
Rhombohedra of middle edges	R	$2R$	$3R$	$4R$
Rhombohedra of shorter polar edges . . .	$4R$	$5R$	$6R$	$7R$
Rhombohedra of longer polar edges . . .	$-5R$	$-7R$	$-9R$	$-11R$

Pyramidal hemihedrism. This, the third case on page 89, corresponds to the extension of alternate faces in the dihexagonal pyramid, to the obliteration of the adjacent ones on the same side of the basal section, half the original edges in that plane being retained, but no others. The result is a regular hexagonal pyramid, indistinguishable geometrically from the holohedral forms, but differing from them in the position of the polar edges, which do not lie in

FIG. 147.

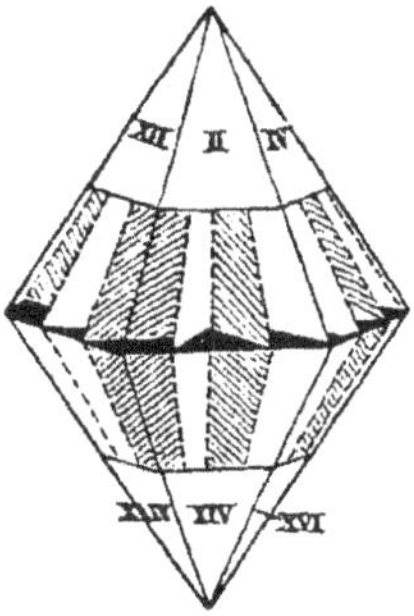

FIG. 148.

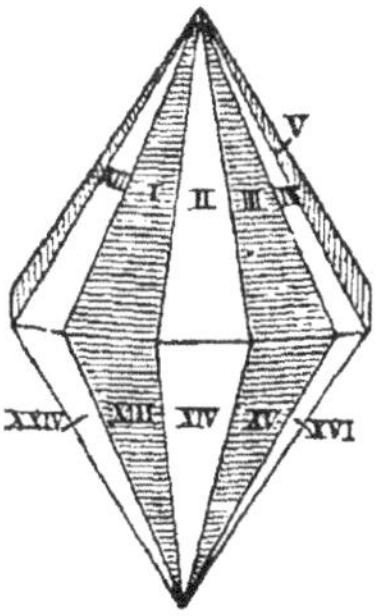

FIG. 149.

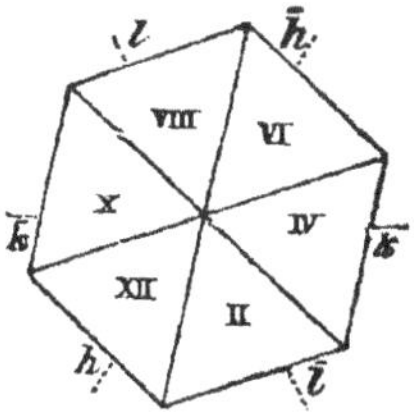

the same planes with the lateral axes or interaxes, but in some intermediate unsymmetrical position. Fig. 147 is the form produced from the left-hand (white or even-numbered)

FIG. 150.

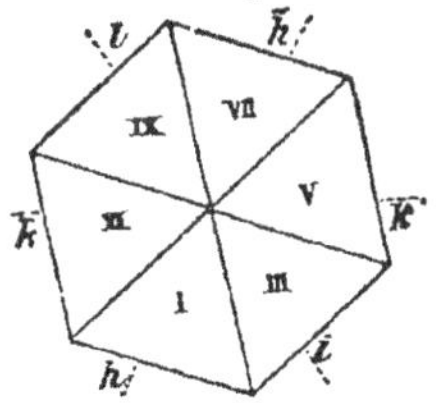

FIG. 151.

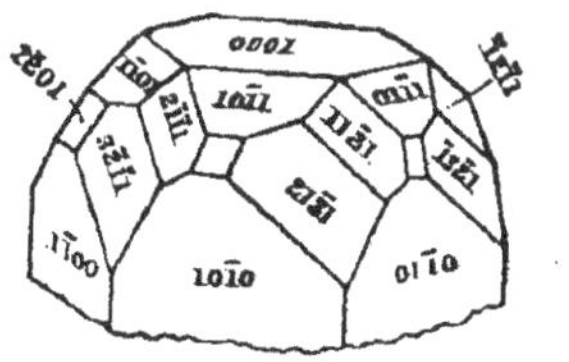

faces in fig. 148, fig. 149 its horizontal projection, and fig. 150 that of the right-hand form, from which it will be seen that in the first case the greatest length of the basal edges is to the left, and in the second to the right of the lateral axes. These are called hexagonal pyramids of the third order; they have one axis and one plane of hexagonal symmetry, but the original binary symmetry is completely lost. By a similar kind of derivation, prisms of the third order are derived from the dihexagonal prism, but none of the remaining holohedral forms are geometrically affected by this class of hemihedrism. The direct and inverse forms are superposable. The symbols are:

$$\pi\,\{h\,k\,\bar{l}\,i\},\ \pi\,\{h\,k\,\bar{l}\,o\},\ \text{and}\ \pm\left[\frac{m\,P\,n}{2}\right],\ \pm\left[\frac{\infty\,P\,n}{2}\right]$$

for the pyramids and prisms respectively. These forms are only known in combination. Fig. 151 is an example in an Apatite crystal, containing ∞P, $2P$, P, oP, $2P2$, $\left[\frac{3P\frac{3}{2}}{2}\right]$. The last form appears only on the right-hand basal solid angles of P, or in the corresponding diagonal zones between P and ∞P.

Tetartohedral hexagonal forms. Supposing a dihexagonal pyramid to be subjected to two different kinds of hemihedrism, the resulting form will have only one-fourth of the full number of faces, or will contain two out of the eight groups in the table on page 89.

As three different kinds of hemihedrism are possible, there should be the same number of kinds of tetartohedrism, but of these only two are possible.

1. *Trapezohedral tetartohedrism*:—By rhombohedral hemihedrism the dihexagonal pyramid is resolved into the two scalenohedra containing the groups $\frac{A}{G}\Big|\frac{B}{H}$ and $\frac{C}{E}\Big|\frac{D}{F}$, which, by a further plagihedral development, divide into the four pairs $\frac{A}{H}$, $\frac{B}{G}$, $\frac{C}{F}$, $\frac{D}{E}$, in which, when the faces of the upper group are even-, those of the lower are odd-numbered, and *vice versâ*. Geometrically this corresponds to taking out the faces of a scalenohedron by pairs adjacent to the alter-

nate middle edges, as in fig. 153. The resulting forms (fig. 154 from the white, and fig. 152 from the shaded faces) are called trigonal trapezohedra; their six faces, which are trapezoids, meet in six polar edges all of the same length,

FIG. 152.

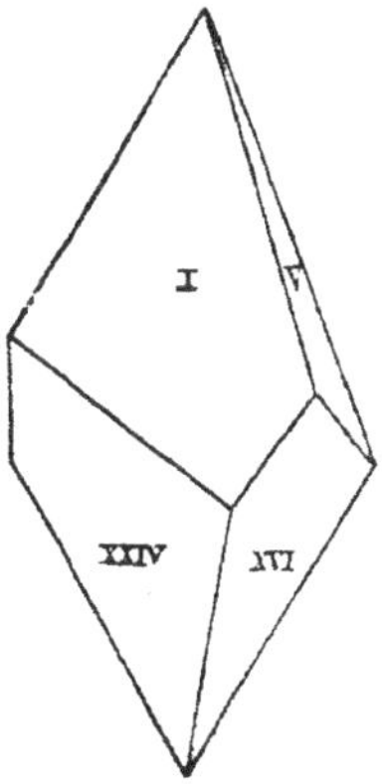

FIG. 153.

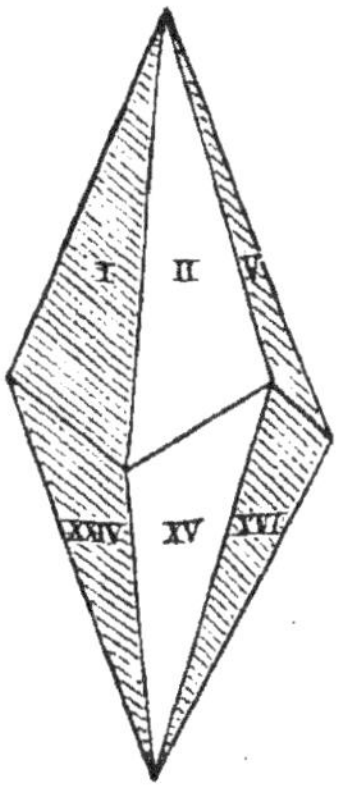

FIG. 154.

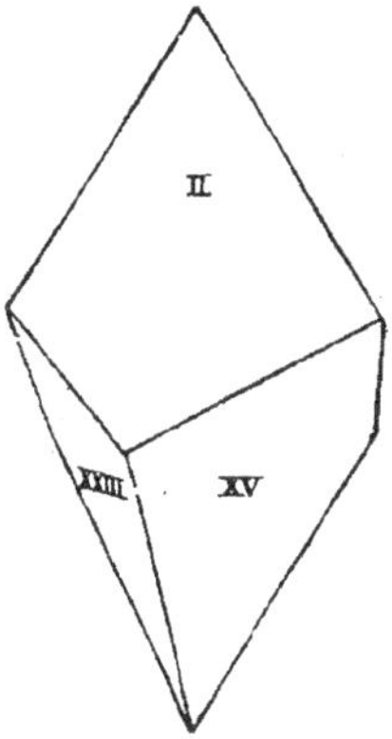

and six middle ones alternately longer and shorter, the longer ones being extensions of edges of the scalenohedron. They have the same axes of symmetry as the scalenohedron—one ternary and three binary—but no planes of symmetry, and are therefore not superposable. They are distinguished as right- and left-handed, positive and negative, forms, the latter having reference to the sign of the originating scalenohedron. Thus, if fig. 153 be considered as positive, fig. 154 will be a right-handed and fig. 152 a left-handed positive trapezohedron. The general symbols are, for all four positions,

$$+\frac{mPn}{4}r, \quad +\frac{mPn}{4}l, \quad -\frac{mPn}{4}r, \quad -\frac{mPn}{4}l,$$

or $\kappa\kappa'\{hk\bar{l}i\}$ if κ' be adopted as indicating plagihedral hemihedrism. The first and third and second and fourth of these—that is, forms of similar direction and opposite signs—are superposable.

The hexagonal pyramid of the second order may be regarded as a special form of scalenohedron, having the

FIG. 155.

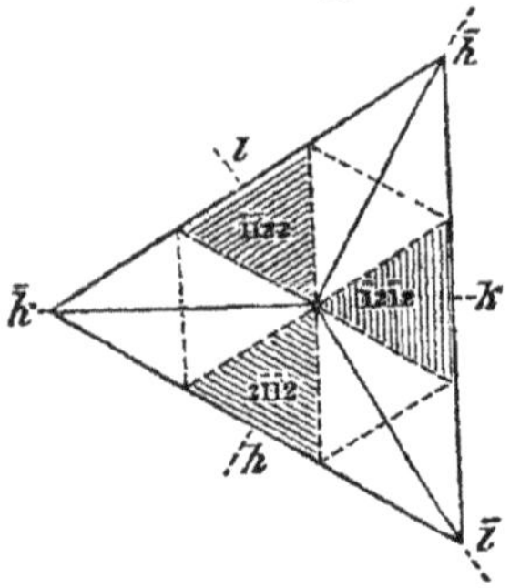

FIG 156.

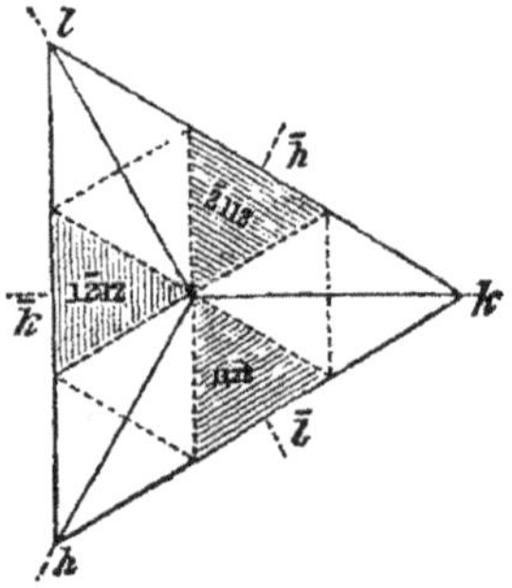

angles of its obtuse, *r*, polar edges = 180°, and its middle edges horizontal; and therefore if half its faces be taken out by alternate pairs above and below, having a middle edge in common, and the remaining ones be extended, it will satisfy this class of tetartohedrism. The form is a trigonal pyramid contained by six isosceles triangles forming a double pyramid, whose base is an equilateral triangle. The symmetry in regard to the axes is therefore the same as in the preceding form: the lateral axial sections as well as the base are planes of symmetry. The derivation of the right-handed form is shown in the horizontal projection (fig. 155), and that of the left-handed one in fig. 156; but there is no distinction required between positive and negative forms, as the first includes both the positive right and negative left trapezohedra, and the second the negative right and positive left ones. The symbols are:

$$\kappa\kappa' \{1\,1\,\bar{2}\,i\} \text{ and } \kappa\kappa' \{\bar{1}\,2\,\bar{1}\,i\}, \text{ or } \frac{m\,P\,2}{4}\,r \text{ and } \frac{m\,P\,2}{4}\,l.$$

These are sometimes written as $\frac{m P 2}{2}$ or $\frac{1}{2} m P 2$, signifying that they contain half the faces of the originating form, which, though convenient as indicating the character, are otherwise misleading, as they are not hemihedral forms; the pyramid of the second order not being susceptible of development into a solid geometrically dissimilar from itself by any of the three possible methods of hemidrism.

FIG. 157.

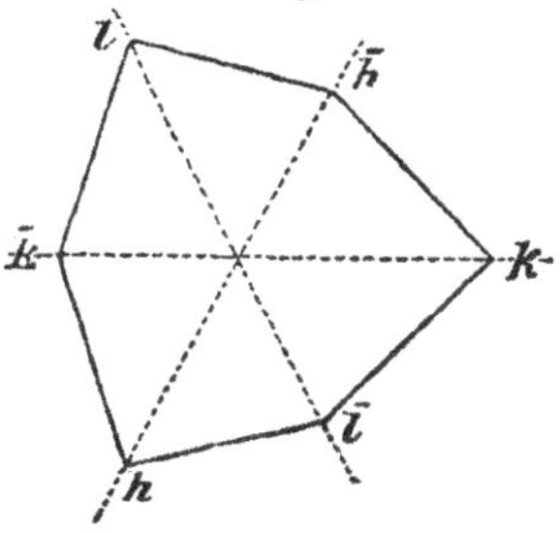

The dihexagonal prism, in the same way, gives rise to two tetartohedral forms called ditrigonal prisms, contained by three pairs of faces making alternately obtuse and acute angles, the former having the same values as those meeting the lateral axes in the holohedral form, as is seen in the horizontal projection (fig. 157). In combination, they bevel the alternate edges of the unit hexagonal prism. The symbols are:

$$\kappa\kappa' (h k \bar{l} o), \text{ or } \frac{\infty P n}{4} r, \frac{\infty P n}{4} l.$$

The prism of the second order produces two exactly analogous forms known as trigonal prisms, contained by three vertical faces making equal angles with each other, corresponding in position to figs. 155, 156. The symbols are:

$$\kappa\kappa' (1 1 \bar{2} 0), \text{ or } \frac{\infty P 2}{4} r, \frac{\infty P 2}{4} l.$$

Pyramids and prisms of the first order do not give any special forms by trapezohedral tetartohedrism.

None of the forms of this class ever occur independently or otherwise than in very subordinate combination, and they are almost exclusively confined to one species; but as that is the most abundant of all minerals, namely quartz, they are of considerable interest, fig. 157 *a* being a characteristic example. It contains R, $-R$, ∞P, and two tetartohedra, $\frac{2P2}{4}r$ (s)[1] and $\frac{6P\frac{6}{5}}{4}r$ (x), which are distinguished as right-handed positive forms on account of their position to the right, and in the case of x, below, the larger rhombohedron R_1, considered as direct or positive. In

FIG. 157 *a*.

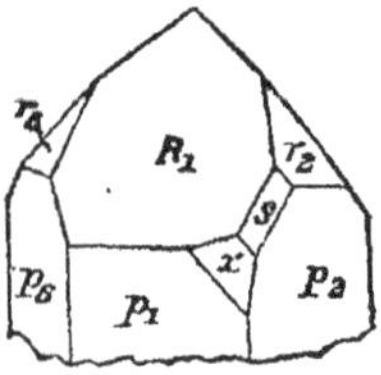

FIG. 157 *b*.

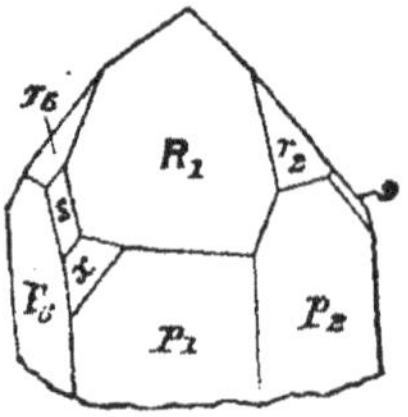

fig. 157 *b* the same faces occur to the left of R_1, and are therefore left-handed positive forms. The trapezohedra are negative when they lie below the faces of the smaller or negative rhombohedron r, or between s and p_2 in fig. 157 *a*, or s and p_6 in fig. 157 *b*, the right- and left-handed character being unchanged. It will be seen from the figures that a face s lies in two diagonal zones of the pyramid or with the faces R_1 p_2 and p_1 r_2 in fig. 157 *a*, and R_1 p_6, and r_6 p_1 in fig. 157 *b*, and that the second of these zones in either case also contains a face, x, of a trapezohedron. No single crystal ever contains both trigonal pyramids, or right- and left-handed trapezohedra of the same direction, as, although a case is known in which both positive and negative trapezohedra of the same kind form an apparent scalenohedron, the crystal has been proved by optical tests to be a compound or twin structure.

1 These, usually known as the rhomb faces in quartz, are remarkable for their brilliancy, whereby they may often be detected in crystals even of microscopic size.

Rhombohedral tetartohedrism. The successive application of pyramidal and rhombohedral hemihedrism to the dihexagonal pyramid corresponds to the extension of alter-

FIG. 158.

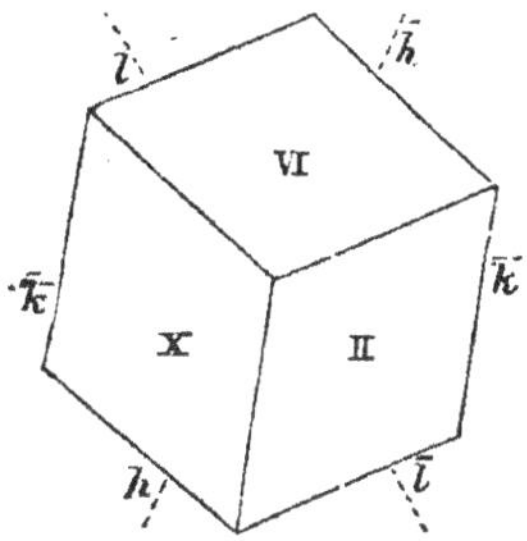

FIG. 159.

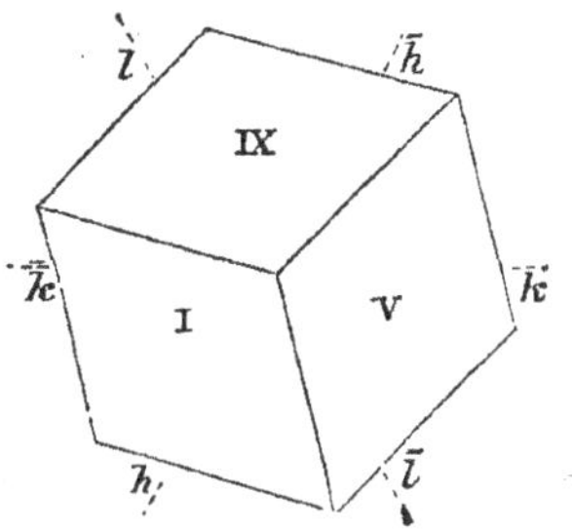

nate faces above and below in the scalenohedron, or the groups $\frac{A\,|\,B}{G\,|\,H}$ and $\frac{C\,|\,D}{E\,|\,F}$ divide into $\frac{A}{G}, \frac{B}{H}, \frac{C}{E}, \frac{D}{F}$, each of which contains six faces, either all even- or all odd-numbered. The resulting form is a rhombohedron, whose edges do not lie in any of the principal crystallographic sections, but are oblique to the lateral axes, as shown in the horizontal projections, figs. 158, 159. This is known as a rhombohedron of intermediate position, or of the third order, its symbol being $\kappa\,\pi\,\{h k \bar{l} i\}$, or the whole series, according to Naumann,

$$+\frac{m\,P\,n}{4}\frac{r}{l},\quad +\frac{m\,P\,n}{4}\frac{l}{r},\quad -\frac{m\,P\,n}{4}\frac{r}{l},\ \text{and}\ -\frac{m\,P\,n}{4}\frac{l}{r};$$

in which positive signs represent the forms derived from the positive scalenohedron, minus signs those from the negative one, and the letters *r l* indicate whether the right- or the left-hand faces are above or below the middle edges. These, like the ordinary rhombohedra, are all superposable.

The hexagonal pyramid of the second order in the same way gives rise to two rhombohedra, whose polar edges lie in the lateral axial planes, or make angles of 30 degrees with those of the ordinary rhombohedron, as shown in figs.

160, 161. These are said to be of the second order. From the analogy of the preceding, their symbols are:

$$\frac{m\,P\,2}{4}\,\frac{r}{l} \quad \text{and} \quad \frac{m\,P\,2}{4}\,\frac{l}{r}.$$

The dihexagonal prism gives rise to two hexagonal prisms of the third order, oblique to the axes, whose hori-

FIG. 160a.

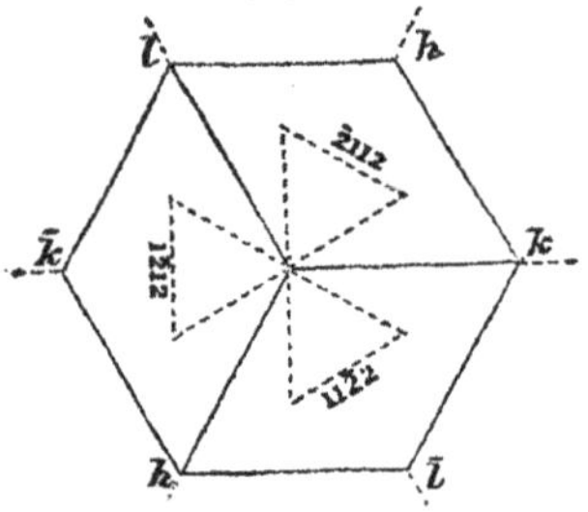

FIG. 161.

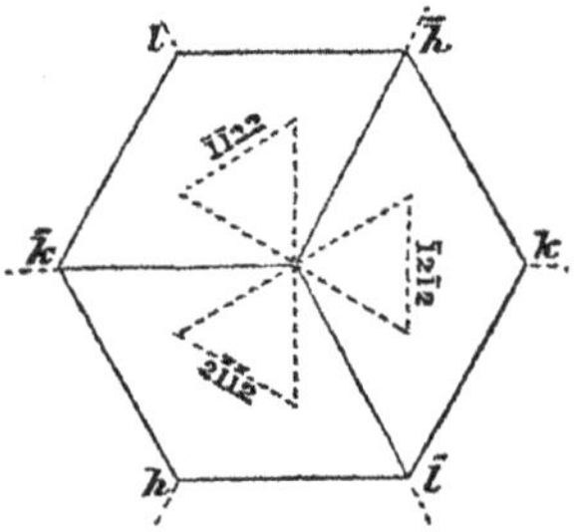

zontal projections will be apparent from the contour of the rhombohedron of the same kind previously given, as either one of these will include the two rhombohedra upon the same base originating from the same scalenohedron. These are exactly similar to the same classes of prisms originated by pyramidal hemihedrism.

The hexagonal pyramid and rhombohedron of the first order, and both kinds of hexagonal prisms, do not give rise to any special tetartohedral forms of this class.

The tetartohedral rhombohedra only occur in combination, and not very frequently. Fig. 161 *a* is a case observed in an Ilmenite crystal $R\,.\,0\,R\,.\,-2\,R\,.\,\frac{1}{4}\,(\frac{4}{3}\,P\,2)$. The faces of the last form appear only on the left-hand polar edges of R.

FIG. 161 *a*.

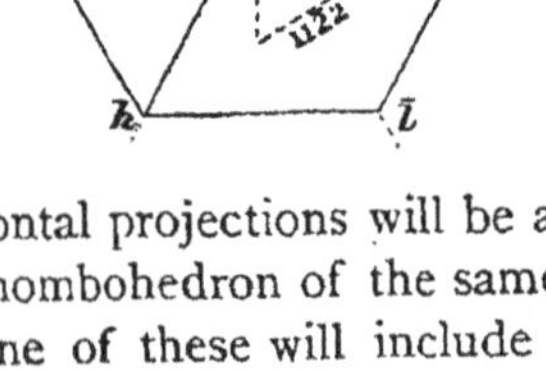

The impossible method of tetartohedrism is that resulting from plagihedral and pyramidal hemihedrism, as the first gives forms with even-numbered faces above and odd below, or *vice versâ*, and the second requires them to be of the same kind above and below. The successive application of the two methods to the dihexagonal pyramid, therefore, leaves only three faces on the same side of the basal section, which is not a symmetrical crystallographic form.

Miller's rhombohedral notation. In this method, the forms are referred to three axes making equal angles with each other, and having equal parameters. These axes are parallel to the polar edges of the unit rhombohedron of the series, and are therefore, as a rule, oblique to one another. The unit form {1 1 1} is the basal pinakoid, and contains two faces (1 1 1) and ($\bar{1}\ \bar{1}\ \bar{1}$). Their normal is called the axis of the rhombohedron or morphological axis, and corresponds to the principal axis of the hexagonal notation. The unit direct rhombohedron includes the three pairs of faces (1 0 0), (0 1 0), (0 0 1), and the corresponding inverse form those with the indices (2 2 1), (1 2 1), and (1 2 2), which together give the hexagonal pyramid as a combination.

The general form *h k l*, is a direct scalenohedron, and the inverse one with which it combines to form the dihexagonal pyramid, is distinguished as *e f g*, the two being related in the following manner :—

$$
\begin{aligned}
e &= 2\,(h + k + l) - 3\,h = -h + 2\,k + 2\,l \\
f &= 2\,(h + k + l) - 3\,k = 2\,h - k + 2\,l \\
g &= 2\,(h + k + l) - 3\,l = 2\,h + 2\,k - l
\end{aligned}
$$

The unit prism has the faces $2\,\bar{1}\,\bar{1}$, $1\,\bar{2}\,1$, $\bar{1}\,\bar{1}\,2$, $\bar{2}\,1\,1$, $\bar{1}\,2\,\bar{1}$, 1 1 2 ; the prism of the second order, $1\,0\,\bar{1}$, $1\,\bar{1}\,0$, $0\,\bar{1}\,1$, $\bar{1}\,0\,1$, 1 1 0, $0\,1\,\bar{1}$; and the dihexagonal prism those of the two forms (*h k* 0) and (*e f* 0).

The hemihedral forms are :

1. Asymmetric α *h k l*, corresponding to the trapezohedral tetartohedral forms ;

2. Inclined κ *h k l*. This is the case not recognised as a symmetrical kind of tetartohedrism in the hexagonal nota-

tion, the faces of the form being all either positive or negative with respect to the principal axis, or the particular class of development subsequently noticed as hemimorphism.

3. Parallel $\pi\, h\, k\, l$, corresponding to rhombohedral tetartohedrism in the hexagonal system.

For practical purposes, in the calculation and determination of crystals this method is generally preferable to the hexagonal notation, as it dispenses with a fourth index in the symbols; but for general descriptive purposes it does not so well express the analogy between the hexagonal and tetragonal systems, and it has therefore not been adopted in this work. The student will, however, do well to become acquainted with Miller's notation, as it may probably supersede the hexagonal form at no very distant date. A simple exposition of it will be found in Gurney's elementary treatise on Crystallography.

CHAPTER V.

TETRAGONAL[1] SYSTEM.

THE complete symmetry of this system is contained in an upright prism upon a square base, which has quaternary symmetry about a principal axis parallel to the vertical edges, and binary about four lateral axes, respectively parallel to the sides and diagonals of the base. These correspond to five planes of symmetry—a basal, or principal plane, and four lateral planes at right angles to the first and at 45° to each other. The reference axes are three, at right angles to each other—namely, the vertical or principal axis, c, and two of the four lateral axes, a_1, a_2—those parallel to the diagonals of the base, their order being similar to that of

[1] Other names are Pyramidal, Dimetric, Quaternary, Quadratic, and Viergliedig or four-membered.

the cubic system. The parameters of the lateral axes are similar, and different from that of the vertical axis, the two being related in the proportion of some arbitrary ratio $a : c$, proper to the species, which has therefore the same signification as in the hexagonal system.

The general symbol of a face having different intercepts upon the three reference axes, corresponding to different inclinations upon three planes of symmetry, none of which is a right angle, is as in the cubic system $(h\,k\,l)$; with the difference that only the first two indices are interchangeable, giving two permutations of letters, $h\,k$, $k\,h$, and four of signs, $++$, $+-$, $-+$, $--$, for each value (positive or negative) of l, or sixteen in all as the maximum number of faces possible in a simple tetragonal form. This, known as a ditetragonal pyramid, represented in elevation with its

FIG. 162.

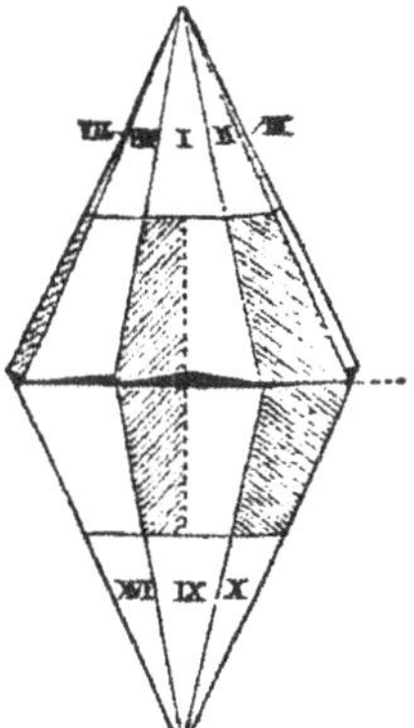

FIG. 163.

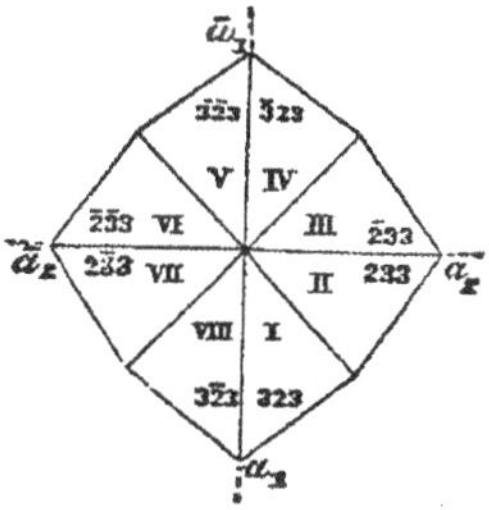

symmetrical sections shaded in the manner described on page 78, in fig. 162, and in plan in fig. 163,[1] is a double pyramid, contained by sixteen faces whose dihedral angles in the eight basal edges are all similar, while those in the polar edges, and the corresponding plane angles of the equi-

[1] This is noted as $P\frac{3}{2}$.

lateral eight-sided base are alternately larger and smaller. The general symbols are :

$$\{h\,k\,l\},\ (a : n\,a : m\,c),\ \text{and}\ m\,P\,n.$$

The pyramid, whose base is a regular octagon, and has the angles of its polar edges all equal, is an impossible form, as requiring for n the irrational value tan. $67\frac{1}{2}°$, or 2·4142, but for any rational value lower than this, the more obtuse polar edges lie in the interaxial planes, and when $n = 1$ or $k = h$, their angle becomes 180°, or the two faces meeting in these planes coincide. This corresponds to the tetragonal or square-based pyramid of the first ord.r, or normal position (figs. 164, 165)—

FIG. 164.

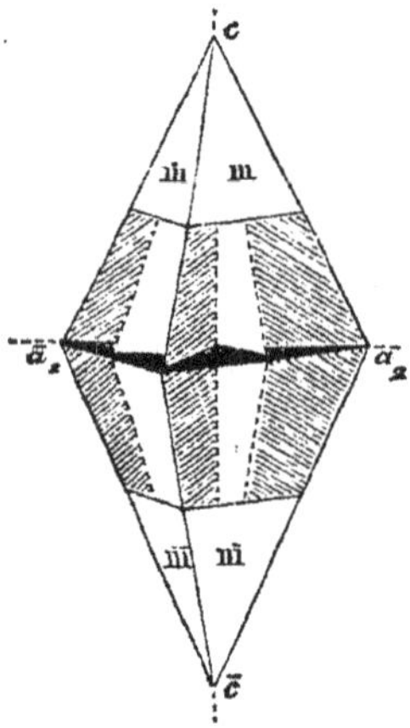

$$\{h\,h\,l\},\ (a : a : m\,c),\ \text{and}\ m\,P,$$

having eight faces meeting at equal angles in the four basal edges and in the eight polar edges which lie in the lateral axial planes, at some other angle whose difference from the first depends upon the disparity in length between the vertical and the lateral axes. The basal angle, β, corresponds to twice the plane angle between the vertical axis and a lateral interaxis a', and as the length of the latter is to that of the adjacent lateral axis a, inclined to it at 45°, as $1 : \frac{1}{\sqrt{2}}$, the fundamental ratio $a : c$ for any pyramid, assumed as the unit of the series, may be determined by the expression—

$$\tan.\frac{\beta}{2} = \frac{c}{a'}\ \text{and}\ c = \frac{\tan.\frac{\beta}{2}}{\sqrt{2}}\ \text{when}\ a = 1.$$

when the measured angle is that over a polar edge $= \pi$, the fundamental parameter of c is found by computing the side p opposite to that angle in a right-angled spherical triangle described about the pole of the principal axis, when—

$$\text{cotan.}\frac{\pi}{2} = \cos. p, \text{ and cotan.} p = c \text{ when } a = 1.$$

When the value of n in the symbol of the ditetragonal

FIG. 165.

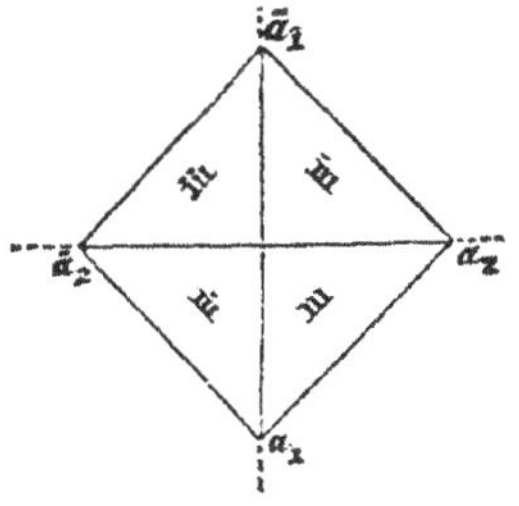

FIG. 166.

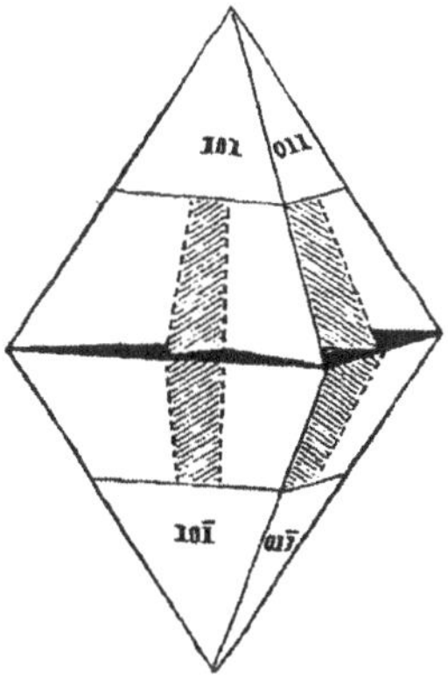

pyramid exceeds tan. $67\frac{1}{2}°$, the more obtuse polar edges lie in the lateral axial planes, and when $n = \infty$ or $k = 0$, the angle becomes 180°, or the adjacent faces meeting in them fall into the same plane, giving the tetragonal pyramid of the second order or diagonal position (figs. 166, 167), whose basal edges are equal and parallel to the lateral crystallographic axes, while its polar edges lie in the lateral interaxial planes. The symbols, as will be apparent from its derivation, are—

FIG. 167.

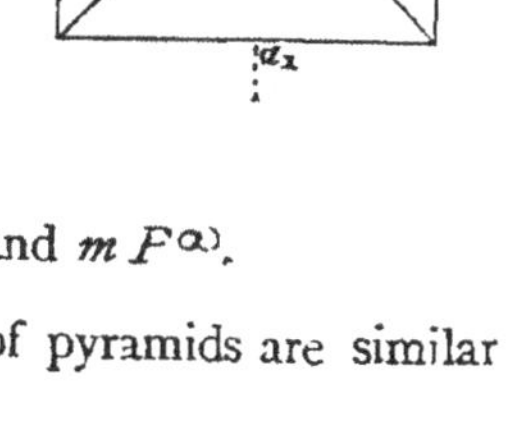

$$\{h\,o\,l\},\ (a : \infty a : mc), \text{ and } mP\infty.$$

The relations of these three classes of pyramids are similar to those subsisting between the allied forms in the hexagonal system, the difference being in the possible value of n, which ranges from 1 to ∞ instead of merely from 1 to 2.

From any pyramid of either kind upon the same base, by multiplying c by any rational quantity m, greater or less than unity, a series of new pyramids of varying altitude is obtained, as in fig. 168, which may be noted as

FIG. 168.

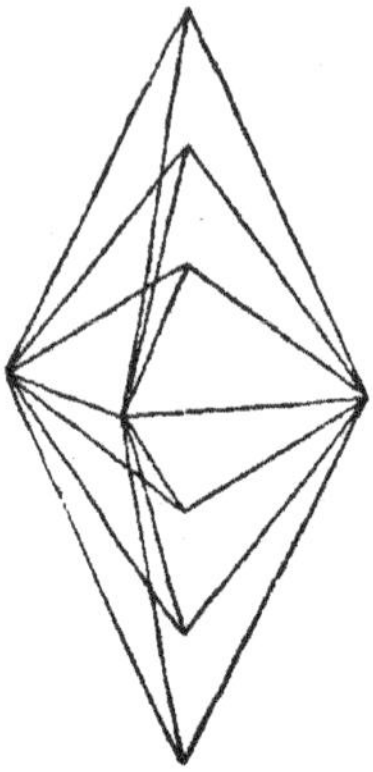

$P, 2P. 3P\{1 1 1\}, \{2 2 1\}, \{3 3 1\}$, or
$\frac{1}{3}P, \frac{2}{3}P. P. \{1 1 3\}, \{2 2 3\}, \{1 1 1\}$, or
$\frac{1}{2}P, P, \frac{3}{2}P, \{1 1 2\}\ \{1 1 1\}\ \{3 3 2\}$,

according as one or other of the three is adopted as the unit form. The basal angle increases with the altitude in these forms, and when $m = \infty$ or $l = 0$ it becomes 180°, or they change to vertical prisms of unlimited height. There are therefore three prisms, one corresponding to each kind of pyramid, namely—

1. Ditetragonal prism, $\{h k 0\}$, $(a : n a : \infty c)$, or $\infty P n$ (fig. 169).

2. Tetragonal prism of the first order, $\{h h 0\} = \{1 1 0\}$, $(a : a : \infty c)$, or ∞P (fig. 170).

FIG. 169.

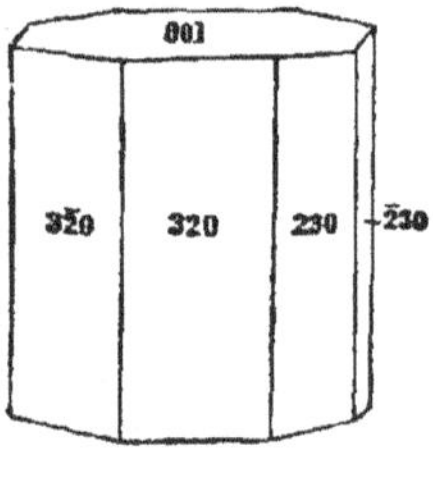

FIG. 170.

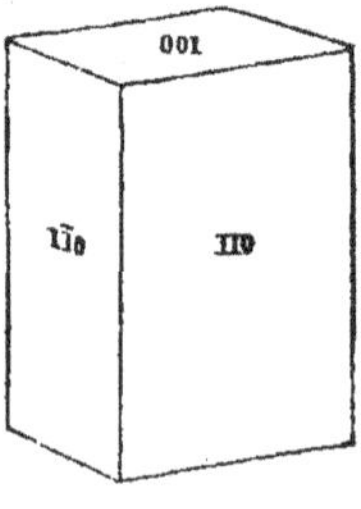

3. Tetragonal prism of the second order, $\{h 0 0\} = \{1 0 0\}$, $\{a : \infty a : \infty c\}$, or $\infty P \infty$ (fig. 171).

These all lie in a zone whose plane is the basal section,

and can only appear in combination. In the first the angles of adjacent faces are alternately larger and smaller, and therefore, from their measure, the characteristic value of n may be deduced; but either kind of tetragonal prism has all its angles right angles, and cannot be used for determining the parameters of the species, except when in combination with some faces not meeting it at right angles. In the other direction, when m is less than 1, the basal edges become sharper, and the polar ones more obtuse, as it diminishes, and when it $= 0$ the whole of the faces fall into the same surface, producing an unlimited plane called, as in the preceding system, the basal or terminal pinakoid, and having the symbols—

FIG. 171.

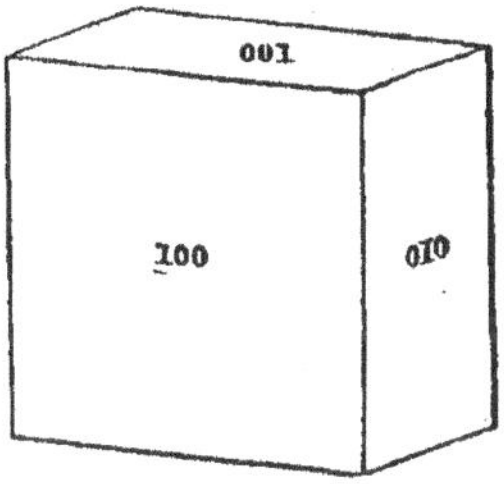

$$\{0\,0\,l\} = \{0\,0\,1\},\ (\infty a : \infty a : c) \text{ and } 0\,P.$$

This can occur only in combination, and is shown as limiting the three prisms in figs. 169–171 when it takes the characteristic shape of their basal sections.

These seven kinds include all the possible holohedral forms: their relations are shown in Naumann's system by a diagram arranged in the same manner as that given for the hexagonal system, and having the same general signification. The only difference between it and the table on page 84 is in the range of the co-efficient n from 1 to ∞, while in the former it is restricted between the limits 1 and 2.

$0\,P$	$0\,P$	$0\,P$
$\frac{1}{m}P$	$\frac{1}{m}Pn$	$\frac{1}{m}P\infty$
P	Pn	$P\infty$
mP	mPn	$mP\infty$
∞P	∞Pn	$\infty P\infty$

In combination, forms of the same order will appear in the succession shown in the vertical lines of the above table, the steepest, the prism, being in the middle, and the flattest, the terminal pinakoid, at the ends: those of intermediate inclination, the pyramids proper, being arranged in regular order towards either of these limiting forms, the steeper truncating the basal edges of the flatter ones, and conversely the latter truncating the polar summits of the former, as in fig. 172, which is noted as ∞P, $\frac{3}{2} P$, P, $\frac{1}{2} P$, $0 P$.

FIG. 172.

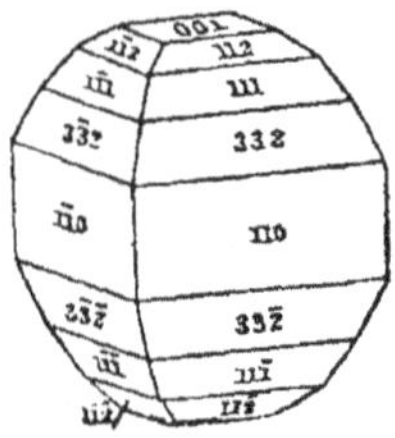

The pyramid of the second order truncates the polar edges of that of the first order, when both are of the same altitude, or have m in common, as seen in elevation and plan in figs. 173, 174, which also illustrate the converse case of the faces of the pyramid of the first order bevelling the solid

FIG. 173.

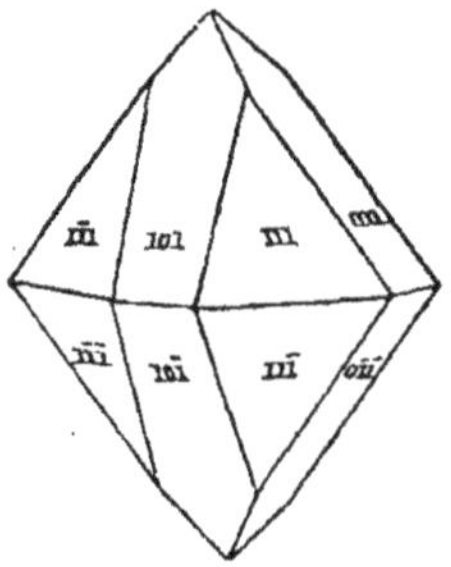

FIG. 174.

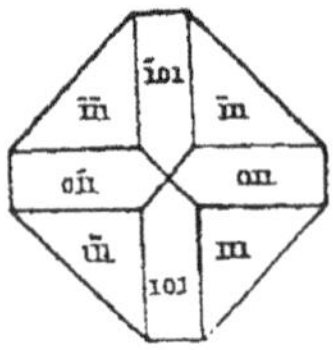

angles of that of the second order. When the two forms differ in the value of m, if $m P\infty$ is steeper it will bevel the basal solid angles of P, as in P, $2 P\infty$, fig. 175; but when $m P\infty$ is the flatter form, it will truncate the polar edges of P obliquely, forming four-faced points upon the polar summits, as in P, $\frac{1}{2} P\infty$, fig. 176. A ditetragonal pyramid, $m P n$, will have its basal solid angles in the lateral interaxial planes bevelled by the faces of a pyramid of the first order of the same altitude, or will bevel the polar edges of the latter, as in P, $P 3$, fig. 177; and similarly, the pyramid of the second order bevels the other basal solid angles, or those in the lateral axial planes of $m\,P\,n$. A ditetragonal pyramid forms eight-faced points upon the principal summits of a steeper tetragonal pyramid of either order, as in $P\frac{1}{3}$, $P\,3$, fig. 178;

FIG. 175.

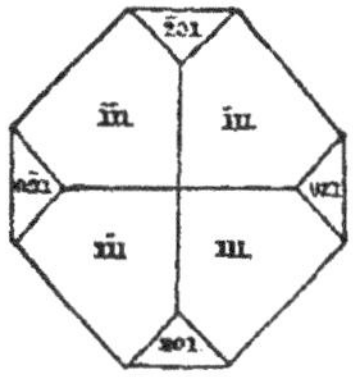

FIG. 176.

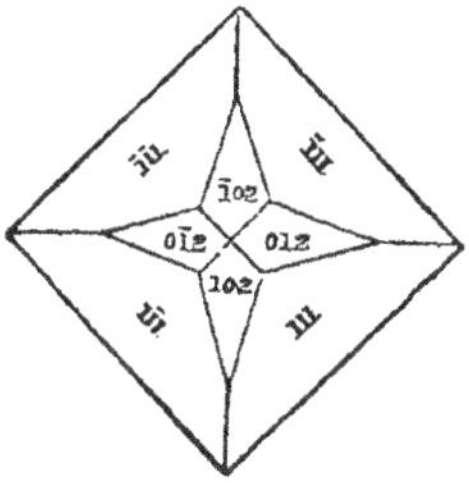

when the latter is the obtuser form it reduces the same summits of $m\,P\,n$ from eight- to four-faced ones, as in $3\,P\,3$, $P\infty$, fig. 179. This class of combination is of considerable interest, as when $a : c$ approximates to $2 : 1$, the solid repre-

FIG. 177.

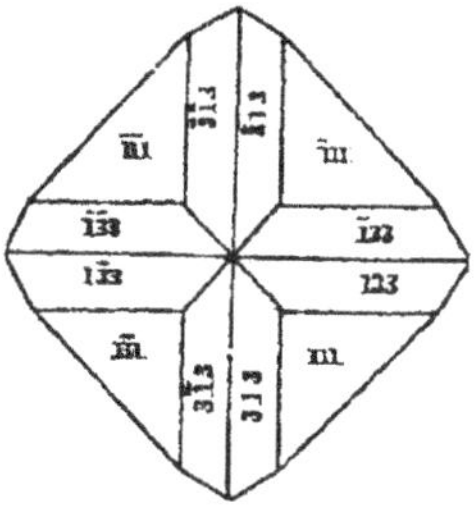

FIG. 178.

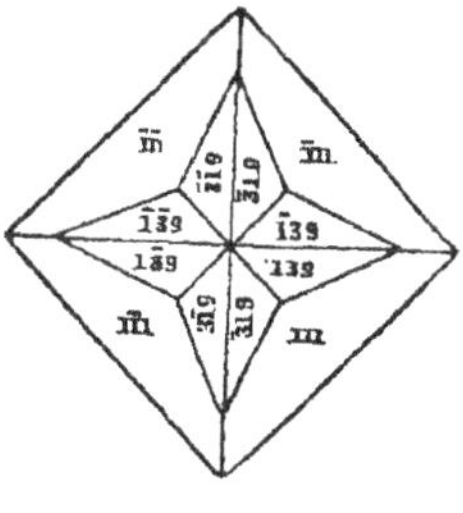

sented by $4\,P\,2$, P or $\{4\,2\,1\}$ $\{1\,1\,1\}$, is almost indistinguishable from the trapezohedron $\{2\,2\,1\}$ of the cubic system. This case actually arises in Leucite, where $a : c = 1 : 0{\cdot}5264$, whose crystals, until recently held to be the most typical examples of the particular trapezohedron in question, have been shown to be probably assignable to a tetragonal combination of the character of fig. 179. One of the simplest cases of a combination of tetragonal prisms and pyramids is shown in figs. 180, 181. It contains ∞P, $\infty P\infty$, P, $0P$, and is a common form of crystal in Apophyllite.

As in the hexagonal system, the number of species with full tetragonal symmetry is comparatively small, but among

FIG. 179.

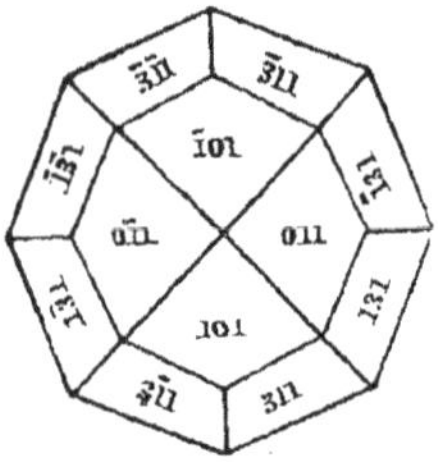

FIG. 180.

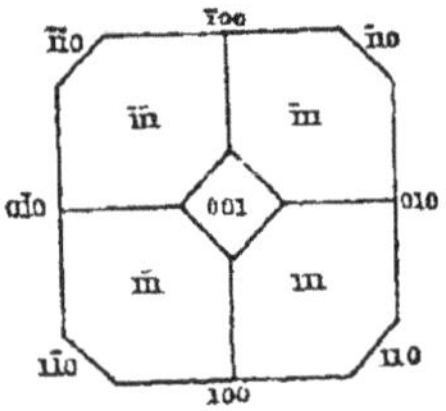

FIG. 181.

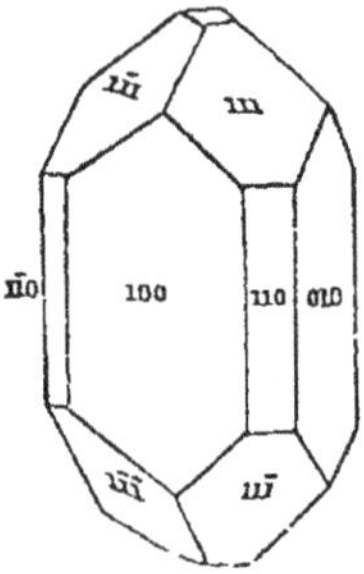

these one, namely Idocrase, is remarkable for the large number of forms that are sometimes combined in single crystals.

Hemihedral tetragonal forms. The faces of the ditetragonal pyramid, when arranged in the following order:

A	B	C	D
I. hkl	II. khl	III. khl	IV. $\bar{h}kl$
V. $h\bar{k}l$	VI. $k\bar{h}l$	VII. $k\bar{h}l$	VIII. $h\bar{k}l$
E	**F**	**G**	**H**
IX. hkl	X. khl	XI. $\bar{k}hl$	XII. $\bar{h}kl$
XIII. $\bar{h}kl$	XIV. $k\bar{h}l$	XV. $k\bar{h}l$	XVI. hkl

—where the Roman numeration corresponds to that in fig. 182, may be symmetrically halved in three ways, giving rise to three kinds of hemihedral forms analogous to those of the hexagonal system, the only difference being that each group contains two instead of three faces.

Trapezohedral hemihedrism. The first method of selection, that by alternate groups both above and below, or the

FIG. 182.

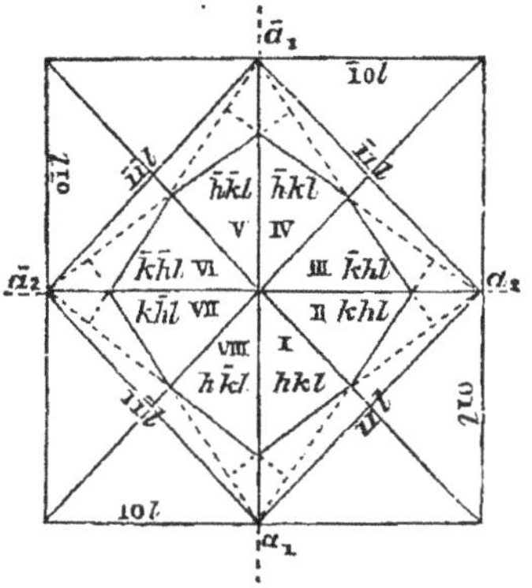

arrangements $\frac{A \mid C}{F \mid H}$ and $\frac{B \mid D}{E \mid G}$, the former containing faces which are all odd-numbered above, and all even-numbered

FIG. 183.

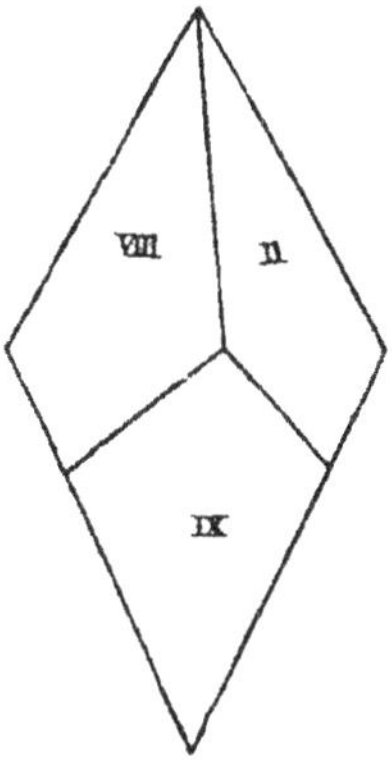

FIG. 184.

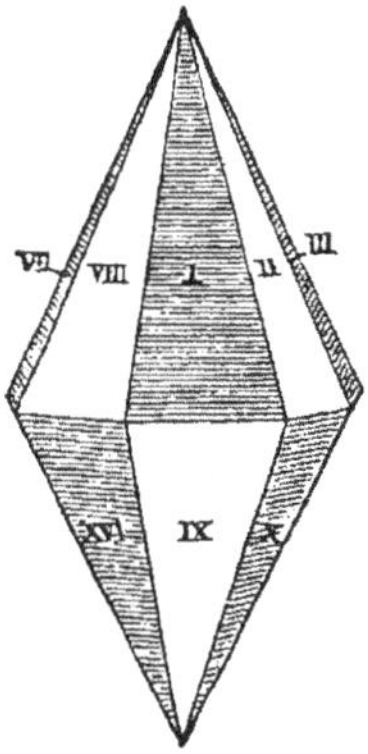

FIG. 185.

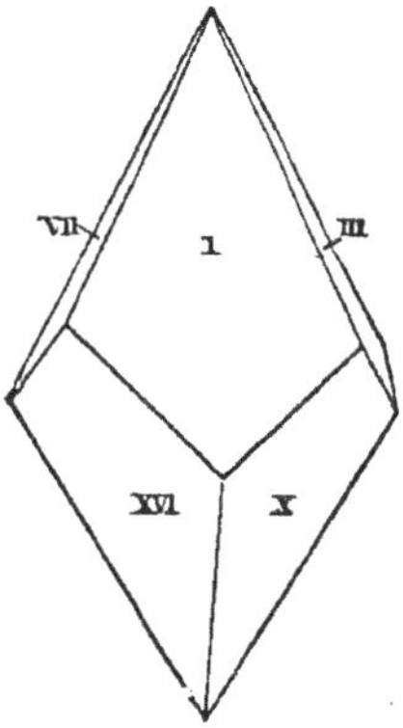

below, while in the latter the order is reversed, gives the forms figs. 183, 185, the first originating from the white, and the second from the shaded, faces of fig. 184. These, known as tetragonal trapezohedra, have eight faces, meeting four equal polar edges, at either end of the principal axis, and eight unequal, alternately longer and shorter, middle ones in zigzag order about the basal section.

From these figures and their horizontal projections (figs. 186, 187), it will be apparent—1st, that as none of these

FIG. 186.

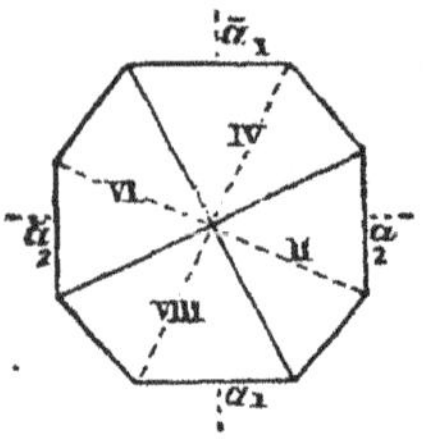

FIG. 187.

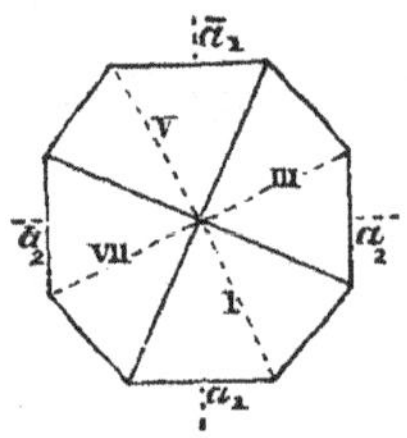

edges lie in planes of symmetry, the forms are plagihedral; and 2nd, that as the principal solid angles are formed by the meeting of four similar edges, and the lateral axes and inter-axes bisect the middle edges, the number and kinds of axes of symmetry are the same as in the holohedral form, namely, one principal or quaternary, and four binary.

The symbols are:

$$a\ \{hkl\}\ \{khl\}, \text{ and } \frac{mPn}{2}r \quad \frac{mPn}{2}l.$$

No other holohedral form than the ditetragonal pyramid gives hemihedra of this class distinguishable from itself, and it is not known whether this particular kind of hemihedrism actually occurs or not. Its existence has been inferred upon physical grounds in a few salts of organic bases, the most pronounced example being a sulphate of strychnine, but no actual plagihedra of the above kind have as yet been observed with certainty.

Sphenoidal hemihedrism. The second method of selection, that by alternate pairs of faces both above and below, gives the arrangements $\frac{A \mid B}{G \mid H}$ and $\frac{C \mid D}{E \mid F}$, the first containing odd-numbered (1st and 3rd) pairs above, and even-numbered ones (6th and 8th) below; and the second the 2nd and 4th pairs above, and the 5th and 7th below, corresponding to the rhombohedral selection of the hexagonal system. When applied to the ditetragonal pyramid it produces the forms figs. 188, 190, the former from the white, and the latter from the shaded, faces of fig. 189. These, known as tetragonal scalenohedra, are contained by eight faces meeting

FIG. 188.

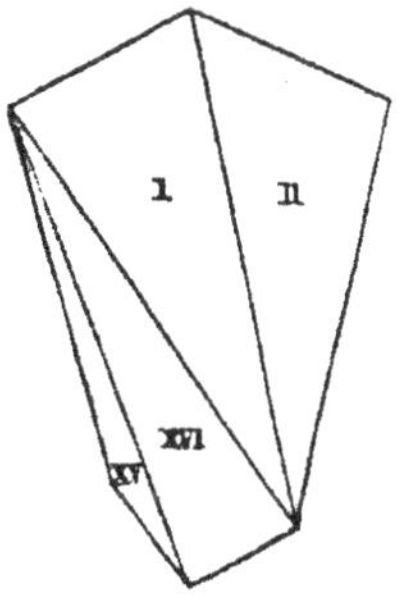

FIG. 189.

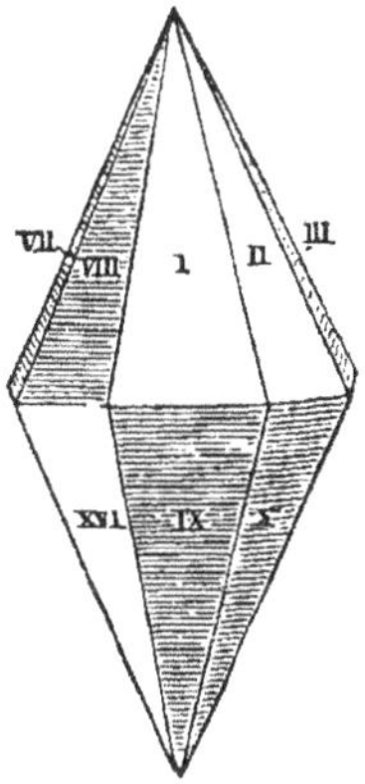

FIG. 190.

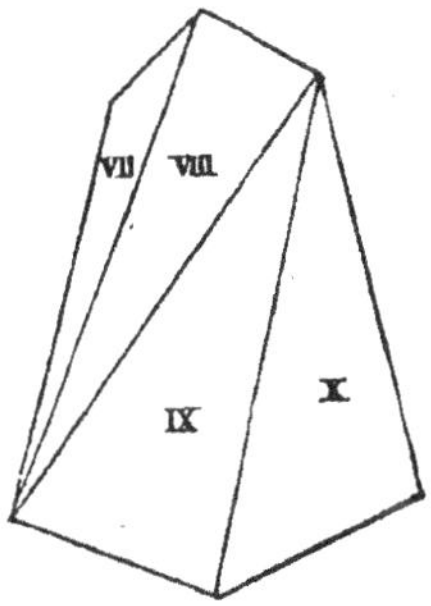

in three prominently dissimilar kinds of edges. Each of the principal solid angles is formed by two longer and two shorter polar edges; the former are those of the holohedral form, and have the same dihedral angles; the third kind, or middle edges, those of the zigzag middle belt, represent alternate middle edges of the tetragonal trapezohedron, and are bisected by the lateral crystallographic axes. On the same side of the base the longer and shorter polar edges lie alternately in planes at right angles to each other; but on opposite sides they are in the same planes, that is, in the lateral interaxial sections, which are therefore the only planes of symmetry. The axial symmetry is binary about the three crystallographic axes.

The symbols are:

$$\kappa\ \{h k l\}\ \{k h l\}, \text{and} + \frac{m P n}{2} - \frac{m P}{2} n;$$

the direct and inverse form of the same origin being superposable by rotation through 90° about the principal axis.

The tetragonal pyramid of the first order in the same way gives rise to figs. 191, 193,[1] the first from the white, and the second from the shaded, faces of fig. 192. These, known

FIG. 191.

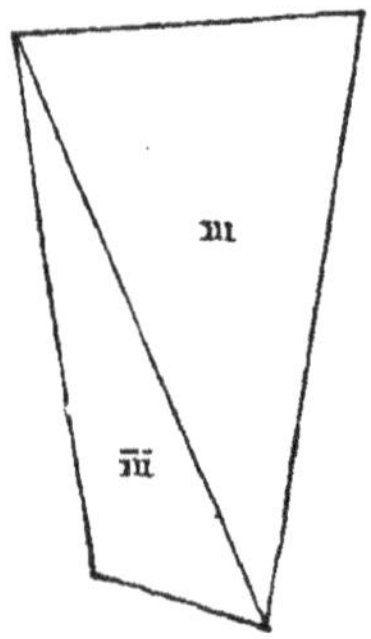

FIG. 192.

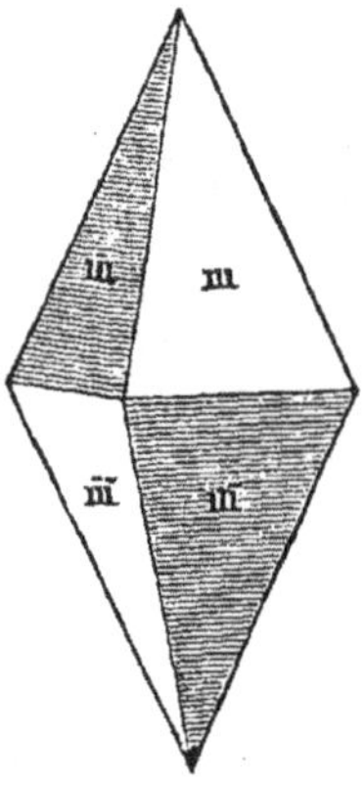

FIG. 193.

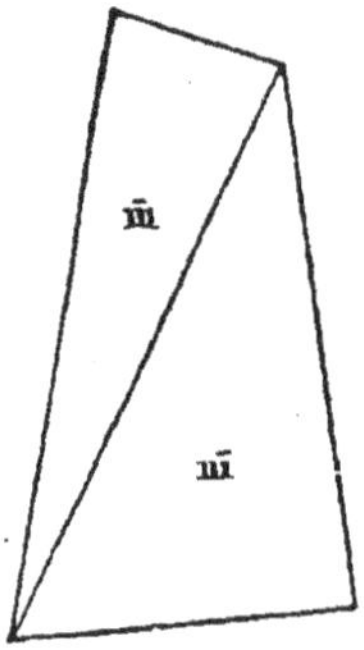

as tetragonal sphenoids, from their wedge-like appearance, are obviously only special cases of tetragonal scalenohedra, having the obtuse polar angles = 180°; but as they are more frequently met with than the latter forms they are considered as most characteristic of this kind of hemihedrism, which is therefore called sphenoidal. The shorter or horizontal edges represent the shorter polar edges of the scalenohedron, and the longer ones, which are parallel to the

[1] These, as well as figs. 188-190, are on a smaller scale than the holohedral form.

axes of the diagonal zone in the pyramid, the middle edges in the same form. The symbols are :

$$\kappa\ \{k h l\}\ \ \{h \bar{h} l\}, \text{ and } +\frac{m P}{2} - \frac{m P}{2}.$$

Like the scalenohedra, they are superposable, and have the same planes and axes of symmetry.

The remaining holohedral forms are not changed by this hemihedrism.

The wedge-like character is not apparent in the forms generally observed, as they mostly originate from pyramids which are either very obtuse or approximate to a regular octahedron, in which latter case the sphenoid is very similar to a regular tetrahedron. This is seen in copper pyrites, which is the only characteristic example of this hemihedrism among minerals, the combination of the two sphenoids $\frac{P}{2}$ $-\frac{P}{2}$ (fig. 194), being very like a slightly distorted regular octahedron. This, however, is only one out of many sphenoids found in the same mineral. The flatter ones have the more characteristic combinations shown in fig. 195, where the faces of one sphenoid truncate the edges between ∞ P and o P alternately above and below. Neither

FIG. 194.

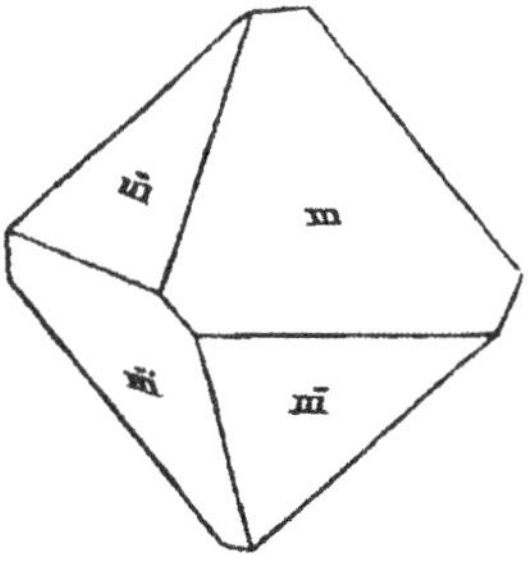

FIG. 195.

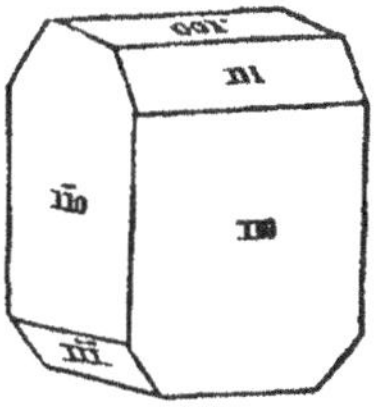

sphenoids nor tetragonal scalenohedra ever occur independently, and the latter when present truncate the solid angles between the prism and sphenoid faces obliquely, but the faces are usually very small.

Pyramidal hemihedrism. The third method of selection, that by alternate pairs of faces adjacent to the same basal edge, gives the arrangements $\frac{A \mid C}{E \mid G}$ and $\frac{B \mid E}{F \mid H}$, the first containing all even-numbered faces, and the second all odd-ones, corresponding to two tetragonal pyramids, which are

FIG. 196.

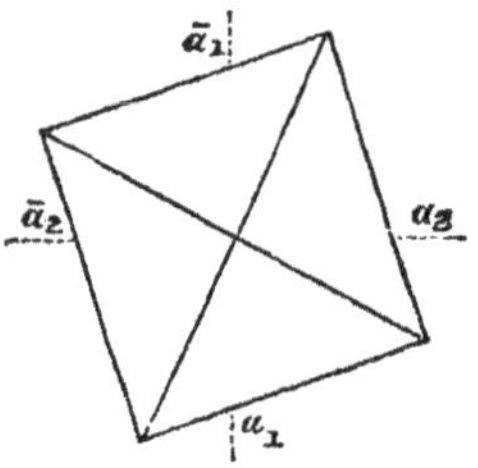

FIG. 197.

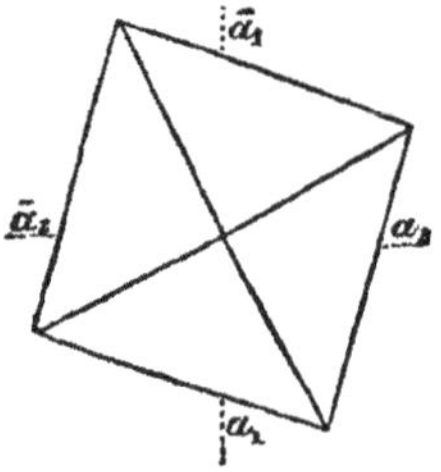

similar in form to the holohedral ones, but not in position; the sections upon planes passing through the polar edges lying in one case to the right, and the other to the left of the lateral axial and interaxial planes of symmetry, as seen in the horizontal projections, figs. 196, 197, which also show that the basal edges are half those of the ditetragonal pyramid; the symmetry is, therefore, to a single plane the base, and quaternary to a single axis the principal one.

These forms, which are superposable, are known as tetragonal pyramids of the third order, or of intermediate position, the symbols being:

$$\pi \{h k l\} \quad \pi \{k h l\}, \text{and} + \left[\frac{m P n}{2}\right] - \left[\frac{m P n}{2}\right].$$

The ditetragonal prism in the same way gives two prisms of the third order, whose basal sections are the same as those of the corresponding pyramid. They have the symbols:

$$\pi \{h k o\} \quad \pi \{k h o\}, \text{and} + \left[\frac{\infty P n}{2}\right] - \left[\frac{\infty P n}{2}\right].$$

The above are the only special geometrical forms produced by this hemihedrism ; they do not occur independently, but only in combinations, and are characteristic of a small but well-defined group of minerals, the Tungstates and Molybdates. One of the simplest examples is given in fig. 198, a crystal of molybdate of lead, in which a tetragonal prism of the third order $\pi\ \{4\,3\,0\}$ or $\left[\frac{\infty P\frac{4}{3}}{2}\right]$ truncates the basal edges of the unit pyramid obliquely to the right hand of the lateral axes.

FIG. 198.

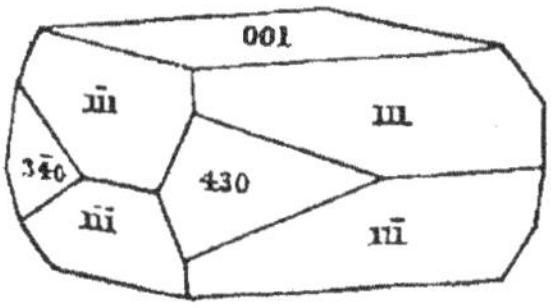

Tetartohedral tetragonal forms. The faces of the ditetragonal pyramid may, by the successive application of two kinds of hemihedral selection, be divided into four symmetrical groups, giving, as in the hexagonal system, two possible classes of tetartohedra, corresponding to the following divisions :—

Hemihedrism . . .	I.	II.	III.	IV.
Plagihedral and sphenoidal	$\frac{A}{H}$,	$\frac{B}{G}$,	$\frac{C}{E}$,	$\frac{D}{F}$
Sphenoidal and pyramidal .	$\frac{A}{G}$,	$\frac{B}{H}$,	$\frac{C}{F}$,	$\frac{D}{E}$

The faces to which these correspond will be seen in the table on page 120.

The forms corresponding to the first of these divisions are sphenoids, differing from those derived by hemihedrism from the tetragonal pyramid of the first order, in the position of their horizontal polar edges, which do not lie in planes of symmetry, but cross each other obliquely, so that the faces are scalene instead of isosceles triangles, and are not symmetrical to any principal section, while preserving the same axes of symmetry as the sphenoid, the relations in the latter respect being similar to those between the hexagonal rhombohedra and plagihedral tetartohedra. The pyramid of the second order in the same way becomes a horizontal prism, whose section is the rhomb, having for its diagonals the vertical and a lateral axis. The ditetragonal prism gives others of rhombic sections, whose diagonals are in the proportion of $a : n a$, and the prism of the second order gives two parallel pairs of faces. By the second method the ditetragonal pyramid produces sphenoids of the same geometrical properties as the hemihedral ones, but differently placed with respect to the axes, and which, like the analogous tetartohedral rhombohedra, may be said to be of the third order. The pyramid of the second order gives sphenoids of the second order, having their horizontal edges parallel to the lateral axes; and the ditetragonal prism gives tetragonal prisms of the third order. None of these tetartohedral forms have as yet been found either in natural or artificial crystals; but they are interesting as geometrical possibilities, and as showing the complete analogy subsisting between the hexagonal and tetragonal systems in all their modifications.

CHAPTER VI.

RHOMBIC[1] SYSTEM.

THE forms of this system are referred to three rectangular axes, whose parameters are all different. The three principal sections are planes of symmetry as in the cubic and tetragonal systems, but, on account of the dissimilarity of the parameters, the symmetry about the axes is only binary. These properties are apparent in a vertical prism of definite height upon an oblong rectangular base, whose length, breadth, and depth are all different, which is only symmetrical to its faces and about its edges, and the lengths of the latter are proportional to the parameters. In the

[1] Other names are, Orthorhombic, Orthoclinic, Prismatic, Trimetric, Terbinary, and Zweigliedrig.

general symbol $h\,k\,l$ the order of the indices is invariable, and the different faces are represented by sign permutations only, giving eight as the largest number of faces that can appear in any simple form. The forms represented by $h\,k\,k$ and $h\,h\,k$ are also of the same geometrical character, and as the crystallographic elements are most readily deduced from these more special kinds, it is convenient to consider them first rather than the general form. The unit of any series is a rhombic pyramid or octahedron, such as fig. 199, which, if represented by $\{1\ 1\ 1\}$ $a : b : c$ or P, has the parameters of $O\,A = a = 6$, $O\,B = b = 10$, and $O\,C = c = 8$.

FIG. 199.

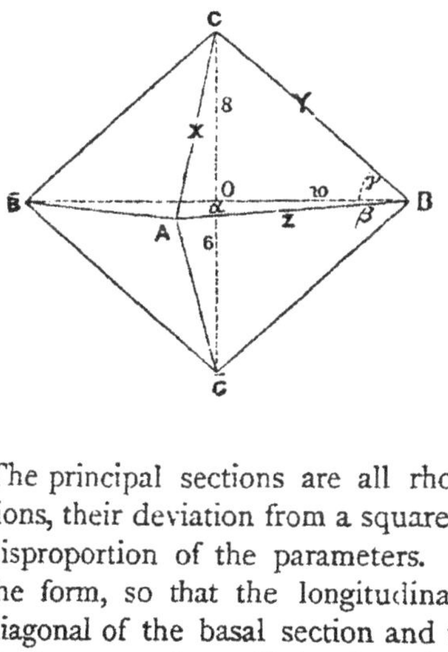

FIG. 200.

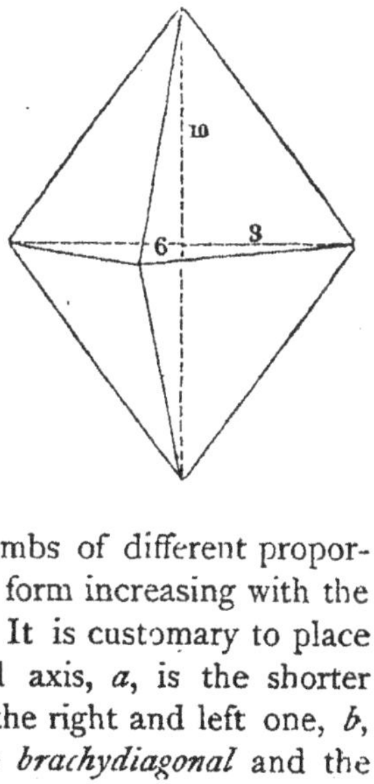

The principal sections are all rhombs of different proportions, their deviation from a square form increasing with the disproportion of the parameters. It is customary to place the form, so that the longitudinal axis, a, is the shorter diagonal of the basal section and the right and left one, b, the longer, and to call the first the ***brachydiagonal*** and the second the ***macrodiagonal*** axis, the third vertical axis being noted by c as in the preceding systems. To express the ratios of the parameters, one of them, usually that of the macrodiagonal, is put $= 1$: thus, in fig. 199, $a : b : c = 0{\cdot}6 : 1{\cdot}0 : 0{\cdot}8$; but if, as some authors prefer, the brachydiagonal is considered as the unit $a : b : c = 1 : 1{\cdot}666 : 1{\cdot}333$.[1] The choice of position in a rhombic pyramid is entirely arbitrary, as there are no peculiarities, either morphological or physical, making any one axis a principal one, so either one may be made vertical or horizontal at pleasure. Thus, turning fig. 199 about a until b is vertical gives fig. 200, when $a : b : c = 6 : 8 : 10 = 0{\cdot}75 : 1 : 1{\cdot}25$; and turning it about b gives the position of fig. 201, where the shortest axis is vertical and $a : b : c = 0{\cdot}8 : 1 : 0{\cdot}6$, all of which ratios have the same value.

FIG. 201.

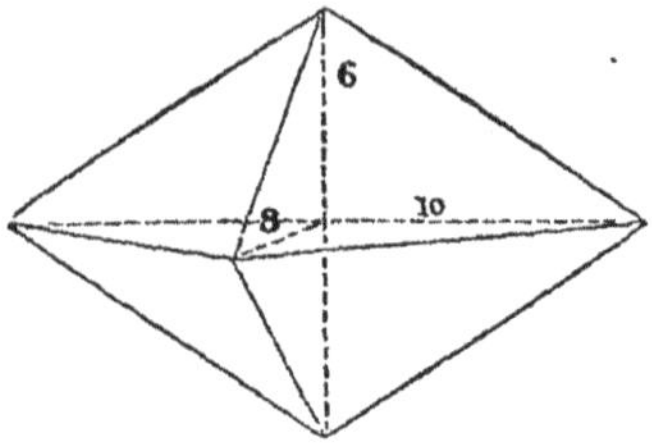

The choice of one of these positions as the normal one may be influenced by considerations based upon structural peculiarities, such as cleavages, analogies derived from similar forms proper to other species of like constitution, or other more arbitrary methods; but no definite rule can be laid down, and indeed the same crystal may be, and often is, differently described by different observers. Schrauf has suggested that the first median line of the optic axes should be taken as the vertical axis, but the suggestion can only be adopted where the crystal is transparent and susceptible of optical examination. The edges of the rhombic pyramid are of three kinds—namely, four basal, four longer, and four shorter polar ones. As the planes bisecting the dihedral angles between the principal sections are not planes of symmetry, the measurement of the angle in the basal edges is not sufficient to determine the parameter of c, one of either polar kind being required in addition. Calling (in fig. 199) Z the angle over a basal edge, Y and X those over a longer and shorter polar edge respectively, and the plane angle of Z upon α, and those of X and Y upon b, β, and γ,

[1] Naumann adopts the first order and Dana the second.

the relation of these angles to the parameters are expressed by the following formulæ:

$$\text{Given X and Y} \quad \cos.\ \alpha = \frac{\cos.\ \frac{1}{2}\ X}{\sin.\ \frac{1}{2}\ Z} \qquad \cos.\ \beta = \frac{\cos\ \frac{1}{2}\ Z}{\sin.\ \frac{1}{2}\ X}$$

$$\text{Y and Z} \quad \sin.\ \alpha = \frac{\cos.\ \frac{1}{2}\ Y}{\sin.\ \frac{1}{2}\ Z} \qquad \cos.\ \gamma = \frac{\cos.\ \frac{1}{2}\ Z}{\sin.\ \frac{1}{2}\ Y}$$

$$\text{X and Y} \quad \sin.\ \beta = \frac{\cos.\ \frac{1}{2}\ Y}{\sin.\ \frac{1}{2}\ X} \qquad \sin.\ \gamma = \frac{\cos.\ \frac{1}{2}\ X}{\sin.\ \frac{1}{2}\ Y}$$

—and when $b = 1$, $a =$ cotan. $\alpha =$ c tan. γ, and $c =$ tan. $\gamma = a$ tan. β. In the case of prismatic forms the angles α, β, and γ are found by direct measurement, and as forms of this kind are, as a rule, of very frequent occurrence, it is customary to give the obtuse angle of the prism as a chief characteristic in describing rhombic mineral species.

From any unit pyramid, by changing the value of c as before, new ones are obtained of varying altitude upon the same base, such as c_2, c_3 (fig. 202), ranging upwards into the prism and downwards into the basal pinakoid as in the preceding systems. These constitute the principal or vertical series, and are distinguished by the symbols—

$\{001\}$	$\{h\,h\,l\}\ h<l$	$\{1\,1\,1\}$	$\{h\,h\,l\}\ h>l$	$\{1\,1\,0\}$
$\infty a : \infty b : c$	$a : b : \frac{1}{m}c$	$a : b : c$	$a : b : mc$	$a : b : \infty c$
$0\,P$	$\frac{1}{m}P$	P	$m\,P$	$\infty\,P$

Any pyramid of a principal series will, by successively lengthening its macrodiagonal axis by any rational multiples n, produce new pyramids all of the same height and length on the axes a and c, but increasing in breadth on b as n becomes greater, as seen for $2b$ and $3b$ in fig. 203; and when $n = \infty$ the form is a rhombic prism, whose edges are horizontal and parallel to the axis b, the axes a and c being the diagonals of its base. Prisms of this kind are called *domes* or *domas*, from their resemblance to house roofs, and the particular one in question is known as the *macrodome*, its zone axis being the macrodiagonal. If this is derived from

FIG. 202.

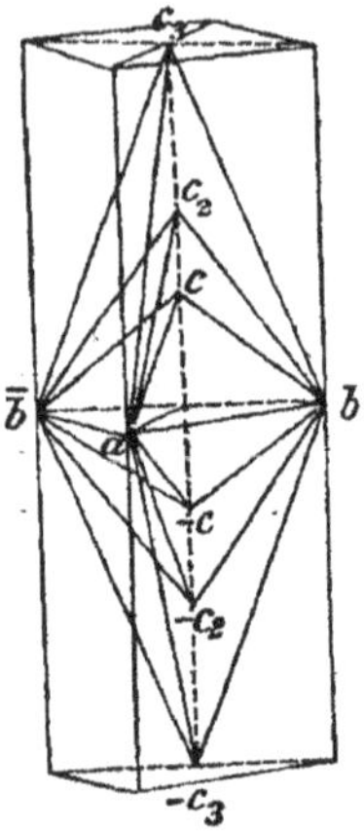

FIG. 203.

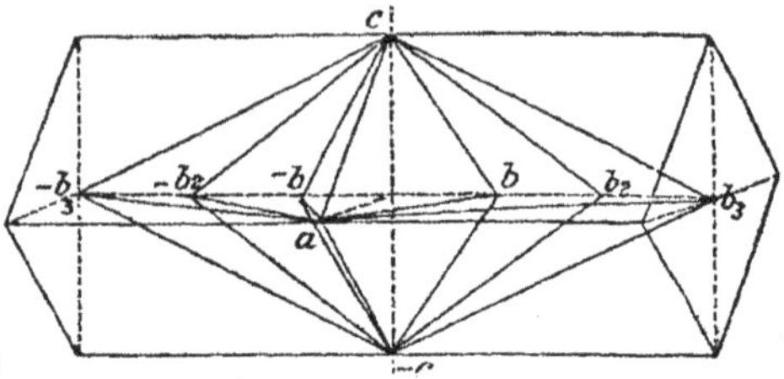

the unit pyramid, it will be the unit macrodome; but as the same development applies equally to any of the pyramids of the principal series, they will give a corresponding succession of macropyramids and limiting macrodomes, the whole constituting the transverse prismatic or macrodiagonal series —represented by Naumann's general symbols $m\bar{P}n$, $m\bar{P}\infty$, where the long sign signifies that n applies to the longer lateral axis. This is more directly indicated in Breithaupt's modification of the symbol, $mP\bar{n}$, which is therefore preferred by some writers, although the first form is most generally used. The order of symbols for the principal types of this series is as follows:—

$\{hkl\}\ h>k\ h<l$	$\{hkh\}\ h>k$	$\{hkl\}\ h>k\ k\gtrless l$
$a:nb:\frac{1}{m}c$	$a:nb:c$	$a:nb:mc$
$\frac{1}{m}\bar{P}n$	$\bar{P}n$	$m\bar{P}n$
$\{h0l\}\ h<l$	$\{h0h\}$	$\{h0l\}\ h>l$
$a:\infty b:\frac{1}{m}c$	$a:\infty b:c$	$a:\infty b:mc$
$\frac{1}{m}\bar{P}\infty$	$\bar{P}\infty$	$m\bar{P}\infty$

The extension of the brachydiagonal of any unit or

other pyramid of the principal series—as in fig. 204, where that axis is successively made $2a$, $3a$, ∞a, while b and c are unchanged—produces a third class of pyramids and domes known as the longitudinal or brachydiagonal series, which are distinguished in their symbols from the transverse series by the sign ˘, indicating the shorter lateral axis, placed over the characteristic P or its coefficient, thus $m\breve{P}n$ or $mP\breve{n}$. The order of the typical symbols is as follows :

$\{hkl\}\ h<k<l$	$\{hkk\}\ h<k$	$\{hkl\}\ h<k\ h\gtrless l$
$na : b : \frac{1}{m}c$	$na : b : c$	$na : b : mc$
$\frac{1}{m}\breve{P}n$	$\breve{P}n$	$m\breve{P}n$
$\{0kl\}\ k<l$	$\{0kk\}$	$\{0kl\}\ k>l$
$\infty a : b : \frac{1}{m}c$	$\infty a : b : c$	$\infty a : b : mc$
$\frac{1}{m}\breve{P}\infty$	$\breve{P}\infty$	$m\breve{P}\infty$

FIG. 204.

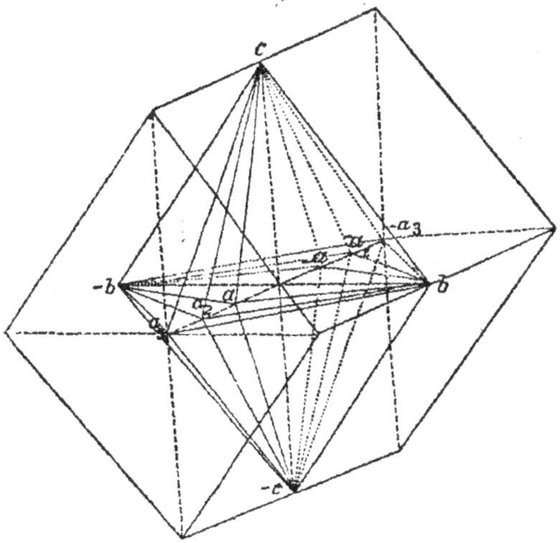

The prism of the principal series having its lateral axes in the unit proportion $a : b$, will, by lengthening either of these axes relatively to the other, give rise to new prisms, whose characteristic angle increases with the value of *n*. These are called macroprisms, when derived from the variation of *b*, and brachyprisms from *a*, which are respectively represented by $\infty \bar{P} n$ and $\infty \breve{P} n$. When *n* in either series becomes ∞, the angle of the prism is 180°, or it is reduced to a single plane parallel to one lateral axis and perpendicular to the other, giving the two forms known as the macropinakoid and brachypinakoid; the former is represented by $\{1\,0\,0\}\ a : \infty b : \infty c$, or $\infty \bar{P} \infty$, and the latter by $\{0\,1\,0\}\ \infty a : b : \infty c$, or $\infty \breve{P} \infty$.

The above include all the simple forms possible in the rhombic system, namely, pyramids of eight faces, the only closed forms, prisms of four faces in three positions, and the three pinakoids of two faces. Their relations to each other may be shown by arranging their symbols in a diagram of the kind given in the preceding systems. Here the unit form *P* in the centre has the obtuser forms of the principal series below and the acuter ones above it in the same vertical line. The lines next to right and left contain the series of the macropyramids and brachypyramids respectively, and the first and last lines, the brachydomes and macrodomes. The top horizontal line contains the symbol of the basal pinakoid as the common limit of all the series; the other pinakoids are at the ends of the lower horizontal line, the intermediate positions being taken by the different prisms. As in the previous system, the forms contained in any line, whether horizontal or vertical, lie in the same zone.

$$\begin{array}{ccccccccc}
0\,P & \ldots & 0\,P & \ldots & 0\,P & \ldots & 0\,P & \ldots & 0\,\dot{P} \\
\vdots & & \vdots & & \vdots & & \vdots & & \vdots \\
\frac{1}{m}\breve{P}\infty & \ldots & \frac{1}{m}\breve{P}n & \ldots & \frac{1}{m}P & \ldots & \frac{1}{m}\bar{P}n & \ldots & \frac{1}{m}\bar{P}\infty \\
\vdots & & \vdots & & \vdots & & \vdots & & \vdots \\
\breve{P}\infty & \ldots & \breve{P}n & \ldots & P & \ldots & \bar{P}n & \ldots & \bar{P}\infty \\
\vdots & & \vdots & & \vdots & & \vdots & & \vdots \\
m\breve{P}\infty & \ldots & m\breve{P}n & \ldots & mP & \ldots & m\bar{P}n & \ldots & m\bar{P}\infty \\
\vdots & & \vdots & & \vdots & & \vdots & & \vdots \\
\infty\breve{P}\infty & \ldots & \infty\breve{P}n & \ldots & \infty P & \ldots & \infty\bar{P}n & \ldots & \infty\bar{P}\infty
\end{array}$$

Series	Symbol
Brachypyramids.	$2a : b : \frac{1}{3}c$, (1 2 6), $\frac{1}{3}\breve{P}2$
	$2a : b : \frac{1}{2}c$, (1 2 4), $\frac{1}{2}\breve{P}2$
	$2a : b : \frac{2}{3}c$, (1 2 3), $\frac{2}{3}\breve{P}2$
	$2a : b : c$, (1 2 2), $\breve{P}2$
	$2a : b : \frac{3}{2}c$, (3 6 4), $\frac{3}{2}\breve{P}2$
	$2a : b : 2c$, (1 2 1), $2\breve{P}2$
	$2a : b : 3c$, (3 6 2), $3\breve{P}2$
Brachyprism.	$2a : b : \infty c$, (1 2 0), $\infty\breve{P}2$
Macropyramids.	$a : \frac{3}{2}b : \frac{1}{3}c$, (6 2 9), $\frac{1}{3}\bar{P}\frac{3}{2}$
	$a : \frac{3}{2}b : \frac{1}{2}c$, (3 2 6), $\frac{1}{2}\bar{P}\frac{3}{2}$
	$a : \frac{3}{2}b : \frac{2}{3}c$, (6 4 9), $\frac{2}{3}\bar{P}\frac{3}{2}$
	$a : \frac{3}{2}b : c$, (3 2 3), $\bar{P}\frac{3}{2}$
	$a : \frac{3}{2}b : \frac{3}{2}c$, (3 2 2), $\frac{3}{2}\bar{P}\frac{3}{2}$
	$a : \frac{3}{2}b : 2c$, (6 4 3), $2\bar{P}\frac{3}{2}$
	$a : \frac{3}{2}b : 3c$, (3 2 1), $3\bar{P}\frac{3}{2}$
Macroprism.	$a : \frac{3}{2}b : \infty c$, (3 2 0), $\infty\bar{P}\frac{3}{2}$
Macropyramids.	$a : \infty b : \frac{1}{3}c$, (1 0 3), $\frac{1}{3}\bar{P}\infty$
	$a : \infty b : \frac{1}{2}c$, (1 0 2), $\frac{1}{2}\bar{P}\infty$
	$a : \infty b : \frac{2}{3}c$, (2 0 3), $\frac{2}{3}\bar{P}\infty$
	$a : \infty b : c$, (1 0 1), $\bar{P}\infty$
	$a : \infty b : \frac{3}{2}c$, (3 0 2), $\frac{3}{2}\bar{P}\infty$
	$a : \infty b : 2c$, (2 0 1), $2\bar{P}\infty$
	$a : \infty b : 3c$, (3 0 1), $3\bar{P}\infty$
Macropinakoid.	$a : \infty b : \infty c$, (1 0 0), $\infty\bar{P}\infty$
Brachypyramids.	$3a : b : \frac{1}{3}c$, (1 3 9), $\frac{1}{3}\breve{P}3$
	$3a : b : \frac{1}{2}c$, (1 3 6), $\frac{1}{2}\breve{P}3$
	$3a : b : \frac{2}{3}c$, (2 6 9), $\frac{2}{3}\breve{P}3$
	$3a : b : c$, (1 3 3), $\breve{P}3$
	$3a : b : \frac{3}{2}c$, (1 3 2), $\frac{3}{2}\breve{P}3$
	$3a : b : 2c$, (2 6 3), $2\breve{P}3$
	$3a : b : 3c$, (1 3 1), $3\breve{P}3$
Brachyprism.	$3a : b : \infty c$, (1 3 0), $\infty\breve{P}3$
Principal-pyramids.	$a : b : \frac{1}{3}c$, (1 1 3), $\frac{1}{3}P$
	$a : b : \frac{1}{2}c$, (1 1 2), $\frac{1}{2}P$
	$a : b : \frac{2}{3}c$, (2 2 3), $\frac{2}{3}P$
	$a : b : c$, (1 1 1), P
	$a : b : \frac{3}{2}c$, (3 3 2), $\frac{3}{2}P$
	$a : b : 2c$, (2 2 1), $2P$
	$a : b : 3c$, (3 3 1), $3P$
Principal-prism.	$a : b : \infty c$, (1 1 0), ∞P
Macropyramids.	$a : 3b : \frac{1}{3}c$, (3 1 9), $\frac{1}{3}\bar{P}3$
	$a : 3b : \frac{1}{2}c$, (3 1 6), $\frac{1}{2}\bar{P}3$
	$a : 3b : \frac{2}{3}c$, (6 2 9), $\frac{2}{3}\bar{P}3$
	$a : 3b : c$, (3 1 3), $\bar{P}3$
	$a : 3b : \frac{3}{2}c$, (3 1 2), $\frac{3}{2}\bar{P}3$
	$a : 3b : 2c$, (6 2 3), $2\bar{P}3$
	$a : 3b : 3c$, (3 1 1), $3\bar{P}3$
Macroprism.	$a : 3b : \infty c$, (3 1 0), $\infty\bar{P}3$
Brachydomes.	$\infty a : b : \frac{1}{3}c$, (0 1 3), $\frac{1}{3}\breve{P}\infty$
	$\infty a : b : \frac{1}{2}c$, (0 1 2), $\frac{1}{2}\breve{P}\infty$
	$\infty a : b : \frac{2}{3}c$, (0 2 3), $\frac{2}{3}\breve{P}\infty$
	$\infty a : b : c$, (0 1 1), $\breve{P}\infty$
	$\infty a : b : \frac{3}{2}c$, (0 3 2), $\frac{3}{2}\breve{P}\infty$
	$\infty a : b : 2c$, (0 2 1), $2\breve{P}\infty$
	$\infty a : b : 3c$, (0 3 1), $3\breve{P}\infty$
Brachypinakoid.	$\infty a : b : \infty c$, (0 1 0), $\infty\breve{P}\infty$
Brachypyramids.	$\frac{3}{2}a : b : \frac{1}{3}c$, (2 3 9), $\frac{1}{3}\breve{P}\frac{3}{2}$
	$\frac{3}{2}a : b : \frac{1}{2}c$, (2 3 9), $\frac{1}{2}\breve{P}\frac{3}{2}$
	$\frac{3}{2}a : b : \frac{2}{3}c$, (4 6 9), $\frac{2}{3}\breve{P}\frac{3}{2}$
	$\frac{3}{2}a : b : c$, (2 3 3), $\breve{P}\frac{3}{2}$
	$\frac{3}{2}a : b : \frac{3}{2}c$, (2 3 2), $\frac{3}{2}\breve{P}\frac{3}{2}$
	$\frac{3}{2}a : b : 2c$, (4 6 3), $2\breve{P}\frac{3}{2}$
	$\frac{3}{2}a : b : 3c$, (2 3 1), $3\breve{P}\frac{3}{2}$
Brachyprism.	$\frac{3}{2}a : b : \infty c$, (2 3 0), $\infty\breve{P}\frac{3}{2}$
Macropyramids.	$a : 2b : \frac{1}{3}c$, (2 1 6), $\frac{1}{3}\bar{P}2$
	$a : 2b : \frac{1}{2}c$, (2 1 4), $\frac{1}{2}\bar{P}2$
	$a : 2b : \frac{2}{3}c$, (2 1 3), $\frac{2}{3}\bar{P}2$
	$a : 2b : c$, (2 1 2), $\bar{P}2$
	$a : 2b : \frac{3}{2}c$, (6 3 4), $\frac{3}{2}\bar{P}2$
	$a : 2b : 2c$, (2 1 1), $2\bar{P}2$
	$a : 2b : 3c$, (6 3 2), $3\bar{P}2$
Macroprism.	$a : 2b : \infty c$, (2 1 0), $\infty\bar{P}2$

If all the columns in this table be placed side by side in the same line, it will form substantially an extension of the diagram on the preceding page. The symbol of the basal pinakoid common to all is omitted.

From the small number of faces possible in any single form, the notation of rhombic crystals, especially in Naumann's method, involves a considerable diversity in the symbols. In illustration of this point, the table on the preceding page gives the notation of a face of $m\,P\,n$, $m\,\breve{P}\,n$, $m\,\bar{P}\,n$, when m is successively made 0, $\frac{1}{3}$, $\frac{1}{2}$, $\frac{2}{3}$, 1, $\frac{3}{2}$, 2, 3, and ∞, and n 1, $\frac{3}{2}$, 2, 3, and ∞, from which it will be seen that these coefficients involve the use as indices of the numbers 0, 1, 2, 3, 4, 6, and 9.

In combinations of rhombic forms, as in those of the preceding systems, any pyramid has its basal edges bevelled by a more acute one, and truncated by the prism; and its polar solid angles blunted by a more obtuse pyramid of the same series, and truncated by the basal pinakoid. The polar edges in the brachydiagonal section are bevelled by any macropyramid and truncated by a macrodome, having a and c in common, and the other polar edges, those in the macrodiagonal section, are similarly modified by the brachypyramid and brachydome, having b and c in common with the pyramid. A prism has its obtuse edges, or those facing the macrodiagonal, bevelled by brachyprisms, and truncated by the brachypinakoid; and its acute edges are similarly modified by macroprisms and the macropinakoid. The observed combinations are exceedingly numerous and diversified in appearance, which diversity is a consequence of there being only lateral symmetry to the axes; and although the edges in all the pinakoid sections may be modified by faces of the same kind, it is more general to find them differently modified in each section. For instance, in the three figs. 205, 206. 207, the first, in which P is truncated longitudinally and vertically by the pinakoids $\infty\,\bar{P}\,\infty$ and $0\,P$, and transversely by the middle edges of the unit brachydome $\breve{P}\infty$, and the

FIG. 205.

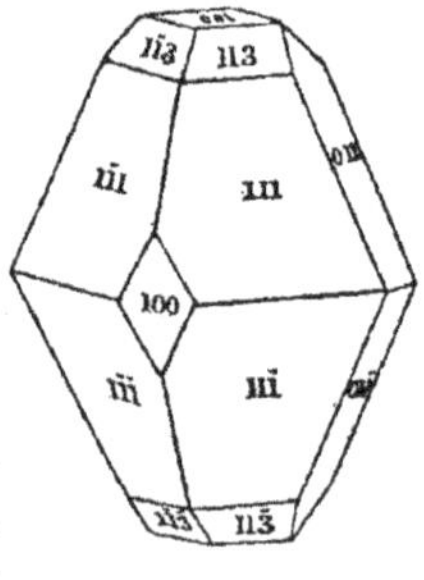

second, where a joins the solid angles formed by the crossing of the edges of the unit prism ∞P and macrodome $\bar{P}\infty$, b, the acute edges of ∞P, and c, the faces of the terminal pinakoid—are more characteristic cases than fig. 207, where all

FIG. 206.

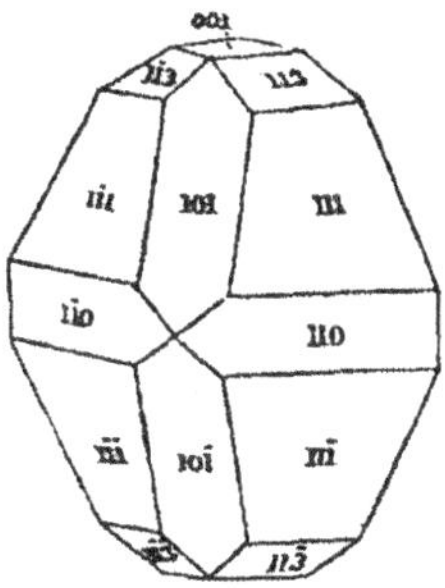

FIG. 207.

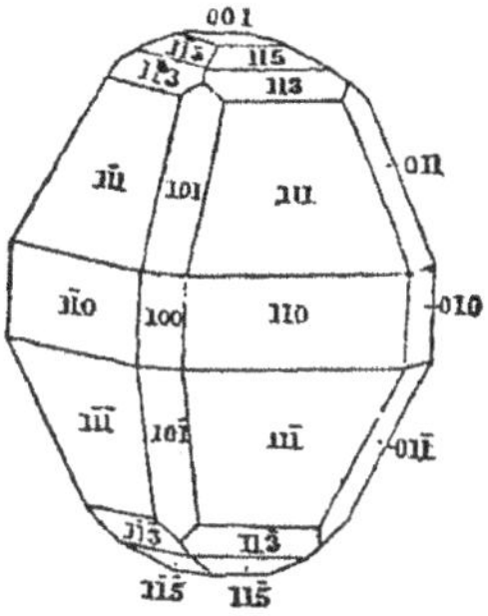

three pinakoids and prismatic forms are present. These examples are taken from the species sulphur, which is one of the few minerals in this system whose crystals have pyramids as the dominant forms. These are noted as P, $\frac{1}{3}P$, and $\frac{1}{5}P$. Fig. 205 is the common form obtained when sulphur

FIG. 208.

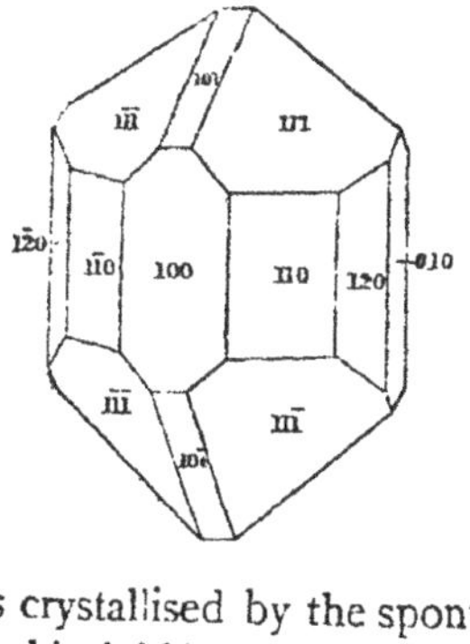

FIG. 209.

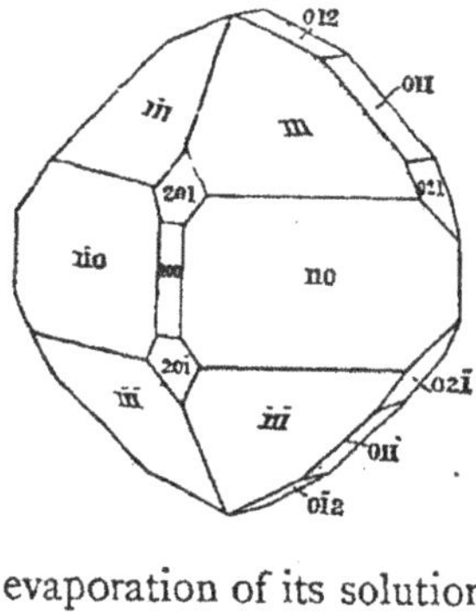

is crystallised by the spontaneous evaporation of its solution in bisulphide of carbon.

Figs. 207, 208 are examples of combinations without the basal pinakoid. The first, drawn to the elements of the species Topaz, contains P, ∞P, $\infty \breve{P} 2$, $\bar{P}\infty$, $\infty\breve{P}\infty$, $\infty\bar{P}\infty$, most of these forms being in the zone of the principal prism; while fig. 209 is chiefly modified in the longitudinal prismatic or brachydiagonal zone, which contains $\frac{1}{2}\breve{P}\infty$, $\breve{P}\infty$, $2\breve{P}\infty$,

FIG. 210.

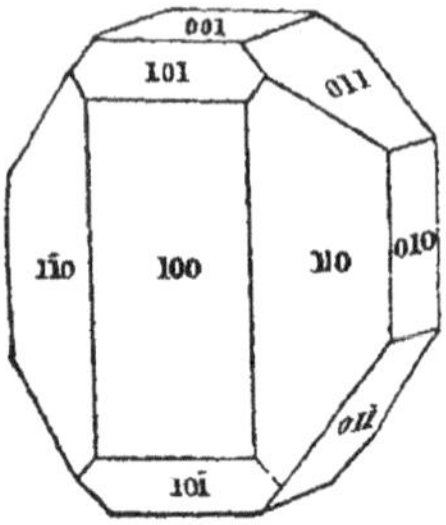

FIG. 211.

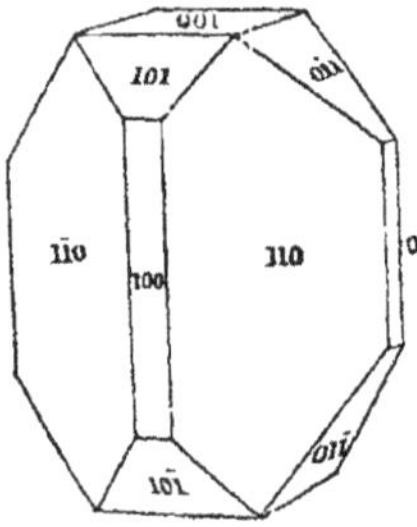

the other forms being P, ∞P, $2\bar{P}\infty$, $\infty\breve{P}\infty$; this is drawn to the parameters of Brookite.

Another consequence of the absence of diagonal axes of symmetry is the tendency to elongation parallel to one axis, producing solids which are essentially prisms, a peculiarity

FIG. 212.

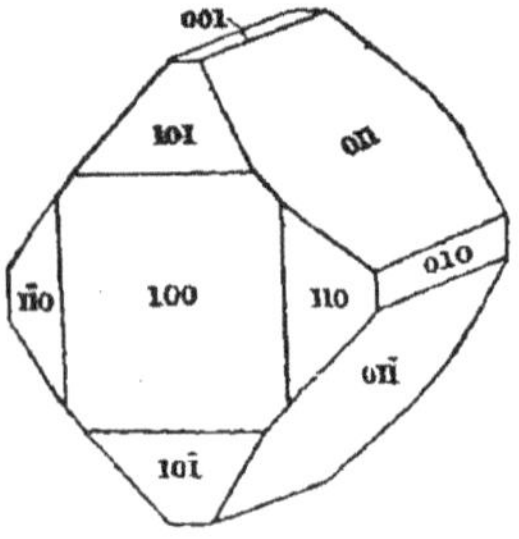

FIG. 213.

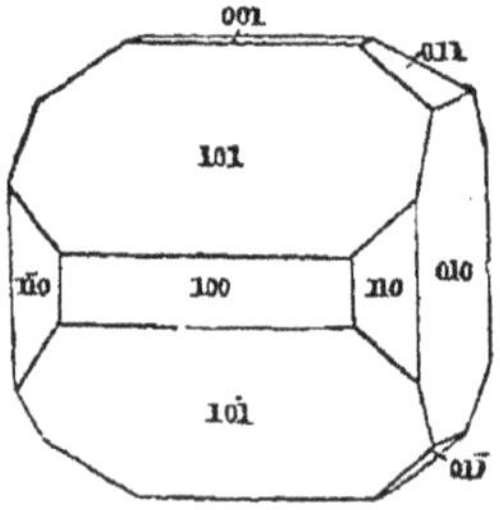

which is indicated in the name 'prismatic' applied by Miller to this system; and as the extension may be along either axis indifferently, the same combination in the same substance may appear in many different shapes, according as one or other form predominates. Thus in fig. 210 the three unit prismatic forms (prism and domes) are combined in about equal dimensions with the three pinakoids; in fig. 211 the solid is essentially prismatic, the vertical edges being the longest. In fig. 212 the character is longitudinal, prismatic or brachydomatic, the greatest length being parallel to *a*; and in fig. 213 it is transverse-prismatic, or macrodomatic, the longest edges being parallel to *b*. These variations are quite possible, and are to some extent represented in the species Barytes.

FIG. 214.

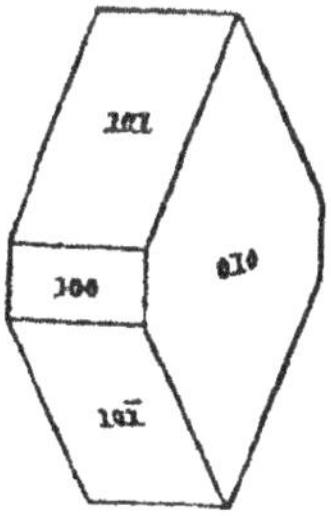

FIG. 215.

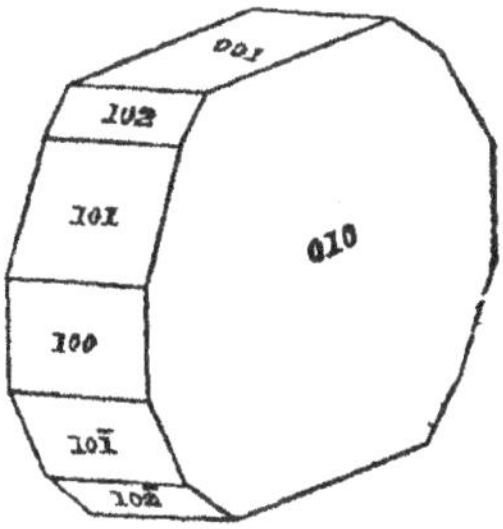

A few more illustrations of the simpler class of combinations are given in figs. 213–223, from the closely allied species, Barytes and Anglesite, which are remarkable for their

FIG. 216.

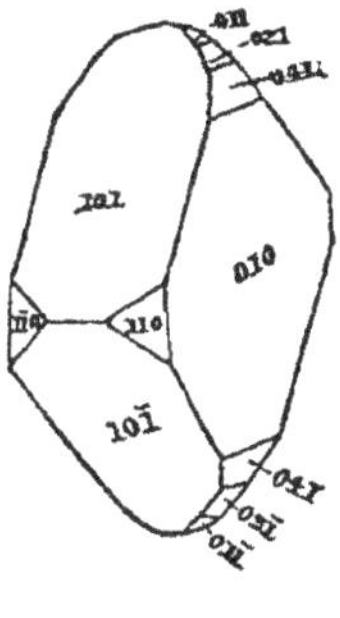

FIG. 217.

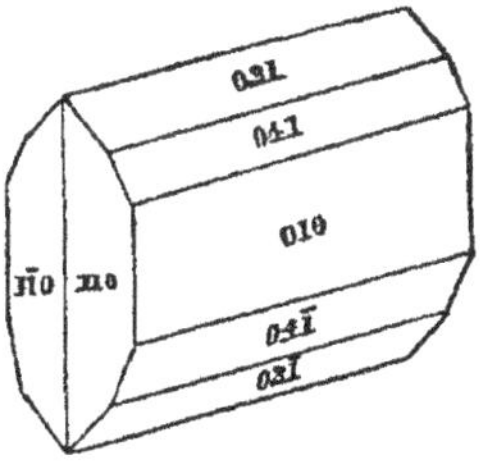

great variety of forms, these being selected from nearly a hundred described crystals of these minerals. Fig. 214 is the unit macrodome of Barytes, $\bar{P}\infty$, shortened in the direction of its axis by $\infty \breve{P} \infty$, and truncated in its middle edges by $\infty \bar{P} \infty$. Fig. 215 is similar, with the addition of $\frac{1}{2} \bar{P} \infty$ and $0 P$. Fig. 216 is $\bar{P} \infty$, with the middle solid angles truncated obliquely by the prism ∞P, and the upper and lower

FIG. 218.

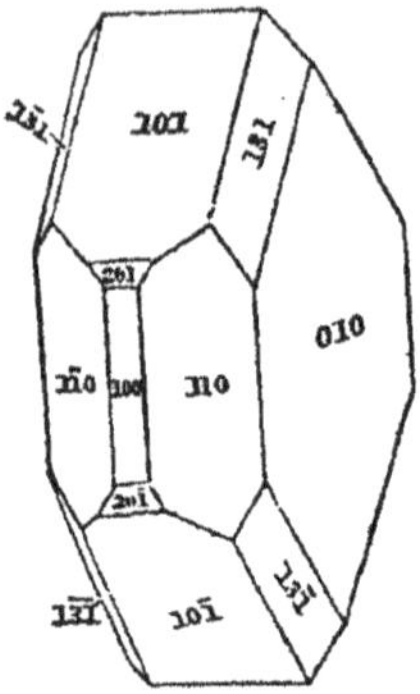

FIG. 219.

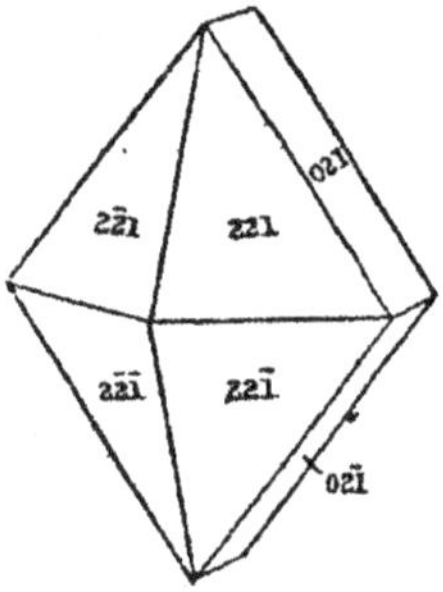

ones by the brachydomes $\breve{P} \infty$, $2 \breve{P} \infty$, $4 \breve{P} \infty$. Fig. 217 contains $2 \breve{P} \infty$, $4 \breve{P} \infty$, $\infty \breve{P} \infty$, elongated parallel to a, and limited by the prism ∞P. In fig. 218, the upper edges between $\bar{P} \infty$ and $\infty \breve{P} \infty$ are truncated obliquely by the acute brachypyramid $3 \breve{P} 3$, and those between ∞P and $\infty \bar{P} \infty$ in front by a more acute macrodome $2 \bar{P} \infty$. Fig. 219 is the

FIG. 220.

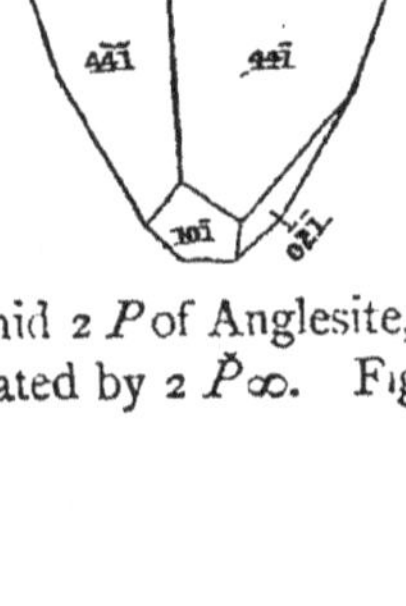

FIG. 221.

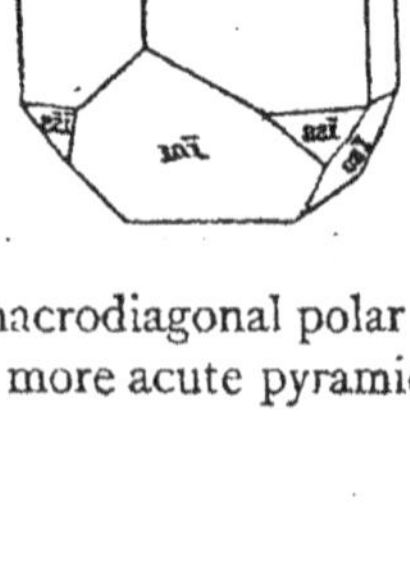

pyramid $2 P$ of Anglesite, with its macrodiagonal polar edges truncated by $2 \breve{P} \infty$. Fig. 220 is a more acute pyramid, $4 P$

with the same brachydome, and the unit macrodome modifying the brachydiagonal polar edges. Fig. 221 contains ∞P, $\bar{P}\infty$, $2\breve{P}\infty$, $\infty\breve{P}\infty$, and $2P$, the latter modifying the solid angles formed by the meeting of the three prismatic forms. That this is one of the pyramids of the principal series is apparent from the horizontality of its edges of combination with the prism.

In fig. 222 the faces of the macrodome $2\bar{P}\infty$ are so

FIG. 222.

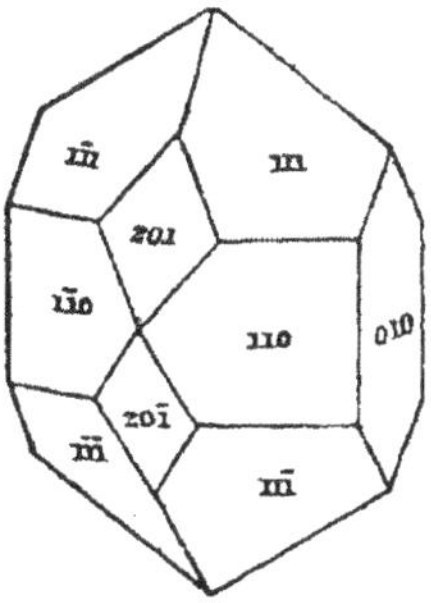

FIG. 223.

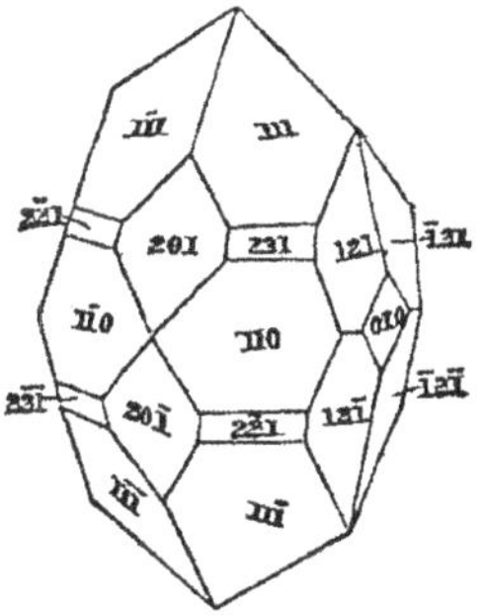

proportioned as to form rhombic planes truncating the front solid angles between ∞P and P. Fig. 223 is of the same general character, but the basal edges of the prism are

FIG. 224.

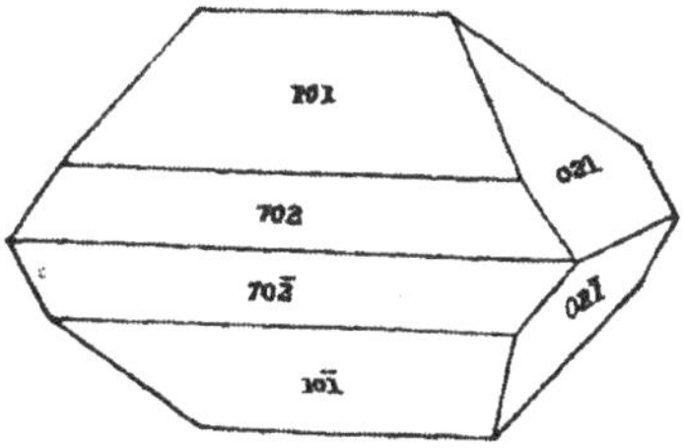

FIG. 225.

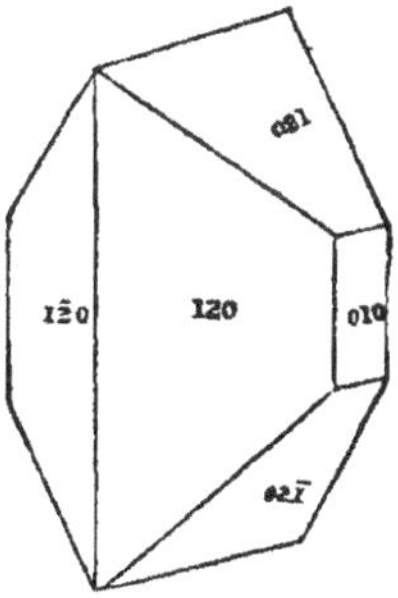

modified by $2P$; and the lateral solid angles formed by P, ∞P, and $\infty\breve{P}\infty$, are replaced by an acute brachypyramid

$2\breve{P}2$, whose edges of combination are parallel to the macrodiagonal polar edges of the pyramid. Fig. 224 is the unit macrodome $\bar{P}\infty$, lengthened parallel to its axis, limited laterally by $2\breve{P}\infty$, and its middle edges bevelled by $\frac{7}{2}\bar{P}\infty$. Figs. 225, 226 are examples of simple combinations of Olivine. The first is $\infty\breve{P}2$, $\infty\breve{P}\infty$, $2\breve{P}\infty$, and the second $\infty\breve{P}2$, $\infty\breve{P}\infty$, $\breve{P}\infty$, the latter being very commonly observed in crystallised slags obtained in puddling and heating furnaces. These examples will suffice to show the general character of the combinations of this system, but they are only of simpler kinds. For those of more complex character the reader is referred to the larger special memoirs and descriptions, especially to Schrauf's atlas of crystalline forms.

FIG. 226.

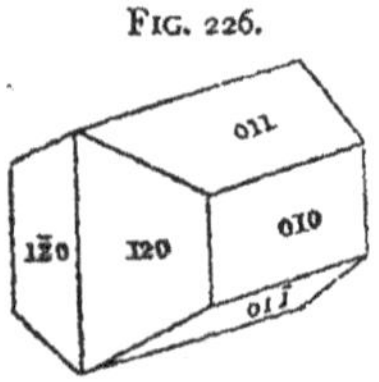

The fundamental parameters of any rhombic series of crystals, being irrational numbers, they may, when two are nearly equal, produce forms approximating in character to those of the tetragonal system, but the true nature will usually be apparent by their modifications. Fig. 221, for instance, would be nearly like the common combination $P\infty P$ in Tinstone but for the rhombic faces of $2\bar{P}\infty$.

FIG. 227.

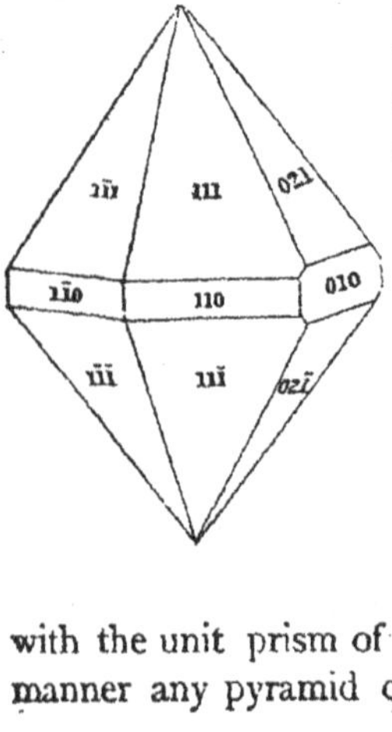

Where the parameters of the axes a and b are related to each other in the proportion of $1 : \sqrt{3}$. the obtuse angle of the prism will be 120°, and as the brachypinakoid truncates its acute angles, the combination of these two forms will be an equal six-sided prism, having all its angles of 120°, or geometrically identical with the unit prism of the hexagonal system; and in like manner any pyramid of the same series combined with a brachydome of twice the height—i.e., P with $2\breve{P}\infty$, or $2P$ with $4\breve{P}\infty$—will produce a regular hexagonal pyramid. Fig. 221 is an example of such a combination in Witherite, which is very similar in appearance to the ordinary form of quartz crystal. Other examples are afforded by Aragonite and Disulphide of Copper, where the prismatic angle approaches very nearly to 120°. Such forms may, however, as a rule, be discriminated without much difficulty by the unequal modification of their edges, peculiarities of cleavage, &c., and, when transparent, by their optical properties. The solid formed by the combination of the three pinakoids may also, in some instances, appear very like a cube; the best example of this is afforded by Anhydrite, or anhydrous Sulphate of Calcium.

Hemihedral rhombic forms. A rhombic pyramid may, by the omission of alternate faces right and left, above and below the base, give rise to hemihedral forms analogous to the tetartohedral sphenoids of the tetragonal system, as

FIG. 228.

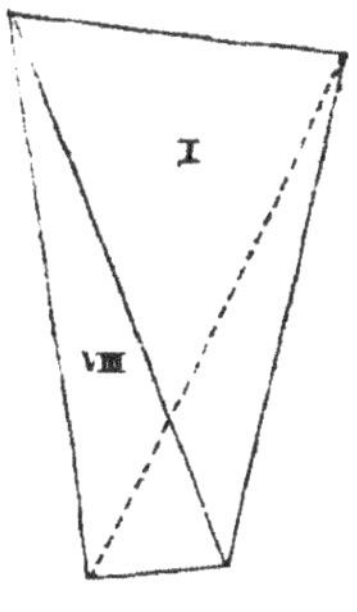

FIG. 229.

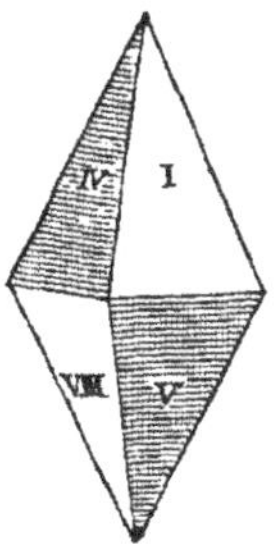

FIG. 230.

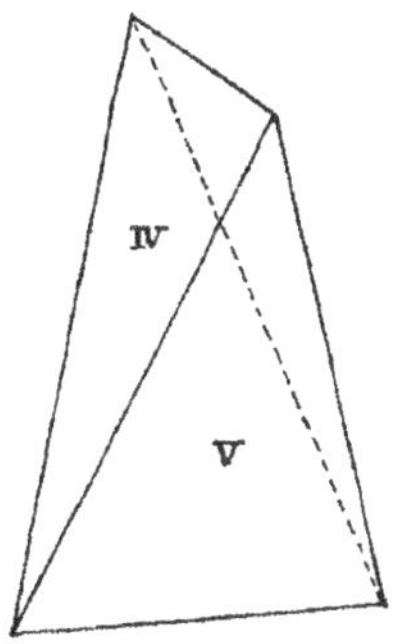

shown in figs. 228 and 230, the first being that derived from the white faces, and the second from the shaded ones in fig. 229. These differ from the tetragonal sphenoid by the inequality of the middle edges, two being obtuse and two acute; and also their polar edges do not lie at right angles to each other, but cross obliquely, the angle between their horizontal projections corresponding to the acute angle of the prism. As in the case of the tetrahedron and tetragonal sphenoids, they are not symmetrical to rectangular axial planes; and these being the only possible planes of symmetry in the system, they are plagihedral, being permanently right- and left-handed, according to their origin. The symbols are:

$$\kappa\ \{h k l\} \text{ and } + \frac{m P}{2} \text{ and } - \frac{m P}{2}.$$

The prismatic forms are not affected geometrically by this kind of hemihedrism, which is not of very frequent occurrence in natural crystals. The principal examples (fig. 231) are found in Sulphate of Magnesium and the isomorphous Sulphate of Zinc.

FIG. 231.

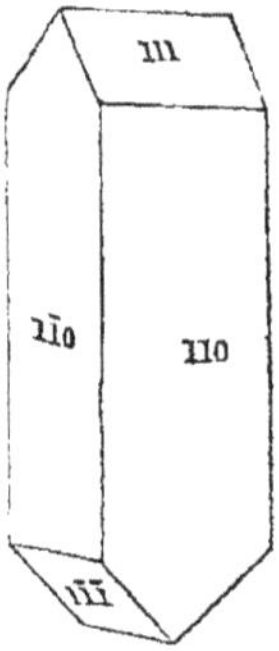

There is another class of hemihedral forms possible in the rhombic system, but only a single example has been demonstrated as existing in an artificially crystallised compound. These are analogous to the scalenohedral or rhombohedral forms of the preceding systems, being produced by the alternate development of pairs of faces preserving one of the original edges. The result is the production of a prism upon the same base as the pyramid, whose edges are inclined instead of being

FIG. 232.

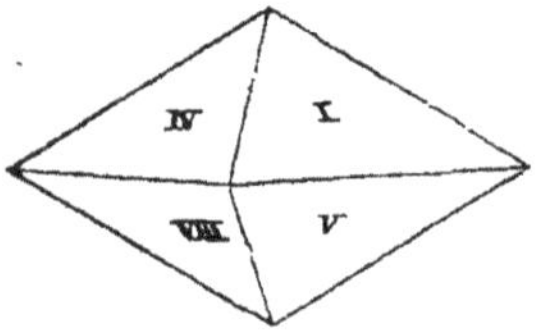

FIG. 233.

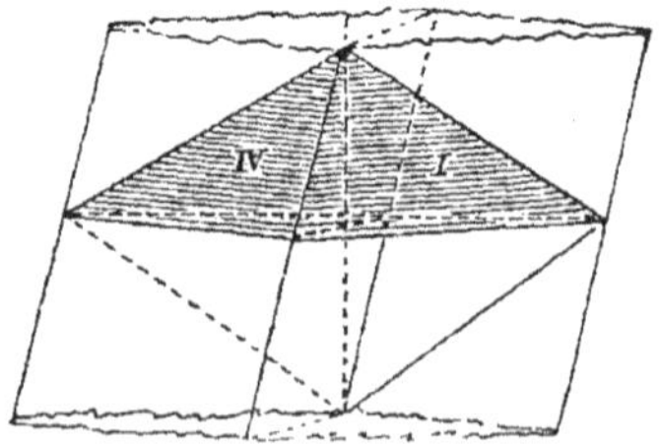

vertical or horizontal, the inclination being similar to that of the edges retained. For instance, fig. 233 is the oblique rhombic prism of this kind produced from the faces I., IV., VI., and VII., in fig. 232; its edges are parallel to the more obtuse polar edges in the pyramid, and its faces are symmetrical to the brachypinakoid alone. The lower front and upper back pairs of faces would produce a similar form, with a forward slope, but symmetrical to the same plane, and in like manner from the extension of alternate pairs of basal and macrodiagonal edges, pairs of similar prismatic forms may be derived, having the same symmetry to one axial plane only. From this latter circumstance, such forms are said to have monosymmetric hemihedrism, a property which they have in common with all other forms of the next system.

Formerly several minerals were referred to this type of hemihedrism, but they are now, upon structural considerations, placed in the oblique system.

CHAPTER VII.

OBLIQUE[1] SYSTEM.

THE forms of this system are referred to three axes having dissimilar parameters, one being at right angles to the other

FIG. 234.

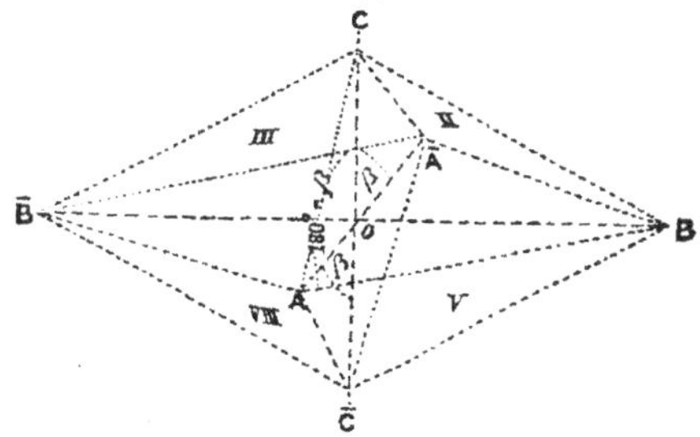

two. If this be considered as the axis of breadth, and placed horizontally, as $\bar{B}$ B in fig. 234, it will be normal to

[1] Other names are clinorhombic, monoclinic, oblique-rhombic, binary, monosymmetric, and zwei und eingliedrig.

an upright longitudinal plane containing the other two axes, which may be oblique to each other, and parallel to two others, in each of which it will be at right angles to one of the remaining axes. The solid whose edges are shown by the fine dotted lines, which may be supposed to represent that derived from the unit parameters, will therefore be rhombic in two of its principal sections, and rhomboidal in the third, as in the three plane projections, figs. 235, 236, 237, which are similarly noted to the preceding one. From these it will be seen that the rhomboid A C $\bar{A}$ $\bar{C}$, fig. 237, divides the solid symmetrically into right and left halves, whether it be looked at from the front, as in fig. 235, or from above, as in fig. 236; while in fig. 237 the division by the

FIG. 235.

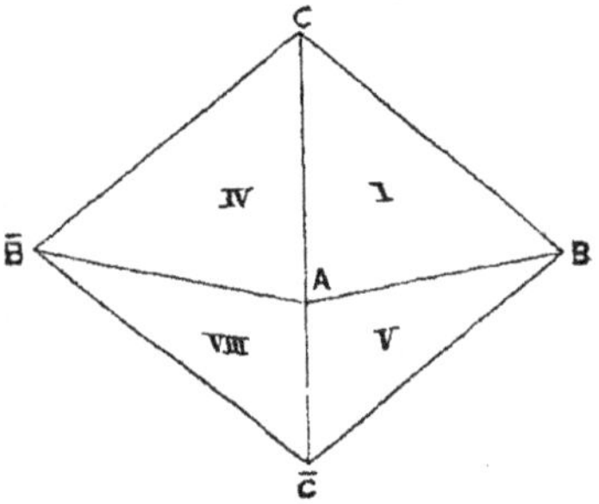

FIG. 236.

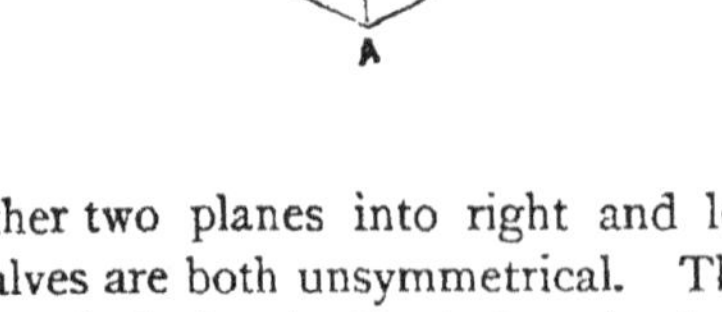

FIG. 237.

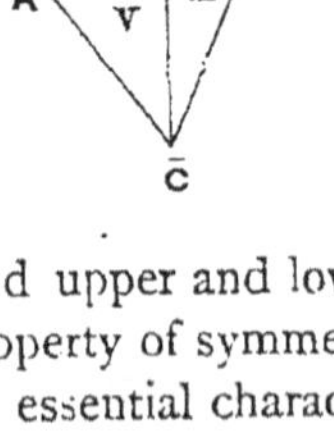

other two planes into right and left and upper and lower halves are both unsymmetrical. This property of symmetry to a single longitudinal plane is the most essential character of the system, and there is usually a marked obliquity between the axes in that plane, which as a necessary crystallographic element, in addition to the parameters *a*, *b*, *c*, is expressed as the angle β. When, as is generally done, one of these axes is placed upright, and the third with a forward inclination, as in fig. 234, β is considered as the acute angle in front below, or on the negative side of the vertical axis; the supplemental obtuse angle $(180° - \beta)$ being on the positive side above, which positions are reversed behind. In this order the axes *a* and *b* are as in the preceding system diagonals of a rhombic section; but the former is inclined, while the latter is horizontal to the third axis *c*. They are therefore distinguished as clinodiagonal and orthodiagonal axes. In the arrangement of the parameters that of *b* is considered as unity, but it is not necessarily longer than that of *a*.

The solid under consideration is geometrically an oblique rhombic pyramid or octahedron, but it is not a simple crystallographic form, being made up of two dissimilar classes of faces marked by longer or shorter edges in the plane of symmetry, according as they face the obtuse or acute angle of the inclined axis, and either set may occur in combination with or without the other. It is therefore to be considered as contained by two half pyramids or hemipyramids, one having the faces II., III., V., and VIII., as in fig. 234, opposite the acute angle, and the other I., IV., VI., VII., facing the obtuse angle of the axes. For the general forms represented by three dissimilar finite indices, the two groups are

$$\frac{\bar{h}\bar{k}l \mid \bar{h}kl}{h\bar{k}\bar{l} \mid hk\bar{l}} \qquad \frac{\mathbf{h\bar{k}l \mid hkl}}{\mathbf{\bar{h}\bar{k}\bar{l} \mid \bar{h}k\bar{l}}}$$

where the stronger letters indicate faces in front of the orthodiagonal section. Naumann calls these positive and negative hemipyramids, or $+P$, and $-P$, and considers the former as that facing the acute angle of the axes, which convention has the inconvenience of throwing the face whose indices are all positive into the negative form; but as it is that most generally followed, it will be adopted in the following pages. The notation of the unit forms is therefore :

$$+P = \begin{array}{c|c} \bar{1}\bar{1}1 & \bar{1}11 \\ \hline 1\bar{1}\bar{1} & 11\bar{1} \end{array} \qquad -P = \begin{array}{c|c} 1\bar{1}1 & 111 \\ \hline \bar{1}\bar{1}\bar{1} & \bar{1}1\bar{1} \end{array}$$

Weiss's symbols are $(a : b : c)$ and $(a' : b : c)$ respectively.

From the unit hemipyramids $\pm P$ we may, by keeping a and b constant and altering the value of c, obtain other forms, which will be flatter or steeper according as the values assigned are greater or less than unity. These are the hemipyramids of the principal series ; the forms are represented by the symbols :

$$\frac{1}{m} \pm P, \left(a : b : \frac{1}{m} c\right), \text{ or } \{h\,h\,l\} \text{ where } (h < l).$$

and the acuter ones by

$$\pm\, m\,P, (a : b : mc) \text{ or } \{h\,h\,l\} \text{ where } (h > l).$$

When in the forms $\pm P$ the axes b and c are kept constant, and a is lengthened, a new series known as clinodiagonal hemipyramids are obtained, represented by the symbols :

$$\pm\, \grave{P}n, (na : b : c), \text{ or } \{h\,k\,l\} \text{ where } (h < k).$$

Here the oblique axis is indicated by the inclined bar in the letter *P*. The same system of derivation applies, however, to any of the forms of the principal series, $\pm P$, so that the general symbols of any hemipyramid of the clinodiagonal series are :

$$\pm\, m\,\grave{P}n, (na : b : mc) \text{ or } \{h\,k\,l\}$$

A third series, as in the rhombic system, is obtained by varying the length of the orthodiagonal, or axis of symmetry b, a and c being unchanged. These are the hemipyramids of the orthodiagonal series, whose symbols are for the forms derived from $\pm P$:

$$\pm\, \bar{P}n, (a : nb : c), \text{ or } \{h\,k\,h\} \text{ where } (h > k),$$

and for those derived from any other of the forms $\pm m P$:

$$\pm m \bar{P} n, (a : nb : mc) \text{ or } \{h k l\}$$

where the straight bar on the stem of the P signifies that the orthodiagonal is the axis modified. The geometrical character of these forms will be generally analogous to those similarly produced in the rhombic system, the difference between the two hemipyramids being remembered. When, however, $m = \infty$ or $l = 0$, the edge lying in the plane of symmetry becomes vertical or parallel to c for either hemipyramid, and a prism is produced, which is known as the primary vertical prism, with the symbols ∞P, $(a : b : \infty c)$, and $\{110\}$. This is only distinguishable from a rhombic prism by the circumstance that the diagonals of the rhomb forming its horizontal section are not the axes b and a, but the orthodiagonal and the horizontal projection of the clinodiagonal, and therefore the fundamental ratio of the axes lying in the basal section cannot, as in the rhombic system, be determined from a measurement of the angle of the prism alone, a knowledge of the characteristic angle β being required in addition. From the clinodiagonal hemipyramid $m \grave{P} n$ by a similar method clinodiagonal prisms, $\infty \grave{P} n = (na : b : \infty c) = \{h k 0\}$ where $(h < k)$ are derived, and from the orthodiagonal series $m \bar{P} n$, orthodiagonal prisms $\infty \bar{P} n = (a : nb : \infty c) = \{h k 0\}$ where $(h > k)$. The first of these in combination modify the right and left edges of the primary prism, and the second those in the front and back plane, the angles being more obtuse as the value of n increases.

In the series of clinodiagonal hemipyramids, $\pm \grave{P} n$, the angle between the edges in the plane of symmetry and the clinodiagonal axis becomes more acute as the value of n is increased, and when it becomes ∞, the four faces of either hemipyramid become parallel, forming an inclined rhombic prism, whose edges are parallel to that axis. This is known as the principal clinodome $\bar{P}\infty = (\infty a : b : c) = \{011\}$. In like manner a more obtuse hemipyramid $\pm \frac{1}{m} \bar{P}$ gives rise to a flatter clinodome $\frac{1}{m}\bar{P}\infty = \left(\infty a : b : \frac{1}{m} c\right) = \{0kl\}$ where $(k < l)$, and a more acute one $\pm m \bar{P}$ to the steeper form $m \bar{P}\infty = (\infty a : b : mc) = (okl)$ where $(k < l)$. In combination the first of these modifies the top and bottom edges, and the second the right and left ones of the principal clinodome $\{011\}$.

In the orthodiagonal series of hemipyramids the angle made by the faces with the plane of symmetry increases with m, becoming 90° when the latter $= \infty$, i.e., the two faces of a hemipyramid right and left of the plane of symmetry fall into one, forming with their corresponding counterplanes pairs of planes parallel to the orthodiagonal axis, distinguished as positive and negative hemidomes, in the same way as the hemipyramids whence they originate. The symbols of the different series of these forms are:

Positive hemidomes,

Principal [1]	$+ \quad \bar{P}\infty = (a' : \infty b : c) = \{\bar{1}01\}$.
Flatter forms $(h < l)$	$+ \frac{1}{m} \bar{P}\infty = \left(a' : \infty b : \frac{1}{m} c\right) = \{\bar{h}0l\}$.
Steeper forms $(h \quad l)$	$+ m \bar{P}\infty = (a' : \infty b : mc) = \{\bar{h}0l\}$.

Negative hemidomes.

Principal [1]	$- \quad \bar{P}\infty = (a : \infty b : c) = \{101\}$.
Flatter forms $(h < l)$	$- \frac{1}{m} \bar{P}\infty = \left(a : \infty b : \frac{1}{m} c\right) = \{h0l\}$.
Steeper forms $(h > l)$	$- m \bar{P}\infty = (a : \infty b : mc) = \{h0l\}$.

[1] In these, as in the hemipyramids, the sign of the first index in Miller's symbol is the reverse of that of Naumann's symbol, being negative in $+ P$ and positive in $- P$. This is a necessary consequence of expressing the obliquity of the axes by the acute angle β, but if the obtuse angle is used the type face of $+ P$ becomes (111), or both are positive. This method of notation is adopted by Schrauf in his great atlas of crystallography.

In combination the principal hemi-orthodome truncates the edges in the plane of symmetry of the corresponding hemipyramid of the same sign, and has its own edges of combination bevelled obliquely, above by the flatter, and below by the steeper forms of the same class.

Pinakoids. When the inclined axis in the clinoprism and the vertical axis in the clinodome are made infinitely long, i.e., when ∞ is substituted for m and n in the symbols of these forms, they are reduced to a pair of planes parallel to the front and back axial plane or plane of symmetry. This is known as the clinopinakoid, and has the symbols $\infty \grave{P} \infty = (\infty a : b : \infty c) = \{010\}$. It truncates all prismatic and domatic edges that lie in the orthodiagonal or transverse axial plane.

The most obtuse kind of orthoprism $\infty \bar{P} n$ in like manner is that having $n = \infty$ or $\infty \bar{P} \infty = (a : \infty b : \infty c) = \{100\}$. This, known as the orthopinakoid, is parallel to the right and left axial plane, and truncates the edges of all prisms lying in the plane of symmetry. The third, or basal pinakoid, parallel to the plane containing the ortho- and clinodiagonal axes, has the symbols, $0 P = (\infty a : \infty b : c) = (001-)$. It is derived from all hemipyramids, hemidomes, and clinodomes when 0 is substituted for m in their symbols.

The solid formed by the combination of the three pinakoids having its edges parallel to the axes represents, when the latter are developed in the ratio $a : b : c$ of the unit form, the fundamental or reticular polyhedron of the Bravais notation, the molecular meshes in the clinopinakoid being rhomboids, whose sides are in the proportion $a : c$; while in the other two planes they are rectangular, having sides in the proportion of $a : b$ and $b : c$ respectively. The angle β is directly obtainable from this combination by measuring the inclination of the face (001) upon ($\bar{1}00$). but no other element can be determined from it, all the other angles being right angles. The appearance of crystals belonging to the oblique system varies very considerably, and is chiefly determined by the obliquity of the axes *a* and *c* or the angle β, which in some cases is as low as 55° or 60°, while in others it may only differ from a right angle by a few minutes, and even this difference is not essential, as there are instances of minerals with the three axes at right angles to each other having the symmetry and physical properties characteristic of this system. The determination of such forms depends, however, not so much on considerations of shape as on physical properties, as will be subsequently shown; but the general principle may be laid down that the essential characteristic of the system is not an oblique axis, but monosymmetry, and therefore the name monosymmetric is more generally appropriate for the system than oblique or monoclinic, and would be preferable were not these latter terms of much wider currency.

As only one of the principal crystallographic lines coincides with an axis or direction of physical equivalence in oblique crystals, any position that satisfies the general symmetry may be adopted for reading them. It is, however, practically inconvenient to adopt faces as base and orthopinakoid that make a very obtuse angle with each other, as the resulting axial obliquity β and one of the angles of the triangles arising in calculation will be so acute that the sides cannot be computed with accuracy, as a small variation in the angle produces a great difference in the length of the side opposite to it. A complete determination of the elements of the crystal is only possible when, in addition to two pairs of faces perpendicular to the plane of symmetry, which are considered as $0P$ and $\infty \grave{P} \infty$, either one hemipyramid or two prismatic forms belonging to different zones, such as ∞P and $\grave{P}\infty$ are present. From a combination containing only a prismatic form with oblique end faces, the ratio $a : b$ and the angle β may be determined, but not the length of the vertical axis. The appearance of monosymmetric combinations varies with their crystallographic elements. When the obliquity of the axis *a* is considerable, and the faces are numerous, the difference from those of systems whose symmetry is more complete is very marked ; while on the other hand, those having a nearly horizontal clinodiagonal axis, and

FIG. 238.

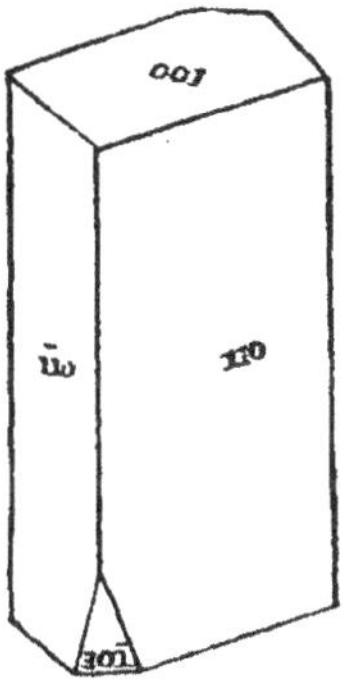

FIG. 239.

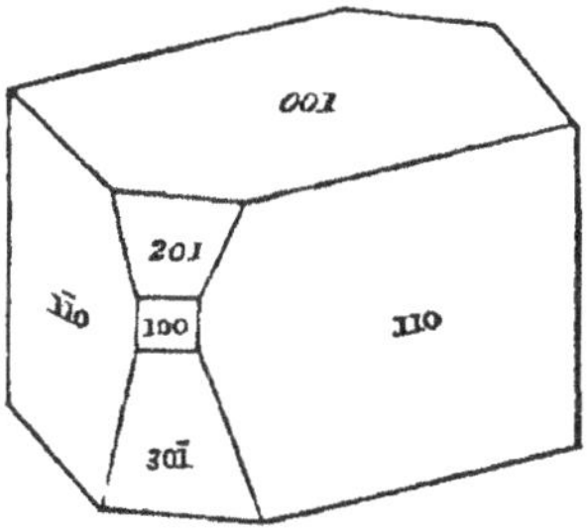

few or slightly developed faces belonging to hemipyramids or hemidomes, are often scarcely distinguishable from rhombic or hexagonal forms. The leading test is, however, in all cases

FIG. 240.

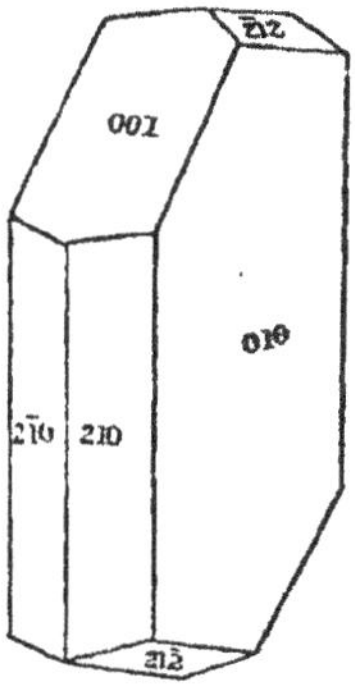

FIG. 241.

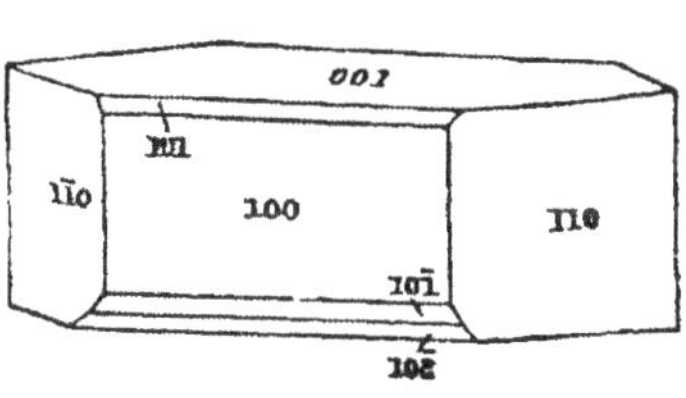

the presence of similar faces right and left of the plane of symmetry, combined with differences in the faces modifying the top and bottom edges parallel to the axis *b*. Some of the simpler cases are represented in figs. 238-247. Fig.

238 has the front lower and back upper solid angles in the plane of symmetry in the combination ∞P, $0 P$, modified by an acute hemidome $+ 3 P\infty$; while in fig. 239 the same

FIG. 242.

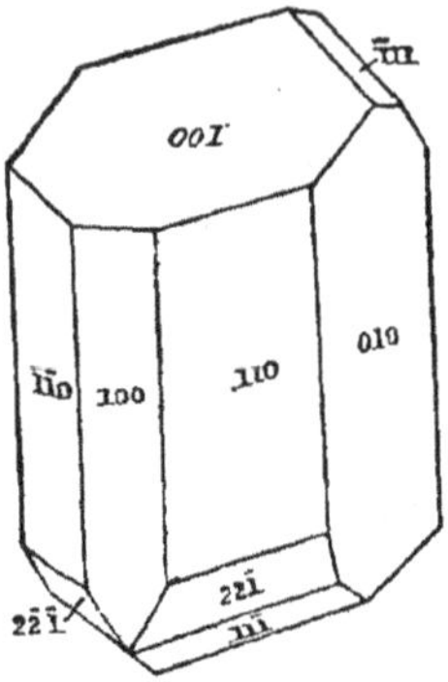

FIG. 243.

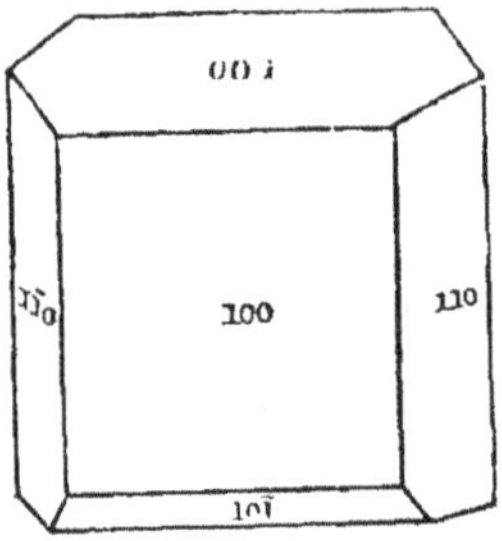

class of combination has in addition a less acute hemidome $- 2 P\infty$ at the top in front and the orthopinakoid $\infty P\infty$. Both are forms of Clinoclase. Fig. 240, a crystal of Brushite,

FIG. 244.

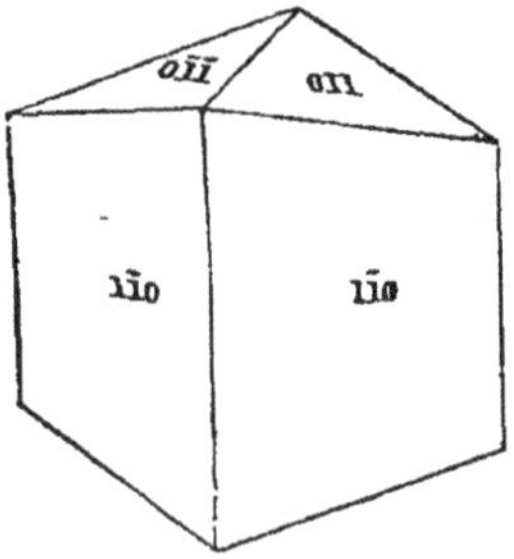

FIG. 245.

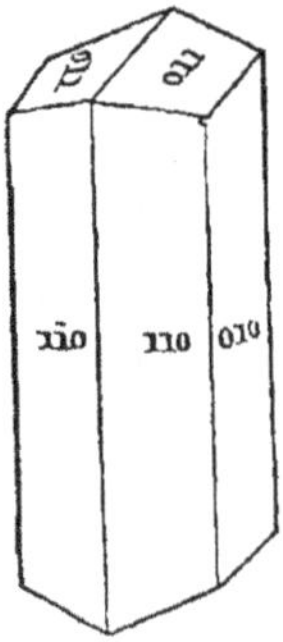

has in addition to the plane of symmetry $\infty P\infty$, the basal pinakoid $0 P$, a hemi-orthopyramid $+ P 2$, and an orthoprism $\infty P 2$. Fig. 241, a combination of a nearly rhombic character observed in Caledonite, is, however, distinguishable by the hemi-orthodome $+\frac{1}{2}\bar{P}\infty$, which appears under the unit hemi orthodome $+\bar{P}\infty$ in front below, but not above. Fig. 242, the common form of Borax crystal, contains the three pinakoids, the principal prism, and both principal hemipyramids, but the oblique character is brought out by the acute hemipyramid $2P$, which occurs only in the positive form. Fig. 243, a crystal of Allanite, contains ∞P, $\infty \bar{P}\infty$, $0P$, and $+\bar{P}\infty$. Fig. 244, one of the simplest forms

FIG. 246.

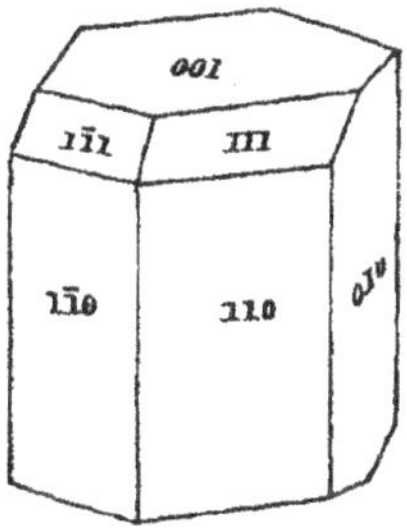

FIG. 247.

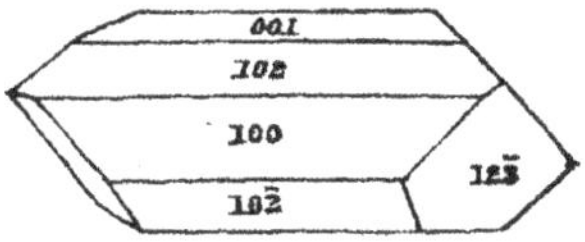

of Hornblende, contains only the unit prism ∞P and clinodome $\grave{P}\infty$; fig. 245 the same, with the acute edges of the prism truncated by the clinopinakoid $\infty\grave{P}\infty$. Fig. 246 has the edges between ∞P and $0P$ modified by the faces of the negative hemipyramid $-P$, together with the clinopinakoid $\infty P\infty$. Fig. 247, a crystal of Azurite, elongated parallel to the orthodiagonal axis, contains $0P$, $-\frac{1}{2}\bar{P}\infty$, $\infty\bar{P}\infty$, $+\frac{1}{2}\bar{P}\infty$, and $+\frac{1}{3}\bar{P}2$. These will be sufficient to indicate the general character of the combinations of this system. A few more complex examples will be given in the descriptive portion of the work.

CHAPTER VIII.

TRICLINIC[1] SYSTEM.

THE forms of this system are referred to three axes all having different parameters and all oblique to each other. The characteristic elements of crystals belonging to it are therefore, in addition to the lengths of the axes, the three angles between them. This gives forms of the most rudimentary character, every face crystallographically possible —that is, having a similar face parallel to itself as required by the general conditions of crystallographic symmetry to a centre—is a complete form, and may combine with any other having similar or dissimilar indices; and as no form can have more than two faces, any actual crystal must be a combination of at least three forms. From the obliquity of the axes there can be neither planes nor axes of symmetry, which property is indicated in the name 'asymmetric.'

The notation of the axes is similar to that in the rhombic system, when one has been selected as the vertical axis *c*,

FIG. 248.

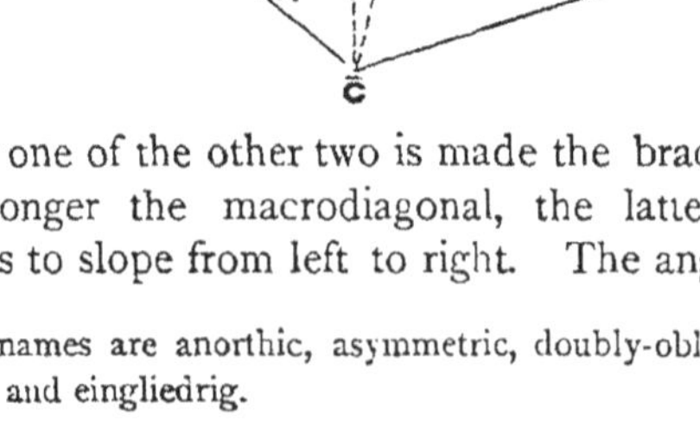

the shorter one of the other two is made the brachydiagonal and the longer the macrodiagonal, the latter being so arranged as to slope from left to right. The angle between

[1] Other names are anorthic, asymmetric, doubly-oblique, oblique-rhomboidal, and eingliedrig.

c and *b* is called α, that between *c* and *a*, β, as in the oblique system, and that between *a* and *b*, γ; these angles being measured upon the positive semi-axes, as shown in perspective projection fig. 248, and in the orthographic projections upon the three principal sections (figs. 249, 250, and

FIG. 249.

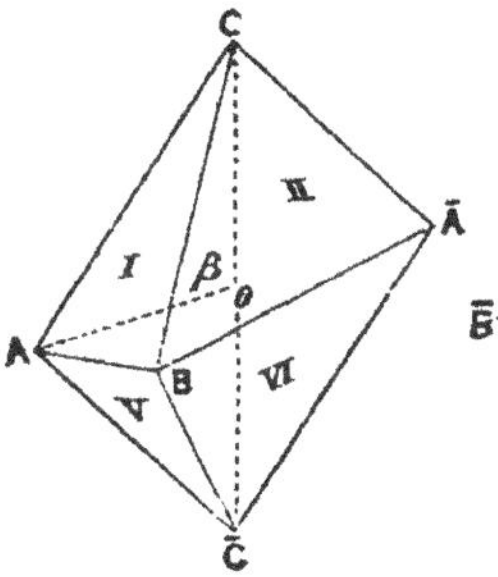

FIG. 250.

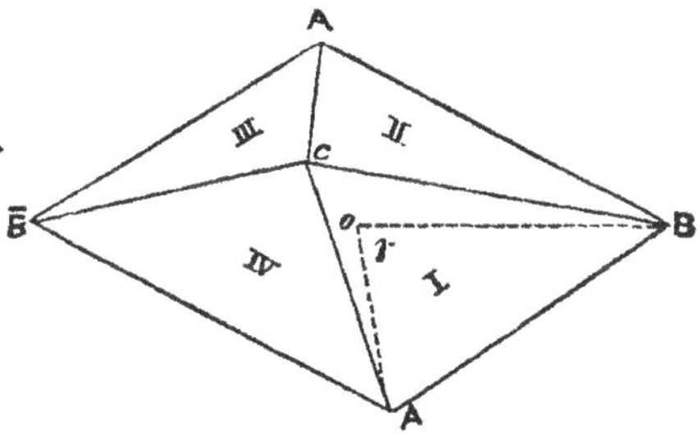

251). Each of these latter shows two of the semi-axes in their true lengths and inclinations, the notation being generally similar to that of the rhombic system. From these it will be seen that the principal sections are all rhomboids,

FIG. 251.

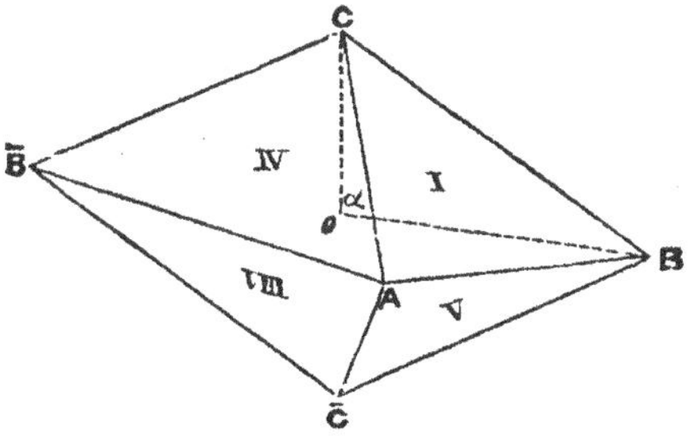

and that the particular solid corresponding to three finite parameters is an oblique rhomboidal pyramid or octahedron contained by four dissimilar pairs of faces, each of which, therefore, represents a different form. These are known as quarter- or tetarto-pyramids, whose positions are in Naumann's method indicated by the letter P differently accented, according as the face indicated belongs to the right or left, upper or lower octants. The complete notation according to the different systems is as follows :—

$$\left.\begin{array}{r} \text{I. } (a : b : c) = (1\,1\,1) \\ \text{VII. } (a' : b' : c') = (\bar{1}\,\bar{1}\,\bar{1}) \end{array}\right\} P' \qquad \left.\begin{array}{r} \text{II. } (a' : b : c) = (\bar{1}\,1\,1) \\ \text{VIII. } (a : b' : c') = (1\,\bar{1}\,\bar{1}) \end{array}\right\} {}_{,}P$$

$$\left.\begin{array}{r} \text{III. } (a' : b' : c) = (\bar{1}\,\bar{1}\,1) \\ \text{V. } (a : b : c') = (1\,1\,\bar{1}) \end{array}\right\} P_{,} \qquad \left.\begin{array}{r} \text{IV. } (a : b' : c) = (1\,\bar{1}\,1) \\ \text{VI. } (a' : b : c') = (\bar{1}\,1\,\bar{1}) \end{array}\right\} {}'P$$

The relations of the observed forms may be developed from the symbols of any of the unit quarter-pyramids in a similar manner to that given for the rhombic and oblique system. Thus, by varying the length of c, while a and b are unchanged, P' gives rise to

$$m\,P' = (a : b : m\,c) = \{h\,h\,l\} \text{ where } h > l$$

when c is lengthened, and to

$$\frac{1}{m}\,P' = \left(a : b : \frac{1}{m}\,c\right) = \{h\,h\,l\} \text{ where } h < l$$

when the vertical axis is shortened.

In like manner a similar group is derived from each of the three other quarter-pyramids, which, together with the preceding, form the principal or vertical series.

By altering the length of the right-and-left axis b, in the principal form $m\,P'$, a and c being unchanged, a macrodiagonal series of quarter-pyramids arises, having the symbols:

$$m\,\bar{P}'\,n = (a : n\,b : m\,c) = \{h\,k\,l\},$$

and when the axis a is similarly varied the result is the brachydiagonal series having the symbols :

$$m\,\breve{P}'\,n = (n\,a : b : m\,c) = \{h\,k\,l\}.$$

In all these series of derived quarter-pyramids, when m becomes ∞, the form changes to a pair of planes parallel to the vertical axis, or a hemiprism, which includes both the upper and lower quarter-pyramids lying on the same side of the centre. This is indicated by the position of the accents in Naumann's symbol. Thus, from P' and $P_,$ is derived the right principal hemiprism:

$$\infty P_,' = (a : b : \infty c) = \{1\,1\,0\},$$

and from $'P$ and $_,P$, the left principal hemiprism,

$$\infty\,_,'P = (a : b' : \infty c) = \{1\,\bar{1}\,0\}.$$

which, like all other triclinic forms, may appear together in the same combination, or independently of each other, there being no true triclinic prism, but only a prismatic combination of the hemiprisms. By increasing the length of the macrodiagonal or brachydiagonal axis respectively in the principal hemiprisms, the other axes being unchanged, macrodiagonal and brachydiagonal hemiprisms are formed. The symbols of the former are

$$\infty \bar{P}_,' n.\ \infty\,_,'\bar{P} n = (a : nb : \infty c)\ (a : nb' : \infty c) =$$
$$\{h\,k\,0\}\ \{h\,k\,0\} \text{ where } h > k,$$

and those of the latter

$$\infty \breve{P}_,' n,\ \infty\,_,'\breve{P} n = (na : b : \infty c)\ (na : b' : \infty c) =$$
$$\{h\,k\,0\}\ \{h\,k\,0\} \text{ where } h < k.$$

In the macrodiagonal quarter-pyramids, $\bar{P}'n$ and $'\bar{P}n$, when $n = \infty$, the angle between the faces meeting in the front and back axial plane becomes 180°, or two fall into one parallel to the macrodiagonal, producing a hemi-macrodome, $'\bar{P}'\infty = (a : \infty b : c) = \{1\,0\,1\}$, which in combination truncates edges parallel to that axis in front of the crystal above the base and below it behind, the correlated form $_,\bar{P}_,\infty$, having the reverse position, or appearing below in front and above behind. Other analogous forms represented by $m\,'\bar{P}'\infty$ and $m\,_,\bar{P}_,\infty$ are derived in the same way from the quarter-pyramids $m\,\bar{P}'n$, $m\,'\bar{P}n$, &c.

Hemiprismatic forms parallel to the axis a, or hemi-brachydomes, are obtained from pairs of the quarter-brachypyramid series by making $n = \infty$ in their symbols, when the two faces meeting in the right and left axial plane fall into one. That derived from $\breve{P}' n$ and $,\breve{P} n$ h s the symbols

$$,\breve{P}' \infty = (\infty a : b : c) = \{0 1 1\}$$

which in combination modifies edges parallel to the brachy-diagonal axis above the centre of the crystal to the right and below it to the left, while the correlated form derived from

$$'\breve{P} n \text{ and } \breve{P}_, n \text{ or } '\breve{P}_, \infty = (\infty a : b' : c) = \{0 \bar{1} 1\}$$

has the opposite position or to the right below and to the left above. As before, the general symbols for the hemi-brachy-domes are

$$m\,,\breve{P}' \infty = (\infty a : b : m c) = \{0 k l\}$$

and

$$m\,'\breve{P}_, \infty = (\infty a : b' : m c) = \{0 \bar{k} l\}.$$

In these, unlike the other prismatic forms (the hemiprisms and hemi macrodomes), the two quarter-pyramid planes included in any face have dissimilar signs. or one is a type or positive plane of one form, and the other the negative or counter-plane of another. Hence the accents in Naumann's symbols lie chequerwise, as in naming the forms the front planes are always meant.

In the hemi-brachydomes, $m\,,\breve{P}' \infty$ and $m\,'\breve{P}_, \infty$, when m is made $= \infty$, the faces become parallel to the front and back axial plane, producing the brachypinakoid, which, like that in the rhombic system, has the symbols

$$\infty \breve{P} \infty = (\infty a : b \,.\, \infty c) = \{0 1 0\}$$

and similarly $m = \infty$ in a hemi-macrodome gives rise to the form parallel to the right and left axial plane or the macro-pinakoid

$$\infty \bar{P} \infty = (a : \infty b : \infty c) = (1 0 0).$$

The third, or basal pinakoid, is the limiting form of the

vertical series of quarter-pyramids $n'\,{}_{,}P'$ when $m = 0$, and is represented by

$$0P = (\infty a : \infty b : c) = \{001\}.$$

This system of development may be represented by the scheme given for the rhombic system on p. 134, if the octants in which the particular quarter-pyramids and hemi-prismatic forms lie, be indicated by properly accentuating the symbol *P*. The lines on which the symbols are arranged will also have the same significance, that is, those in any horizontal and vertical lines will be in the corresponding zones, except that the axes of the principal zones, instead of being at right angles, will be oblique to each other.

The appearance of triclinic combinations is chiefly dependent upon the obliquity of the axes. When the three angles differ but slightly from right angles, as in the mineral Cryolite, the crystals have a general resemblance to cubic forms, while, on the other hand, in Axinite and Sulphate of Copper, they are marked by extreme obliquity and apparent want of symmetry. In other species, notably in the felspar group, triclinic crystals occur, which in Albite are closely allied morphologically to those of the analogous species Orthoclase, in the oblique system. In this latter case the resemblance is often so close that the system to which the crystals belong cannot always be determined by consideration of forms alone. The combination of the three pinakoid planes, also called the doubly-oblique prism, is the primitive solid of the system according to the French notation, the faces being oblique parallelograms whose sides represent the meshes of the molecular network, each being dissimilar from the other two.

The determination of the elements of a triclinic form requires at least five independent observations, and involves calculations which cannot be described in few words. The student is therefore referred for information on this subject to the larger works on determinative mineralogy and practical crystallography. As there is no direct relation between form and other physical properties, the choice of position is quite arbitrary, so that there may be and often is considerable diversity of opinion as to the symbols to be assigned to the faces by different authors.

The general characters of the simpler triclinic combinations will be seen in figs. 252–254. Fig. 252 is one of the most unsymmetrical kinds, a crystal of Axinite,[1] containing $'P.\ P'.\ \frac{1}{2}P'.\ o\,P'_{,}.\ '\bar{P}'\infty.\ 2'\bar{P}'\infty$.

FIG. 252.

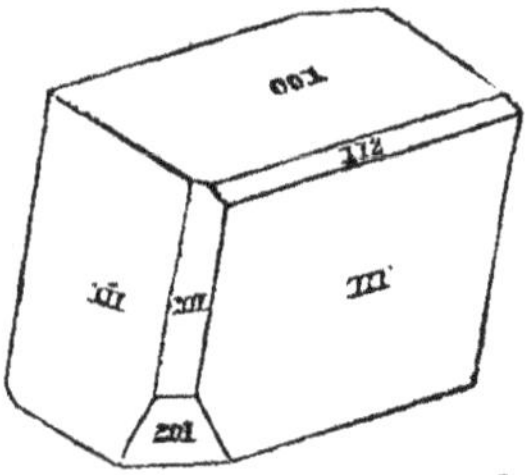

FIG. 253.

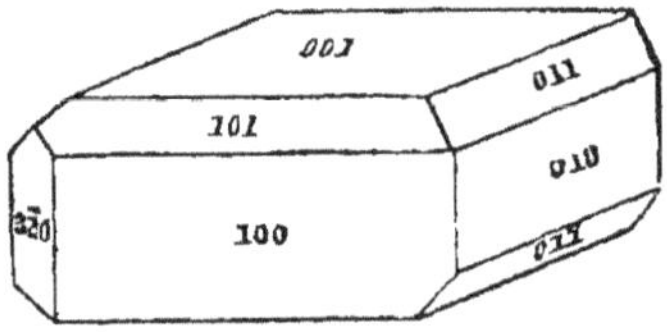

Fig. 253, a crystal of Babingtonite, contains $\infty\breve{P}\infty.\ \infty\bar{P}\infty.\ o\,P.\ \infty{}'\bar{P}\frac{3}{2}.\ {}_{,}\breve{P}'\infty.\ '\bar{P}_{,}\infty$. Fig. 254 is similar, with

FIG. 254.

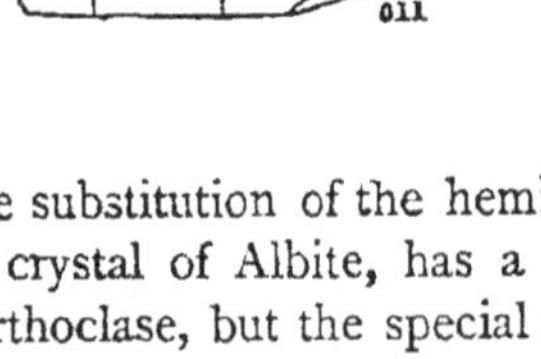

FIG. 255.

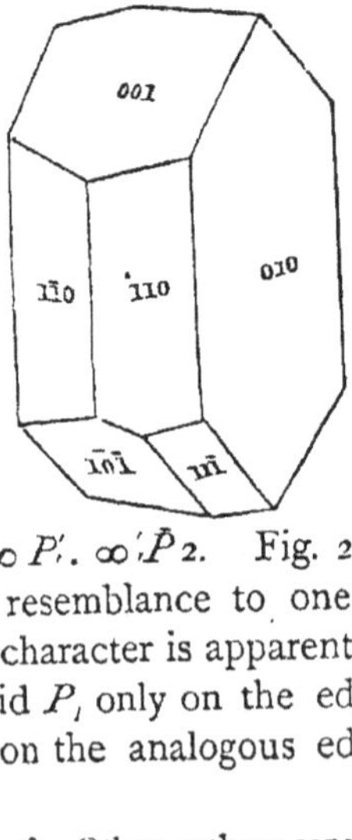

the substitution of the hemiprisms, $\infty P'_{,}.\ \infty{}'\breve{P}2$. Fig. 255, a crystal of Albite, has a general resemblance to one of Orthoclase, but the special triclinic character is apparent by the presence of the quarter-pyramid $P_{,}$ only on the edges between $\infty\breve{P}\infty$ and ${}_{,}\bar{P}_{,}\infty$ and not on the analogous edges between $\infty\breve{P}\infty$ and $o\,P$.

[1] This is the position adopted by Schrauf. Other authors consider the quarter-pyramid faces as belonging to the zone of the prism.

CHAPTER IX.

COMPOUND OR MULTIPLE CRYSTALS.

IN demonstrating the geometrical characteristics of the different systems in the preceding chapters, the solids illustrated have been assumed to be of the most regular character, every face of the same form being similarly placed in regard to the symmetrical centre or origin of the axes. Such crystals, however, without being absolutely unknown, are of comparative rarity, at any rate in individuals of any size, and in by far the larger number of instances one or more faces of any form may be largely developed, with a corresponding reduction or even entire suppression of the remainder, as, for example, in the common case of a prism terminated by pyramids or domes, the faces of the latter forms appear only at one end of the prism, because the other forms the surface of attachment to the rock. In such cases the missing faces have to be assumed in reasoning out the character of the completed from the observed form. Besides this, there are other cases in which two or more crystals are united into a mass having a particular regularity of arrangement, the component crystals preserving, to a great extent, their individuality. Such multiple crystals are of two principal kinds known as parallel and twinned groups, but before considering these it is necessary to notice a third special kind of development which is, to some extent, of a compound character.

Hemimorphism. There are a few minerals and artificial products, whose crystals are dissimilarly ended, the faces limiting a prismatic zone at one end of its axis belonging to different forms from those in the corresponding position at the other end. Such crystals are not properly hemihedral, as, although they contain but half the full number of faces possible in their constituent forms, these faces are not, as they should be, uniformly distributed about the axes, but are so grouped that we may have all the faces whose indices are positive to an axis, while the corresponding negative ones are entirely absent, their places being occupied by some totally different form. This arrangement is incompatible with regular hemihedrism,[1] and it is therefore distinguished by the name of hemimorphism. The most conspicuous examples are afforded by Tourmaline, the Ruby, Silver Ores, and Greenockite in the hexagonal, Struvite and Electric Calamine in the rhombic, and Cane Sugar in the oblique system.

Fig. 256 represents a crystal of Tourmaline contained above by $R\,\kappa\,\{1\,0\,\bar{1}\,1\}$ (*a*), and $-R\,\kappa\,\{0\,1\,\bar{1}\,1\}$ (*b*); below, by R (*a*) and $-\frac{1}{2}R\,\kappa\,\{1\,0\,\bar{1}\,2\}$ (*c*); and in the zone of the prism by $\infty P2\,\{1\,1\,\bar{2}\,0\}$ (*d*), $\infty P\,\{0\,1\,\bar{1}\,0\}$ (*e*), and $\infty P\frac{4}{3}$ $\{1\,3\,\bar{4}\,0\}$ (*f*). Of the latter three forms the first appears with

FIG. 256.

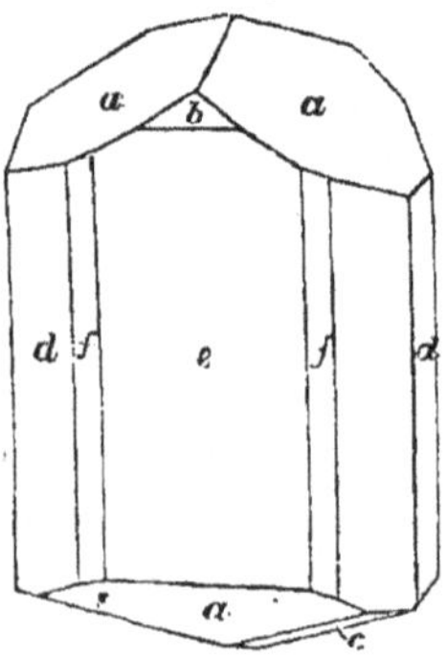

its full number of faces, and the others with only one-half, as trigonal and ditrigonal prisms respectively. The reason of this is, that the prism of the first order, considered as a rhombohedron of infinite altitude, falls into two groups of three faces, one of which belongs to the upper and the other to the lower end of the crystal, either of which may be present to the exclusion of the other in a hemimorphic group, and the dihexagonal prism in the same way as an unlimited scalenohedron divides into an upper and a lower ditrigonal prism; but a face of the hexagonal prism of the second order includes both upper and lower faces of the scalenohedron, of which either may be omitted without changing its geometrical character. The occurrence of trigonal prisms of

[1] This is, however, distinguished by Müller as asymmetric hemihedrism in the rhombohedral system.

this kind is therefore evidence of hemimorphic development, even when both ends of the crystal are not available for observation, as is generally the case in the ruby silver ores. In Greenockite the dissimilarity of the ends is much more marked, the crystals showing numerous pyramids at one end of the prism, squared off by the basal plane at the other. Fig. 257, a crystal of Struvite, has the upper faces of $\bar{P}\infty$ {1 0 1}, combined with the lower ones of $\frac{1}{3}\bar{P}\infty$ {1 0 $\bar{3}$} and o P. {0 0 $\bar{1}$}, which are limited transversely by $\breve{P}\infty$ {0 1 1}, 4 $\breve{P}\infty$ {0 4 1}, and $\infty\breve{P}\infty$ {0 1 0}; the latter may be con-

FIG. 257.

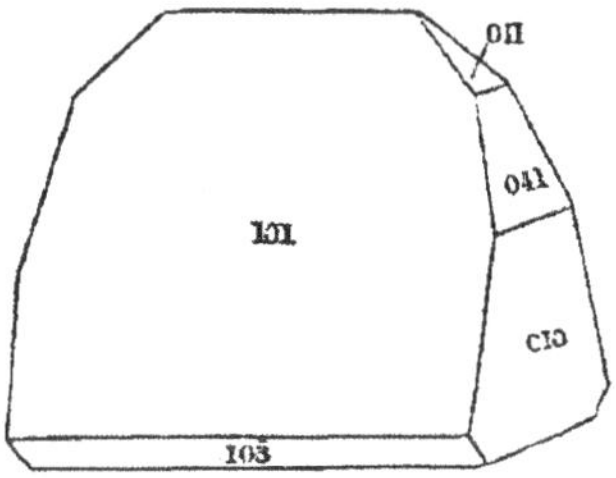

FIG. 258.

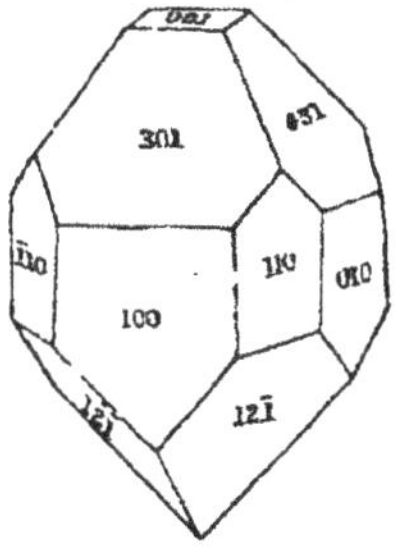

sidered as common to both sides of the base, while the former two are only represented by their upper faces. Fig. 258 is a crystal of Electric Calamine contained by ∞P {1 1 0}, $\infty\bar{P}\infty$ {1 0 0}, $\infty\breve{P}\infty$ {0 1 0}, limited above by 3 $\bar{P}\infty$ {3 0 1}, 3 $\breve{P}\infty$ {0 3 1}, and o P {0 0 1}, and below by the brachypyramid 2 $\breve{P}$2 {1 2 $\bar{1}$}. These are some of the more striking examples of this class of crystals, which, as a rule, are distinguished by the property of pyroelectricity, the opposite developing dissimilar polarity when heated.

Parallel grouping. In the simplest case of the aggregation of two similar crystals, the individuals are so arranged that a line joining their centres is either on the prolongation of a crystallographic axis or parallel to it, as in fig. 259, representing two octahedra having a common vertical axis where the surface of contact represented by the shaded plane is obviously equivalent to a face of a cube, and no alteration of character is effected by mere rotation of either crystal through one or more right angles about the line $O\,T\,O'$. The compound nature of such growth is evidenced by the re-entering angles of the faces adjacent to the plane of contact which will be more apparent as the distance between the centres $O\,O'$ is increased. This kind of grouping, often many times repeated, is commonly seen in crystals of alum, and also in native silver and other cubic minerals. If we suppose two cubes to be united in the same way, there will be a mere shifting of the top and bottom faces, giving a cube drawn out in height, but otherwise indistinguishable from a single crystal. The same remark holds good if either cubes or octahedra are in contact parallel to a face of the rhombic dodecahedron, their aspect not being changed by a half turn about the normal to that face.

FIG. 259.

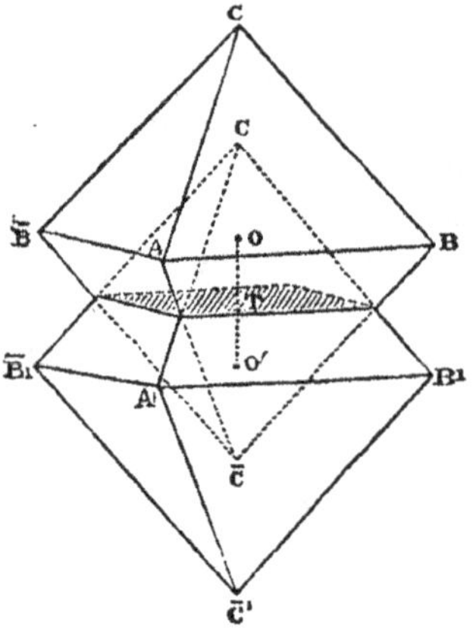

FIG. 260.

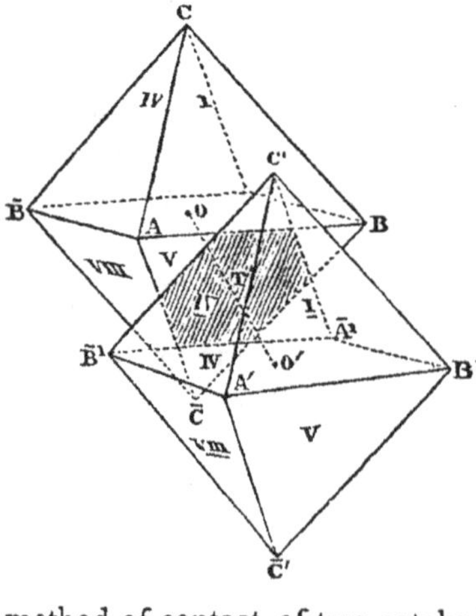

Twin grouping. Fig. 260 represents another method of contact of two octahedra, namely, on a face common to both, in which the axes are parallel as long as the particular position is retained, but if one crystal, say the front one, be turned 180° about the line $O\,T\,O'$, the result shown in fig. 261 is obtained, where the axes are no longer parallel, the original positive extremities in the movable crystal coinciding with the negative ones in the fixed one, or A with $\bar{A}'$, B with $\bar{B}'$, and C with $\bar{C}'$, while their opposite extremities make large angles with each other, the individual crystals being symmetrical to their common face, the surface of contact, which, however, as we have previously seen, is not a plane of symmetry of the form. This arrangement is known as a twin structure, or *twin crystal*, the common plane of symmetry is the *twin plane*, its normal the *twin axis*, and the surface joining the two crystals, the *plane of contact* or *composition*. In this, as in many other simple cases, the planes of twinning and composition coincide, but it is not always so, and the distinction between them must be carefully borne in mind, especially in dealing with the twin forms in the systems of lower symmetry where only the composition face is apparent, and the position of the crystals must often be shifted to arrive at the true twin plane. As a rule, the twin plane may be any actual or possible face of a form proper to the series, other than a plane of symmetry, and it is generally one having low indices, such as 1 1 0, 1 0 0, 1 1 1, &c. If we suppose two octahedra in the position of fig. 261 to be freely penetrable, and the line $O\,T\,O'$ to be shortened until O and O' coincide, we obtain the solid fig. 262, where the faces of

FIG. 261.

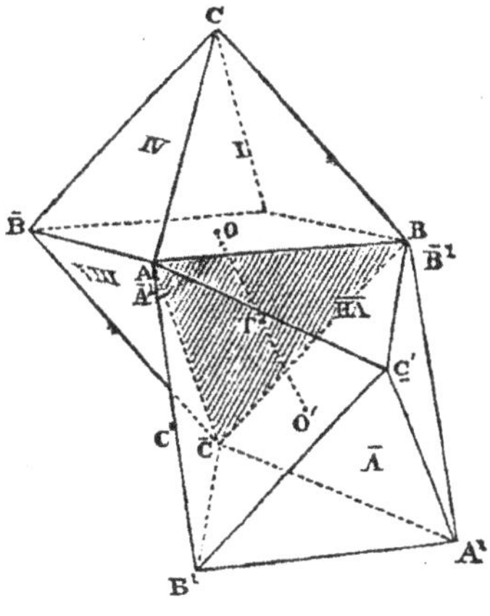

contact lie in the same planes in front and behind, but all the others meet in re-entering angles, the points and edges of both crystals being fully developed. This is known as a *penetration twin*, the shaded parts belonging to the inverted, and the white to the direct or fixed crystal, the faces being numbered according to the original positions in the preceding figures.

If the individual crystals, instead of being regularly developed, are supposed to be flattened to one half of their

FIG. 262.

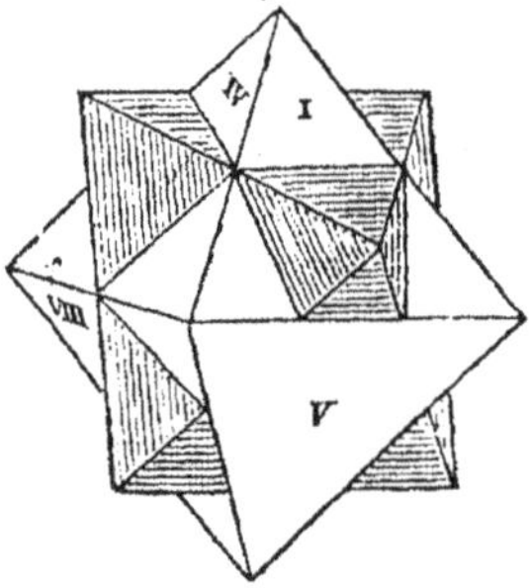

FIG. 263.

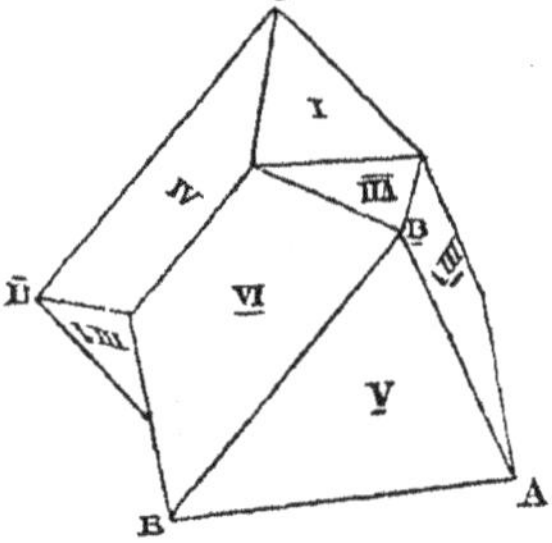

normal thickness upon the twin axis, the group will resemble fig. 263, where there is no penetration, and only those faces that are parallel to the twin plane appear of their full size. This is exactly what happens when a single crystal is divided by a twin plane passing through the centre as in fig. 264, and one half turned through 180°, the other remaining stationary. This is one of the most convenient methods of explaining twin structure, and is that most generally used, the resulting forms are called *contact-twins* as well as *macles* and *hemitrope* crystals. The latter terms, which were formerly

FIG. 264.

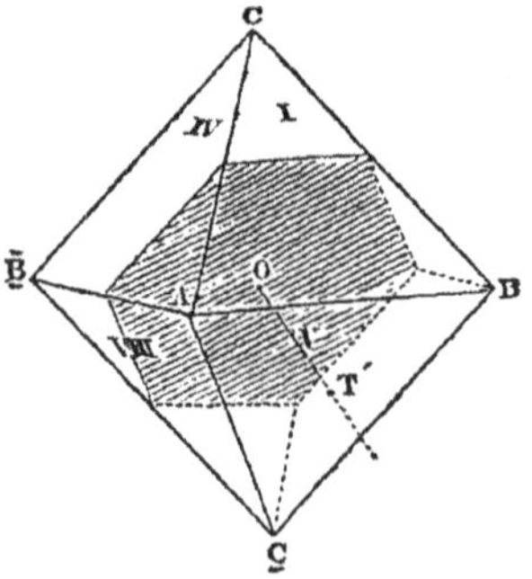

in general use, are now mainly confined to the works of French authors. German writers describe twin crystals, *zwilling*, *drilling*, *vielling*, &c., according as two, three, or more individuals are apparent in the group. Fig. 265 is a contact twin of two rhombic dodecahedra upon a face of the octahedron. Here there are no re-entering angles, the section upon the twin plane being a regular hexagon. A

FIG. 265.

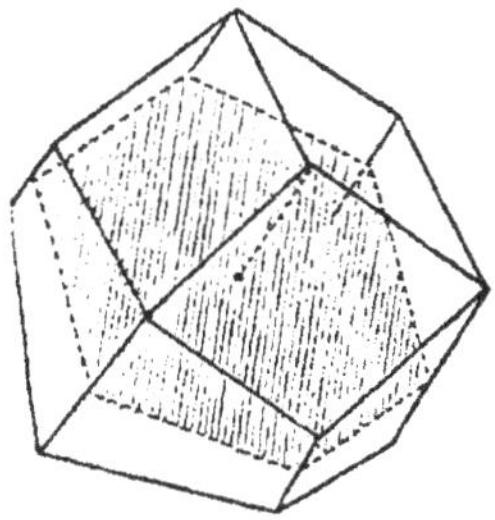

FIG. 266.

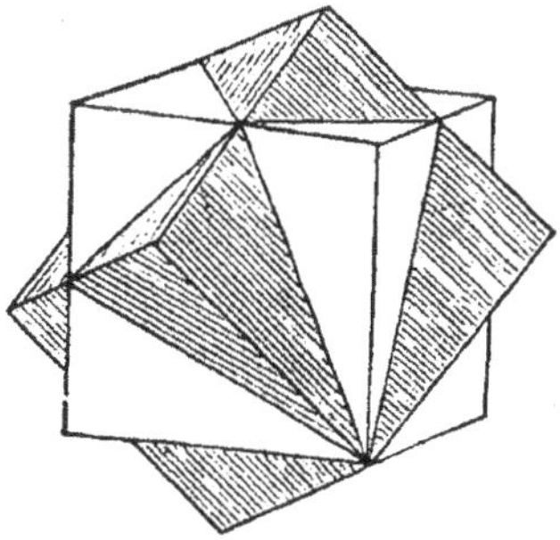

complete penetration twin of the same kind has also been observed in crystals of Sodalite. Fig. 266 is a penetration twin of two cubes, which, being exactly centred, have their points on the twin axis in common. This is a common twin form of Fluorspar, but the observed crystals are not generally quite regular, so that the projecting portions of the shaded crystal above the faces of the white one are alternately of different sizes instead of being all exactly alike.

FIG. 267.

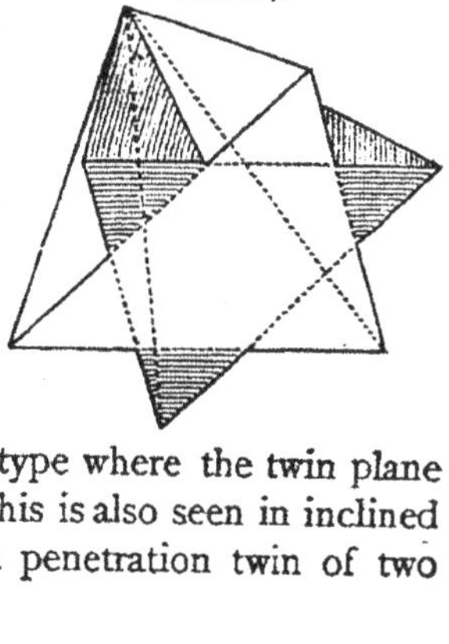

The greater number of cases of twin crystals among holohedral cubic forms are upon the above type where the twin plane is the face of an octahedron, and this is also seen in inclined hemihedral forms, as in fig. 267, a penetration twin of two tetrahedra, two of whose faces, parallel to the twin plane, lie in the same surface at one end of the twin axis, while the other six meet in the same point at the other end.

Fig. 268 is a more common case of penetration-twinning of tetrahedra, the twin-plane shown by oblique shading being a face of the cube. When the individuals in such a

FIG. 268.

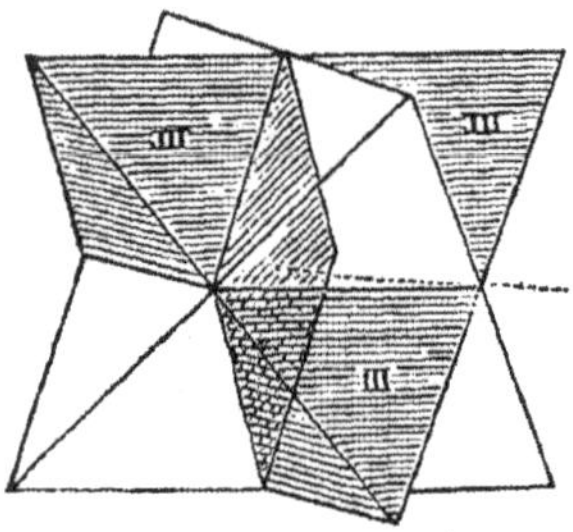

FIG. 269.

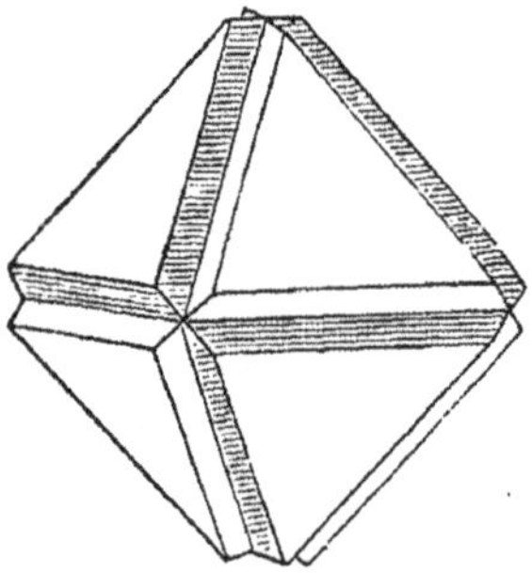

group instead of being simple tetrahedra are unequally developed combinations of both positive and negative ones, the appearance is similar to that of fig. 269, or an octahedron with a V-shaped groove along each of its edges; and the combination of $\frac{O}{2} \infty O$ (fig. 95), twinned in the same way, resembles a rhombic dodecahedron grooved parallel to the longer diagonals of its faces.

In the parallel hemihedral forms one of the most frequently observed cases is the penetration twin of two pentagonal dodecahedra $\left[\frac{\infty O 2}{2}\right]$ (fig. 270), the twin plane being a face of the rhombic dodecahedron. This is especially characteristic of iron pyrites, and the similarly constituted sulphides and arsenides of nickel and cobalt.

The above are the principal kinds of twin-crystals in the cubic system, in their simplest and most regular development; other and more complex cases arise when the component crystals are of different sizes, or the plane of composition is not central, when the groups are often considerably distorted.

The same structure may also be repeated with three or more individual crystals, producing multiple or polysynthetic

FIG. 270.

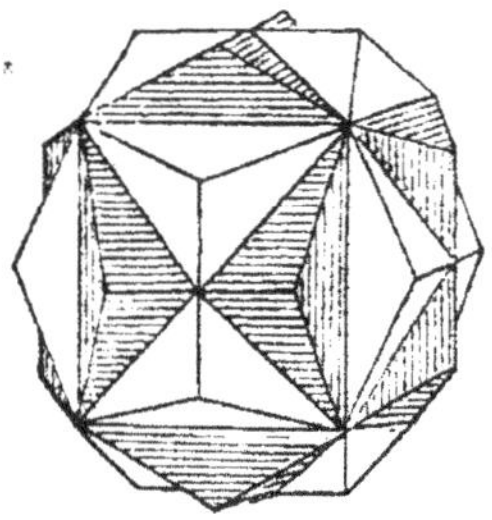

FIG. 271.

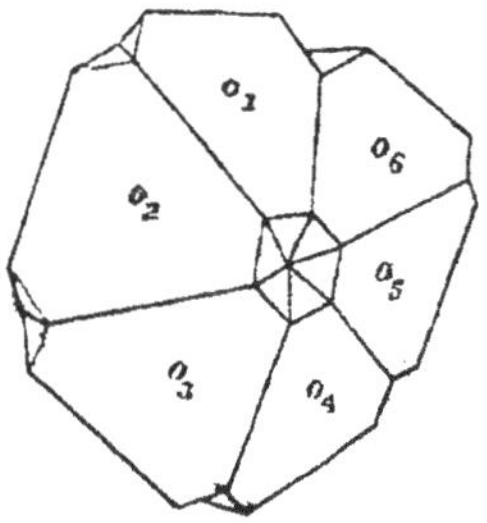

twin groups. Fig. 271 is an example of a peculiar polysynthetic twin of Spinel recently described by Struver. It is made up of six tetrahedral combinations, o_1 to o_6, the first four being repeated contact twins, on an octahedral face, while the fifth and sixth are parallel to the second and third; and as all their twin axes lie in the same face of a rhombic dodecahedron, whose axis, the line joining the hollow six-faced angles in the centre, is the edge common to all the individuals, there is complete lateral symmetry to that face.

In many instances the structure of a twin group may be explained in more than one way, or the twin axis may be exchanged for another line at right angles to itself, rotation about which produces a similar geometrical form, although the position of individual faces may be different. Thus, in fig. 264 the axis *O T* may be exchanged for a line in the twin plane joining the middle points of opposite edges, and normal to a new plane, cutting the edges at one-half and one-third of their lengths alternately, which has the properties of a face of the icositetrahedron 2 *O* 2 and gives forms exactly similar to figs. 262 and 263, but with this difference, that the faces brought opposite to each other belong to different tetrahedra, instead of to the same one as they do with the octahedron face in the twinning plane.

Twin crystals of the hexagonal system. The faces of di-hexagonal pyramids and prisms and hexagonal pyramids are possible twin planes in the holohedral forms of this system, but the only observed groups are twinned upon the latter form, and they are not of very common occurrence. In the rhombohedral hemihedral forms, on the other hand, twin structure is extremely common, the twin plane being most frequently either the face of the same or some other rhombohedron or the basal pinakoid, the latter not being one of their planes of symmetry. Fig. 272 is a common twin group of Calcite, in which two rhombohedra of the same

FIG. 272.

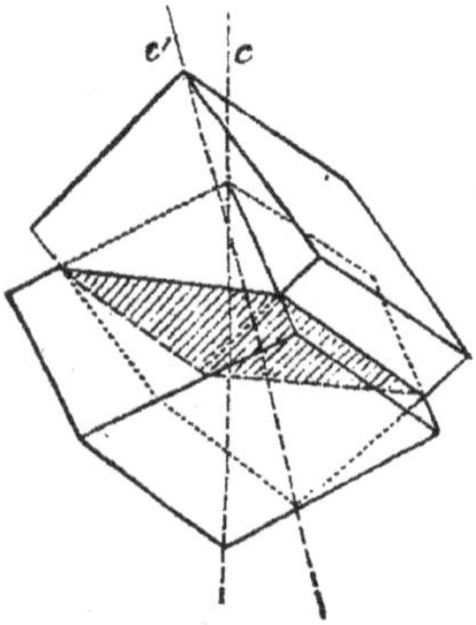

FIG. 273.

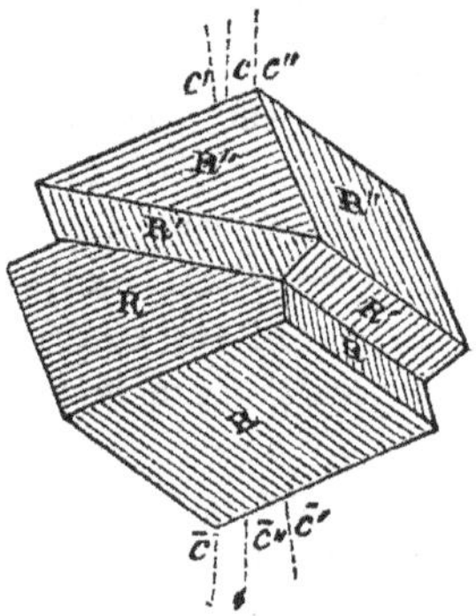

sign are twinned upon a face of the more obtuse rhombohedron $-\frac{1}{2}R$. The faces in front meet in a re-entering angle, and those below in a parallel salient one, the twin edges in both cases being parallel to the longer diagonals of the faces, while those of the side faces are parallel to the middle edges of the rhombohedron. When this structure is repeated by the addition of a third crystal, as in fig. 273, the middle member of the group R′ is often reduced to a thin parallel plate, the third one R″ being parallel in position to the first R; and when the number of individuals is much greater, and the intermediate ones are very thin, the group is scarcely distinguishable from a simple crystal, the twin structure being only apparent in the numerous fine striations covering two of the faces parallel to their horizontal diagonals, and the other four parallel to their middle edges. Fig. 274 is a contact twin

FIG. 274.

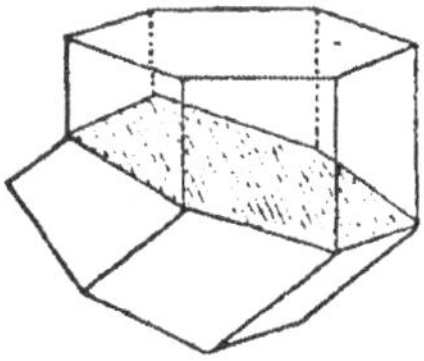

FIG. 275.

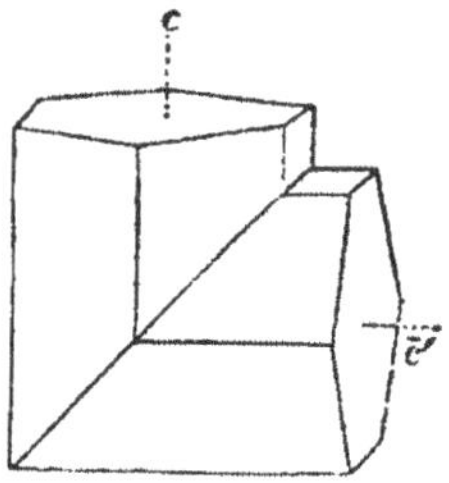

of the hexagonal prism upon a face of the same rhombohedron $-\frac{1}{2}R$, and fig. 275, another having a face of the unit rhombohedron as a twin plane, the two crystals making a nearly

FIG. 276.

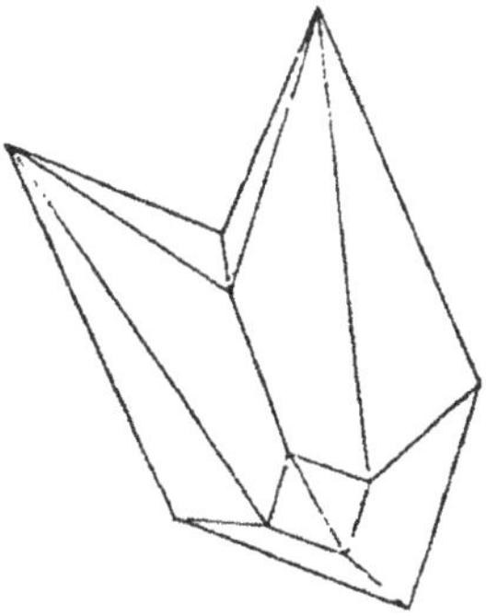

FIG. 277.

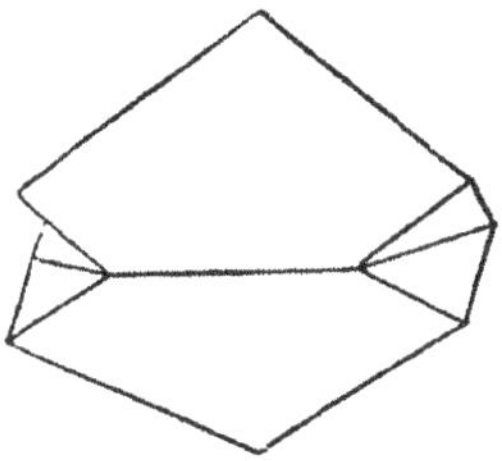

right-angled group, the inclination of the principal axes to each other being 89°.04′. Fig. 276 is a contact group of the common scalenohedron R^3 of Calcite twinned upon a face of the acute rhombohedron $2R$.

In the second case, where the basal pinakoid is the twin plane, the axes of the component crystals are parallel. Fig. 277 is the contact twin or hemitrope of a single rhom-

FIG. 278.

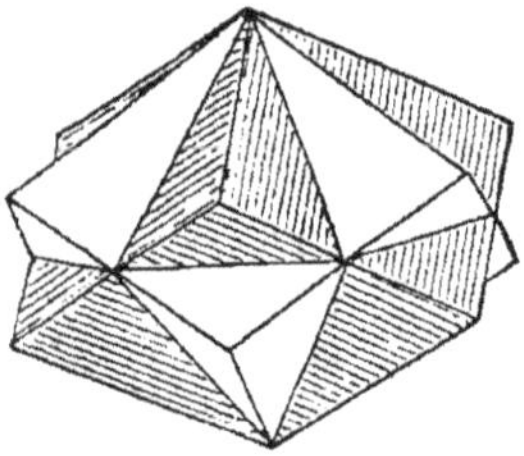

FIG. 279.

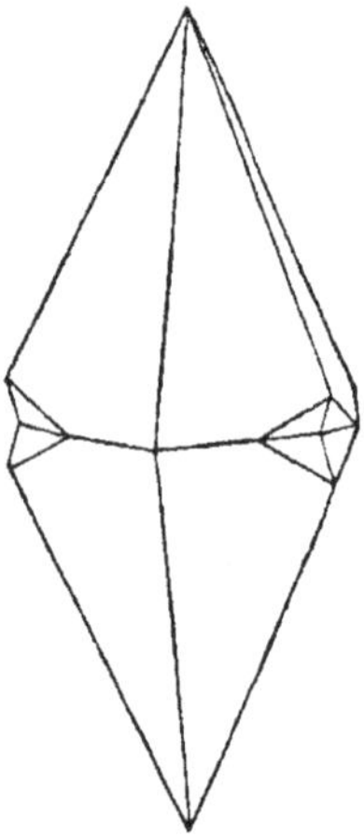

bohedron, and fig. 278 the same completely developed as a penetration twin of two. Fig. 279 is a hemitrope of the common scalenohedron R^3 of Calcite, also upon the basal plane; and fig. 280 a similar twin of the combination $\infty P. - \frac{1}{2} R$ in the same mineral. Here the two prisms are in contact, and their section being a regular hexagon, there are no re-entering angles, but the compound character is apparent from the shape of the prism faces, which are alternately rectangular and six-sided, instead of all being irregular five-sided figures, as in the single crystal, fig. 144.

FIG. 280.

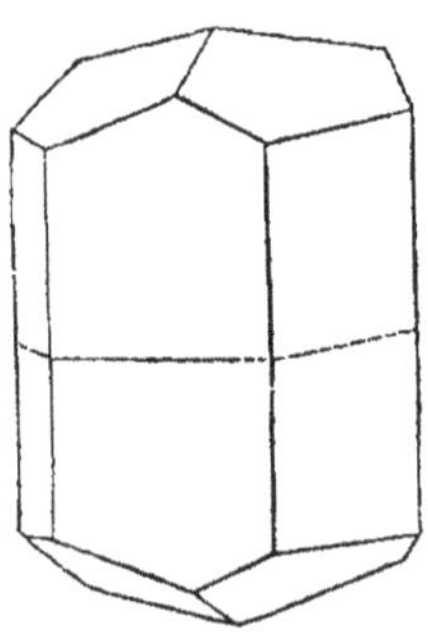

Fig. 281 is a contact twin of the common combination $\infty P. R. - R$ of Quartz, the faces of the positive rhombohedron in one crystal lying parallel with those of the negative one in the other. This is a case of partial penetration, the faces adjacent to the surface of contact, which is not the twin plane, meeting in re-entering angles; but when, as very frequently happens, there is more complete penetration, these angles are convex, and the group can only be distinguished from a simple crystal by the irregular character of the faces, which rarely have even surfaces, portions of one rhombohedron being irregularly distributed through the other in a manner which shows the crystals to be combined, and that their contact is not in a plane surface. A complete description of twin structures of this kind will be found in Descloizeaux's memoir on Quartz.

FIG. 281.

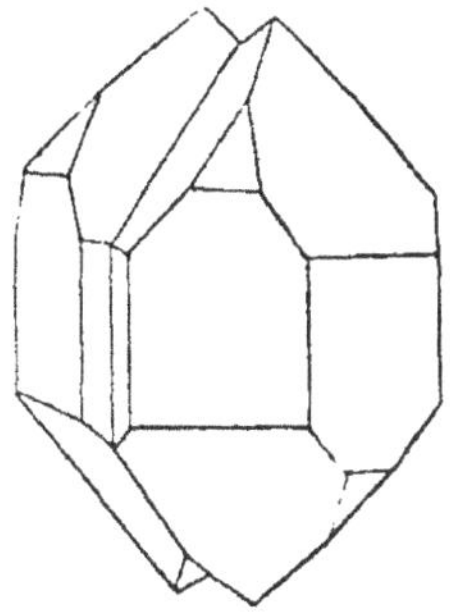

Twin crystals of the tetragonal system. In the type of twin structure most frequently observed in this system, the twin plane is a face of the pyramid of the second order,

FIG. 282.

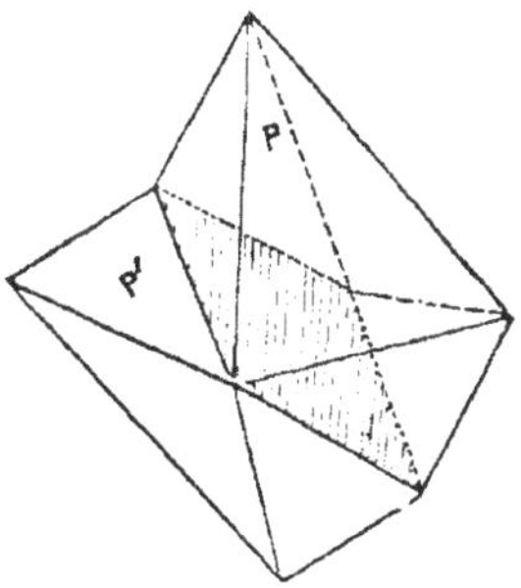

FIG. 283

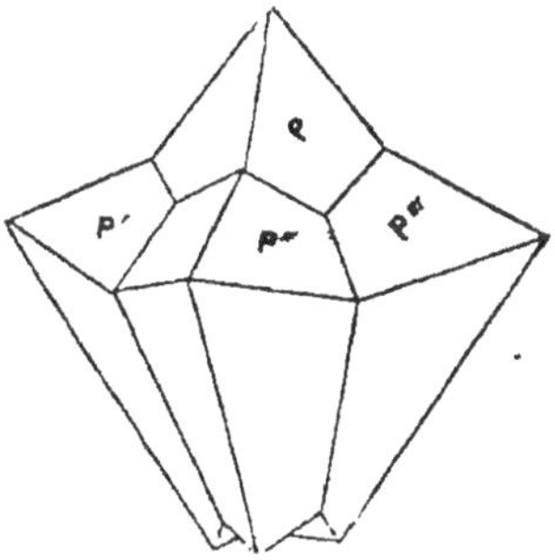

$m P \infty$. Fig. 282, one of the simplest examples, is a hemitrope of the tetragonal pyramid in Hausmannite, the lower half being rotated to the left on the lower plane $(0\bar{1}\bar{1})$. This, when repeated symmetrically upon all four sides of the pyramid, gives the group of five individuals, fig. 283.

Another very common example of the same kind (fig. 284) occurs in the combination *P*. ∞ *P* of Tinstone, and this, when repeated with a third individual, the middle one being shortened to a parallel plate, gives the bent-kneed or geni-

FIG. 284.

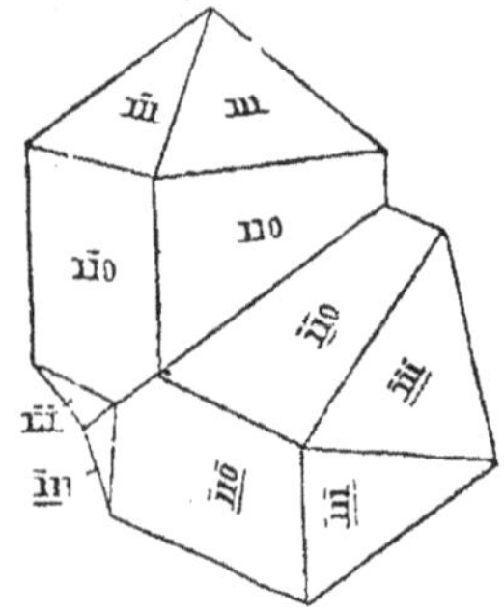

FIG. 285.

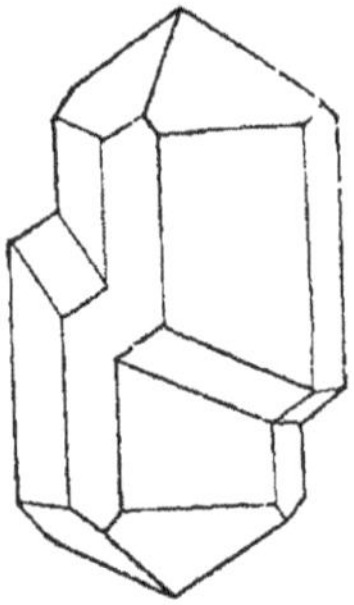

culated group (fig. 285) whose ends are both in the direct position, the middle one alone being reversed. Fig. 286 is another example of a triple group, very characteristic of the allied species Rutile. Here the ends are bent away from the middle, the twin planes being different faces of *P*∞.

FIG. 286.

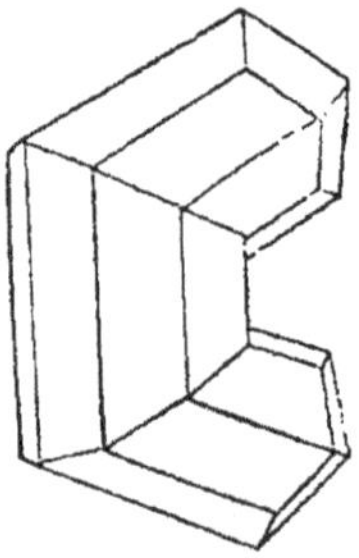

FIG. 287.

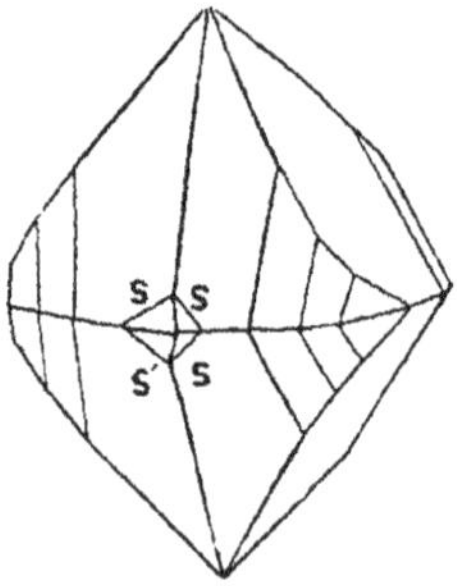

In sphenoidal hemihedral forms, the twin plane is commonly a face of a pyramid, and when the proportions of the latter differ but slightly from those of a regular octahedron, the twin groups, whether contact or penetration, are very like tetrahedral twins in the cubic system. This is especially observed in Copper Pyrites. In the pyramidal hemihedral forms the twin plane is usually a face of the diagonal prism $\infty P \infty$, or the twin axis is one of the lateral axes. This in crystals like fig. 287 has the effect of bringing the faces $S S'$ of the pyramids of the third order $\frac{1}{2}$ $(3 P 3)$ into a re-entering or negative solid angle in the basal section of the pyramid of the second order $P\infty$, but when these planes are less completely developed, they usually appear as contrasted diagonal striations upon the faces of $P\infty$.

Twin crystals of the rhombic system. Crystals twinned upon the faces of pyramids or prisms are of frequent occurrence in this system. Some of the most familiar examples

FIG. 288.

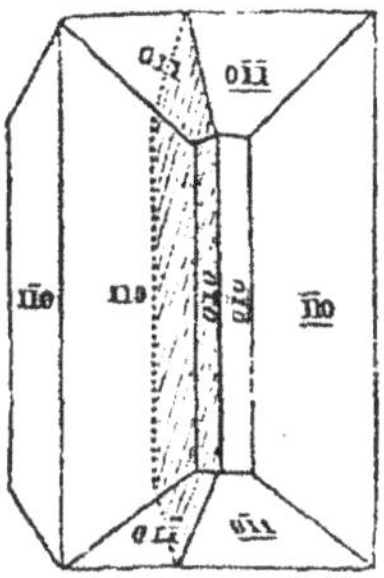

FIG. 289.

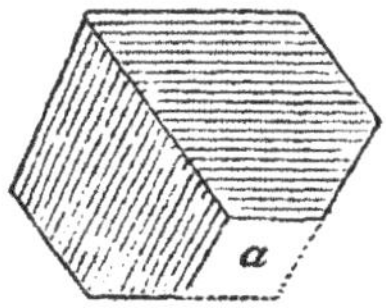

are afforded by the allied species, Aragonite and White-lead Ore. Fig. 288 is a contact twin of Aragonite on the face $(\bar{1}\ 1\ 0)$ of the prism ∞P, and fig. 289 its section on the basal plane, the shading lines being parallel to the brachypinakoid in each individual, which faces meet on one side in a projecting, and on the other in a re-entering, angle. If, as very commonly happens, the hollow space *a* is filled up by parallel elongation of the two crystals, the group may resemble a single crystal if the development is confined to the zone of the prism. Figs. 290*a*, 290*b* give the same group with a third individual, the twinning being repeated on the same face; this brings the prism faces in I. and III. into similar position, the middle crystal II. being reduced to a parallel plate.

This is also a very common case, the middle individual being many times repeated, and appearing as a series of fine

FIG. 290*a*.

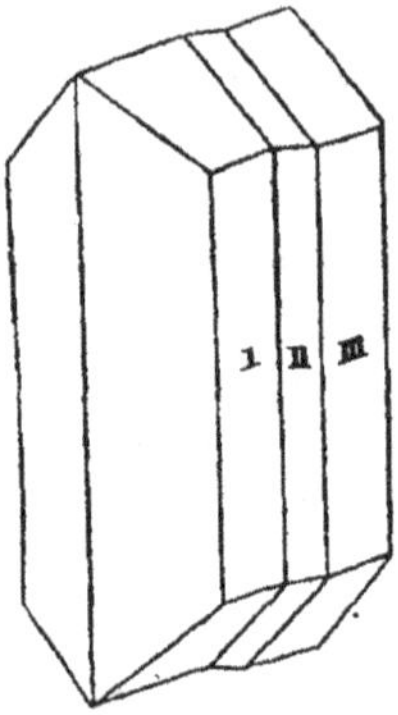

FIG. 290*b*.

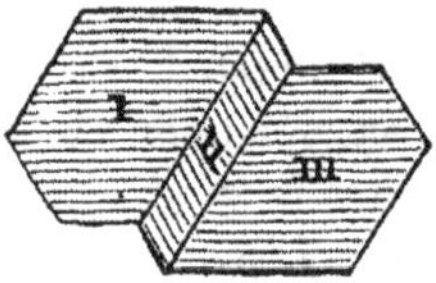

striations parallel to the twin plane. Fig. 291 is another group, in which the ends I. and III. are twinned upon adjoining faces of the prism in the middle crystal II. Fig. 292 is the

FIG. 291.

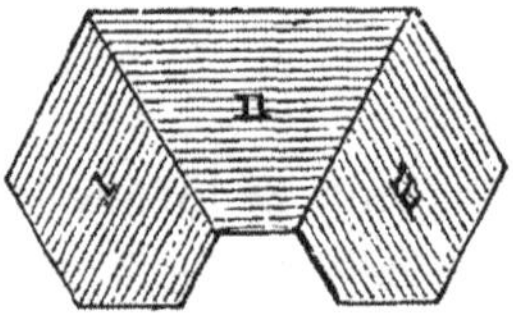

FIG. 292.

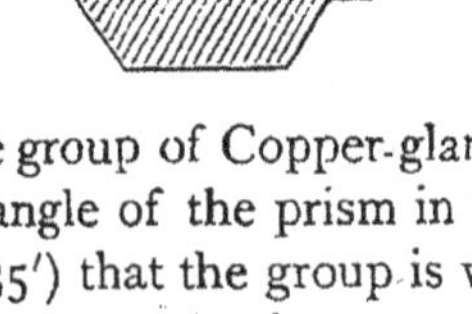

horizontal section of a similar triple group of Copper-glance, twinned upon faces of ∞P. The angle of the prism in this species is so nearly = 120° (119° 35′) that the group is very similar in appearance to a regular hexagonal prism.

Figs. 293, 294 are examples of cruciform twins produced by the penetration of two crystals of the combination $\infty P . \infty \breve{P} \infty . 0 P$ in Staurolite. In fig. 293 the twin plane is a face $(0 3 \bar{2})$ of $\frac{3}{2} \breve{P} \infty$, the vertical arms of the cross being in direct, and the transverse ones in the inverted position.

The latter are nearly, but not quite, horizontal, the angles between the vertical axes of the two crystals being alternately

FIG. 293.

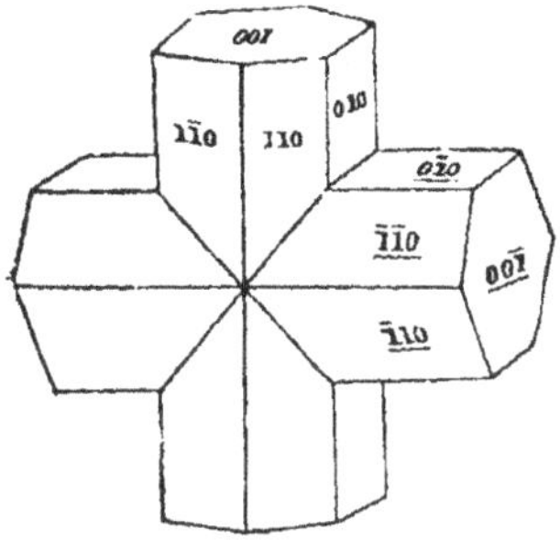

FIG. 294

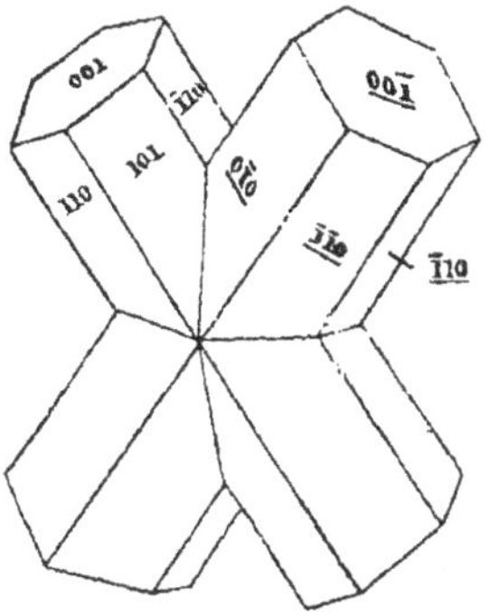

91° 36′ and 88° 24′. In fig. 294 the twin plane is the face $(\bar{2}\,3\,\bar{2})$ of $\frac{3}{2}\breve{P}\frac{3}{2}$, the left-hand crystal being placed in direct and the right-hand one in reversed position. Here the axes of the two prisms cross at 58° 46′, and the re-entering angle between the brachypinakoid faces is 119° 34′, or nearly 120°. These peculiarities are caused by the fundamental ratios $a : b : c$ of this species being very nearly as $\frac{1}{2} : 1 : \frac{2}{3}$; the actual values are 0.48 : 1 : 0.67, which give a twin axis inclined to the vertical at nearly 45° in the first case, and 60° in the second.

In the sphenoidal hemihedral forms of this system, the three pinakoids are possible twin planes; but examples of such twinning are not common. The best known case occurs in Manganite where a combination containing $\frac{6}{5}\breve{P}\infty$ or $\{365\}$ as a sphenoid is twinned upon the brachypinakoid.

FIG. 295.

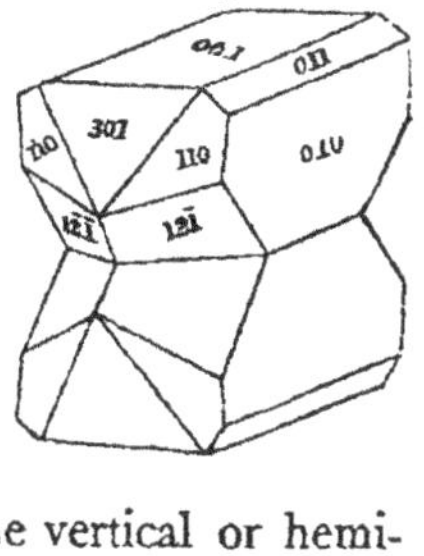

Fig. 295 is a compound structure observed in the hemimorphic species, Electric Calamine; the twin plane is the basal pinakoid, whose normal is the vertical or hemimorphic axis. As, however, the faces of $2\,P\,2$ in these crystals only appear at the negative end of the vertical axis, it is necessary to consider the lower component as inverted about one of the lateral axes as well as about the twin axis, and the notation of the faces of the inverted form will differ according as this reversal takes place about *a* or *b*. In the first case the front transverse faces $3\breve{P}\infty$, ∞P, $2\breve{P}2$ will all be positive to the axis *a*, while the longitudinal or side faces, $\infty\breve{P}\infty$ and $\breve{P}\infty$, on the same side will be positive above and negative below to the axis *b*; and in the second the longitudinal faces on the same side will have similar signs, while the transverse ones will be negative to the axis *a* in the lower crystal. This difference is only geometrical, and there is no reason to consider either reading as preferable to the other. The same structure may also be explained by supposing the vertically reversed crystals as essentially penetrating the direct one, which, however, requires either the brachypinakoid or macropinakoid to be the twin face, an assumption which is not compatible with the exclusion of planes of symmetry from possible twin planes.

Twin crystals of the oblique system. In this system any face is a possible twin plane except the clinopinakoid, or plane of symmetry, and of the three crystallographic axes only the vertical is a possible twin axis. The type of most frequent occurrence is that having the orthopinakoid as the twin plane, the twin axis being horizontal. When the same face is the plane of contact, the group resembles fig. 296, a hemitrope of the combination $-P.\ \infty P.\ \infty P\!\!\!/\infty$, common in Gypsum, the front half of the crystal being supposed to be the reversed one. The faces of the hemipyramids in the two crystals meet in re-entering angles above and parallel salient ones below, and their clinopinakoids in the same surface at either end. When the individuals are differently proportioned, and have their greatest length parallel to the inclined axis, they

FIG. 296.

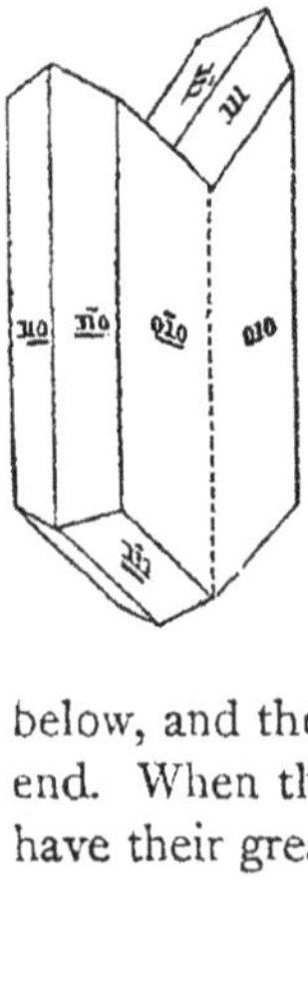

may form a complete cruciform penetration twin, the re-entering angles of the faces of the hemipyramid breaking the lines of the vertical edges in front and behind. These may be also considered as twinned by rotation upon the vertical axis, in which case the twin plane is horizontal and not a possible crystallographic face; while on the former view the normal to the twin plane, being horizontal, is not a possible crystallographic axis.

The same type of twin crystal is very common in the

FIG. 297.

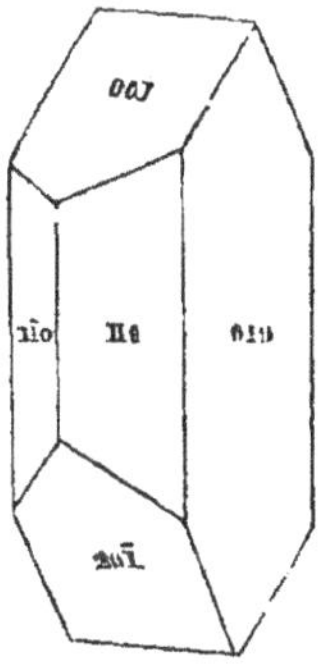

FIG. 298.

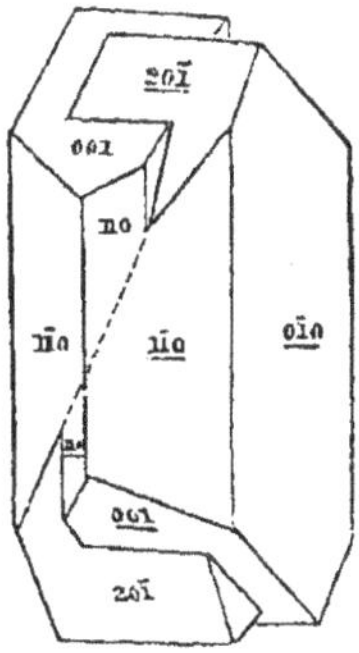

species Orthoclase, two individuals of the combination ∞P, $0 P$, $2 \not{P} \infty$, $\infty \not{P} \infty$ (fig. 297), being combined with partial penetration upon the clinopinakoid in the groups, figs. 298, 299, each component preserving its individuality, except in the case of the prism faces, which, though apparently simple, are made up of parts of different crystals, joined along the diagonal dotted lines in the figures. This is known as the Carlsbad type of twin crystal, the groups being further distinguished as right- and left-handed, according as the reversed crystal is to the right (fig. 298) or left (fig. 299) of the direct one, a geometrical distinction which is only

FIG. 299.

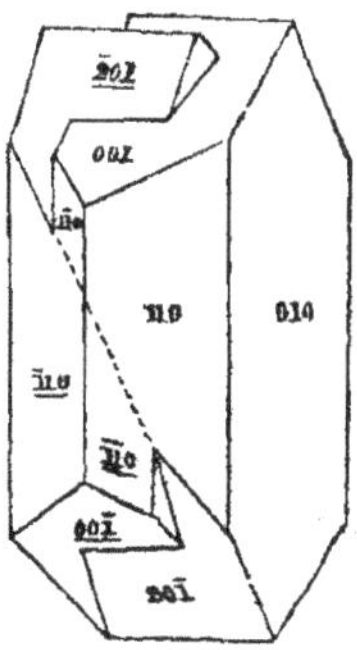

recognisable so long as the penetration is incomplete. Fig. 298 is noted as having a horizontal twin axis, and fig. 299 a vertical one.

Fig. 300 is a similar combination to fig. 296, the positive hemi-orthodome $\bar{P}\infty$ being substituted for the more acute one

FIG. 300.

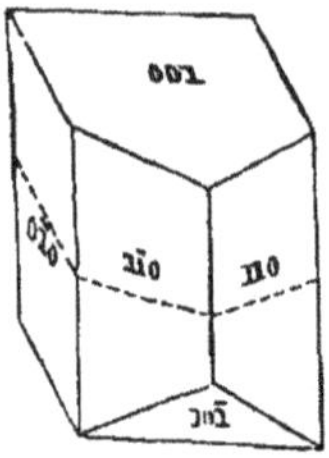

FIG. 301.

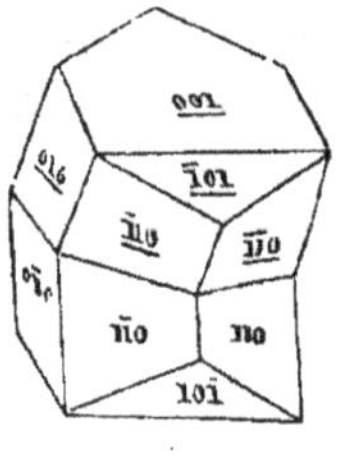

$2\,\bar{P}\infty$. This, when twinned by rotation upon a plane parallel to the base, as shown in the dotted line, gives fig. 301, where the faces of the prism meet in re-entering angles in front and projecting ones behind. The twin axis in this case is inclined to the vertical at the angle $180° - (90 + \beta^c)$. This is sometimes called the Manebach type of twin in Felspar,

Fig. 302 is a crystal of the same combination as the last, but of a nearly square section on the orthopinakoid ; this,

FIG. 302.

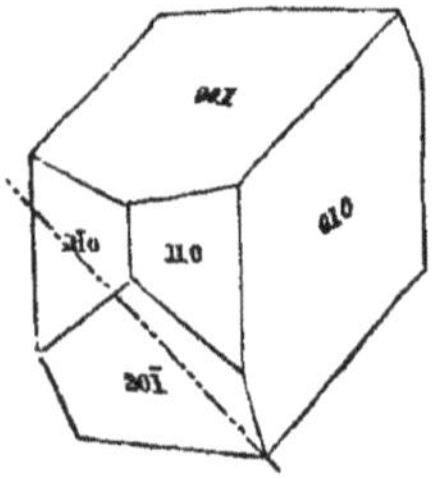

FIG. 303.

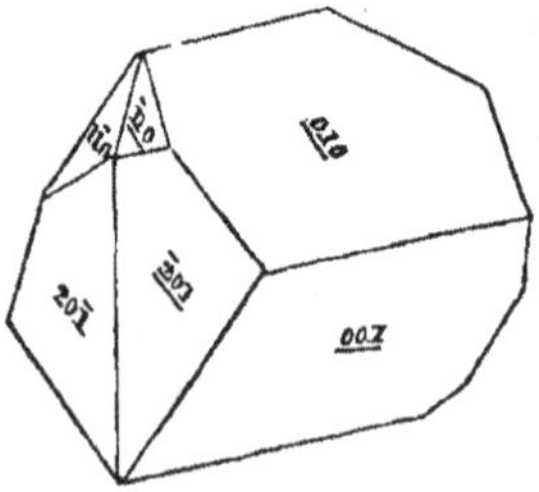

when twinned on the clinodome $2\,\grave{P}\infty$, indicated by the dotted line, produces fig. 303, in which there are no re-entering angles. This is known as the Baveno type of twin in Felspar.

Twin crystals of the triclinic system. As there are no planes of symmetry in this system, any face is a possible twin plane. The observed cases are generally similar to those of the oblique system, with an additional one special to the system where the brachypinakoid is the twin plane. This is represented in fig. 304, the right-hand half of the divided crystal being supposed to be rotated upon a normal to the brachypinakoid, which brings the opposite halves of the basal pinakoids together in a re-entering angle at the top, as indi-

FIG. 304.

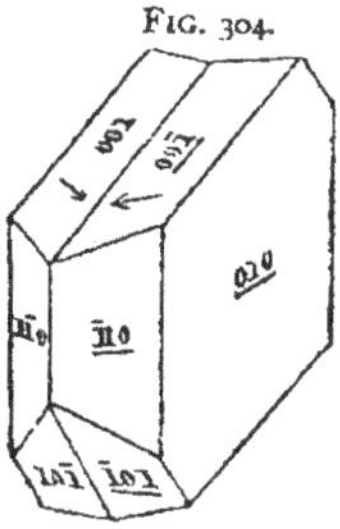

FIG. 305.

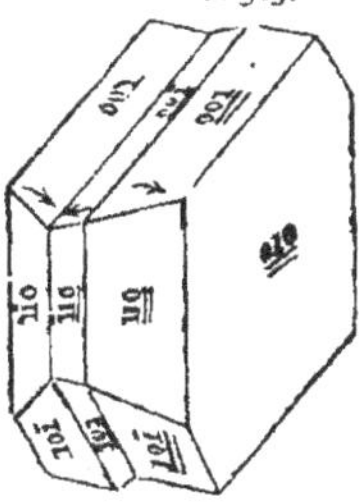

cated by the arrows, and in a corresponding salient one at the bottom; and those of the hemi-macrodome in the same way are convex in front and concave behind. Supposing such a group to have only its lower faces developed, it would be impossible to distinguish it from a simple crystal by considerations of form alone. If this structure is repeated with a third individual, as in fig. 305, the exterior components resemble the halves of crystals in their normal positions, divided by an intermediate parallel plate. This middle individual may, however, be replaced by a large number of much thinner plates, in which case the re-entering angles of the basal pinakoids appear as fine striations parallel to the edge between $\infty \breve{P} \infty$ and $0 P$ upon the latter faces. These, known as polysynthetic striations, are especially characteristic of the Felspars crystallising in this system, there being no apparent limit to the number and fineness of such twin lamellæ, which are, as a rule, easily recognised by the microscope, even when not apparent to the naked eye. In many cases, however, they are perfectly visible without being magnified, as, for instance, in Labradorite.

Fig. 306 is an example of complex twin grouping of triclinic crystals, the individuals I. II. and III. IV. being twinned upon the brachypinakoid, as in the preceding example, and these are further compounded according to the Carlsbad type. This, though generally similar in appearance to fig. 305, differs from it by having similar faces arranged in pairs instead of alternating singly. The above are the principal types of twin crystals in the different systems. Other and more special cases will be noticed in treating of the minerals in detail.

FIG. 306.

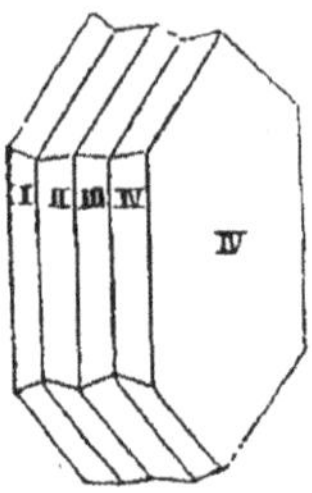

Irregularly developed crystals. In the artificial preparation of crystals in the laboratory, it is a well-known practice to select some of the most regular individuals from those first deposited, and place them in such a manner in fresh portions of the solution that they may increase as equally as possible in all directions, for which purpose it is necessary to alter their positions from time to time. If, on the other hand, the growth upon any part is hindered, as, for example, upon the face in contact with the surface of the vessel, the deposition of the additional material will increase the faces remaining free to such an extent that their true character may not be readily seen. This condition prevails to a great extent in natural crystals, and in fact one of the chief objects of determinative crystallography is the reduction of such distorted forms to their theoretical regularity. Only a few of the simpler cases can be given in this place. Fig. 296 is an octahedron with two faces $(1 1 1)$ $(\bar{1}\bar{1}\bar{1})$ prominently larger than the other six, in which the solid angles are no longer

FIG. 307.

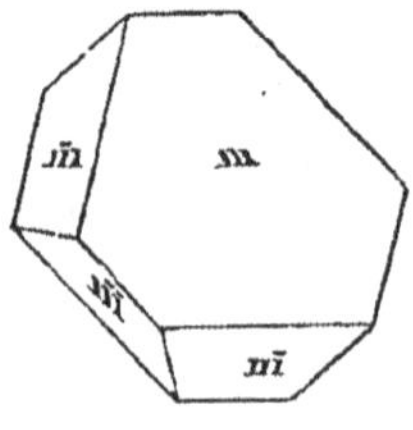

apparent, as the four faces similarly placed with respect to any principal axis do not meet in the same point, but in an edge parallel to the axis of a dodecahedral zone, or. in other words, the crystal is elongated in the zones $[\bar{1}10]$, $[01\bar{1}]$, and $[10\bar{1}]$, and shortened on $[110]$, or shortened on the ternary axis (111). This is common in crystals of Nitrate of Barium deposited from solution upon a flat

FIG. 308.

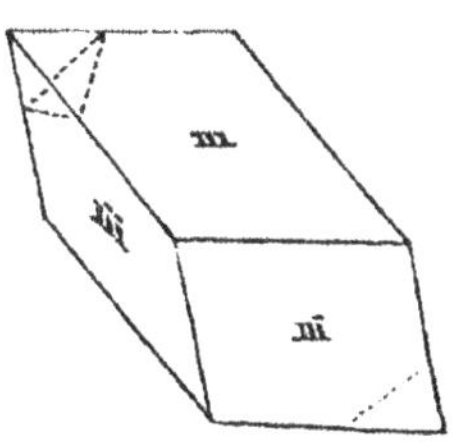

FIG. 309.

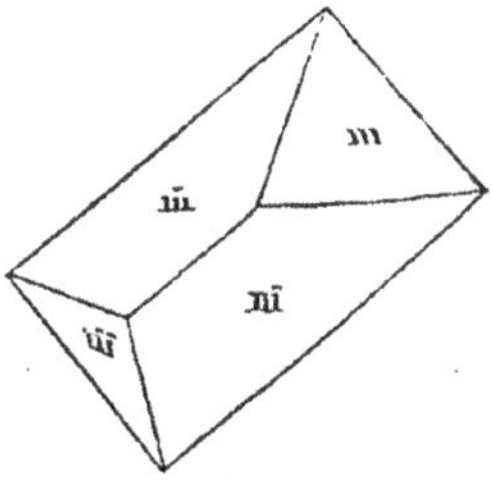

surface, the enlarged faces appearing as nearly regular hexagons; and it is also readily obtained by cleavage from octahedra of Fluorspar. When the octahedron is elongated on a ternary axis the faces perpendicular to that axis may be completely obliterated, producing an acute rhombohedron, as in fig. 308. This is easily obtained by cleavage in Fluorspar, and its octahedral character is as easily restored by cleaving off two tetrahedra in the directions shown in the dotted lines. Fig. 309, an octahedron distorted by elongation on a binary axis, has a general resemblance to a combination of two prismatic forms in the rhombic system.

FIG. 310.

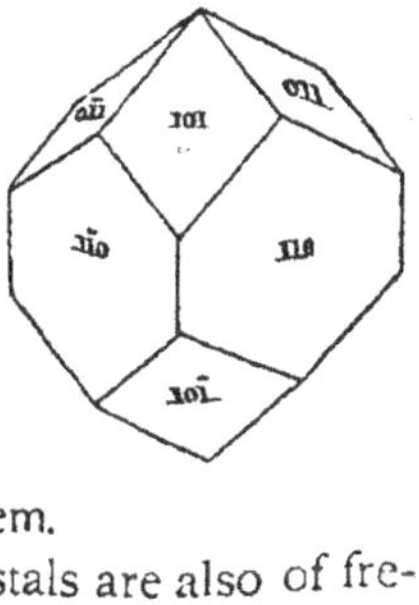

Fig 310 is a rhombic dodecahedron elongated vertically, which converts the upright faces into a square prism, and the inclined ones being unchanged, the general effect is that of the combination of a prism and pyramid of diagonal position $\infty P . P\infty$ in the tetragonal system.

In the hexagonal system distorted crystals are also of frequent occurrence, producing great variation in shape. Some of the most familiar examples are afforded by the common combination $R - R \infty R$ in Quartz, three of which are represented in the following figures. In fig. 311 the four rhombohedral and two prismatic faces in the zone $[2\bar{1}\bar{1}2]$ are

FIG. 311.

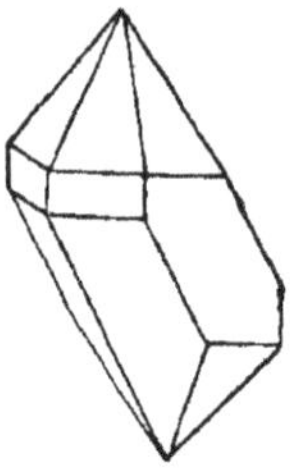

FIG. 312.

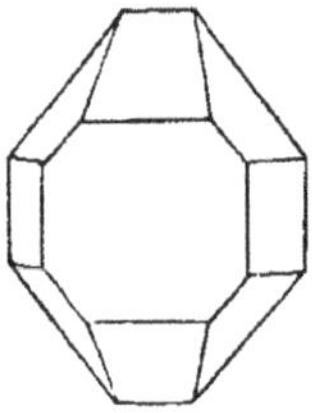

lengthened parallel to its axis; in fig. 312 the crystal is elongated on the axes of the zones $[\bar{1}2\bar{1}0]$, $[\bar{1}2\bar{1}2]$, and $[1\bar{2}12]$, giving an apparent rhombic character to the combination; and in fig. 313 one face of one rhombohedron is prominently longer than the others, giving a kind of chisel-edged termination. Crystals of this kind generally with only one end developed are common among the brilliant groups of rock crystal found in the Western Alps of Dauphiné and Piedmont.

FIG. 313.

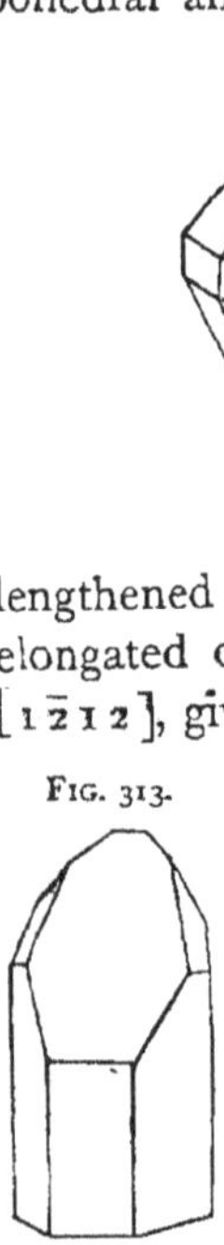

In the rhombic and oblique systems the commonest case of distortion is that of the unequal development of the faces of the prism, one pair being broader than the other, giving a rhomboidal instead of a rhombic basal section.

Imperfections in the faces of crystals. In addition to the irregularities arising from distortion and compound structure, crystals, when of large size, often appear with roughened, striated, or even partially hollow faces. The latter imperfection is common in substances that crystallise easily, whether from solution, as salt and alum, from sublimed vapours as arsenious acid, or from molten masses, as lead, silver, bismuth, and silicates produced in furnace slags. In all these cases crystals are often observed having their edges perfectly defined, while the faces themselves are hollowed out and reduced to very narrow surfaces adjacent to the edges, giving skeleton structures, in which the general elements of the form are, however, usually recognisable without difficulty. In Fluorspar, octahedra with roughened faces are occasionally found, which are made up of minute cubes piled up like courses of masonry, the side of each successive course being diminished by the breadth of two cubes. The roughness of the face is, therefore, due to the step-shaped section of the pile.

In the hexagonal system, instances of irregular single crystals, built up from smaller individuals of the same or different kinds, are very common in Calcite. In the north of England lead mines the obtuse rhombohedron known as Nailhead Spar, and the combination shown in fig. 144, are often aggregated in such a manner as to produce rough-faced rhombohedra and scalenohedra often of considerable size. Sometimes this step-shaped outline is apparent on some of the faces, while the others are comparatively smooth and regular. The same kind of structure may often be brought out in the most regularly developed crystals by the action of solvents, which produce the so-called *corrosion figures*, which show that in many cases the smallest crystals are fully as complex, or even more so, than those of larger size. The characteristic habit of particular crystals is often apparent in the most minute individuals; for instance, the peculiar geniculated groups of Tinstone and Rutile (fig. 284), are developed when these oxides are crystallised from solution in melted borax or phosphate of soda before the blowpipe, even when the crystals must be magnified from 400 to 500 diameters to render them visible.

The prismatic faces of Quartz crystals are very generally covered with horizontal striations, representing very minute portions of the faces of an acute rhombohedron. This is generally called oscillatory combination, a tendency towards the formation of rhombohedral ends being supposed to have alternated with another towards prismatic elongation. In Beryl the prisms are striated vertically, as is also the case in Tourmaline, the continued repetition of the prisms of the first and second order producing nearly cylindrical forms.

Another class of imperfection, where the faces of crystals are curved instead of plane surfaces, is characteristic of certain minerals, the most striking examples being afforded by Siderite, which occurs in rhombohedra having strongly curved faces; Gypsum and Diamond: the latter, when in the form of the hexakisoctahedron, are often nearly spherical in shape.

Habit of crystals. In describing minerals it is usual to speak of their crystals as affecting particular types, according to the character of the dominant or principal faces; thus, in the cubical system, they may be cubic, octahedral, dodecahedral, &c., as one or other of the principal forms prevail in the combination. In the other systems the types are pyramidal, sphenoidal, rhombohedral, or scalenohedral when the closed forms are most apparent, and prismatic when the development is mainly in the direction of the open forms. In the latter case several further distinctions are founded upon the relation of the height of the prism to the breadth of its base or closing pinakoid. The shorter forms, or those having their principal dimensions in the direction of the lateral axis, are said to be tabular, or, if very thin, platy or scaly: when the height is only a few times the breadth, they are short columnar, and as the relative length of the prismatic axis increases they become columnar, slender-prismatic, and acicular, or needle-shaped. In these terms the expression prismatic is not restricted to the forms assumed as vertical prisms, but is used with the general signification of any zone of prismatic planes, whether vertical or horizontal. The habit of considering any prominently defined axis as a prismatic one in describing the appearance of crystals is very general and convenient, but care must be taken not to confound such 'columnar' forms with the more exactly determined crystallographic prisms.

Irregular grouping of crystals. Masses of crystals, when not arranged as symmetrically twinned forms, are spoken of as groups or crystalline aggregates. These are commonly found in hollow spaces or druses in the containing rock, attached at one end, with the faces terminating the opposite end, freely developed, the individuals of the group having a more or less radial arrangement diverging from the point of attachment. This, in general terms, may be considered as the most typical kind of grouping of well individualised crystals. When the aggregates are of a more compact kind, the individuals are rarely recognisable with anything like their full number of faces, but appear, as a rule, as columnar or fibrous masses arranged in parallel or divergent forms. The latter, when in sufficient numbers, make up more or less spheroidal masses, which, according to the size of the spheroids, are spoken of as *mamillary*, *reniform*, or kidney shaped, and *botryoidal*, or grape-like mas es or concretions. Other smooth spheroidal masses of substances, having no apparent definite structure, are generally called *nodules*.

Parallel aggregates of a fibrous structure, such as those of Calcite and Gypsum, often form regular beds of a silky character on the face. These are known as Satin Spar; the same structure is common in salt and alum, where the fibres are often bent or contorted. Aggregates resembling corals, mosses, and other organised forms, are common in Arragonite, the so-called *flos-ferri*, or flowers of iron, and native metals; the latter are usually called dendritic forms, the same name being also applied to the plant-like stains of Brown Iron Ore and Peroxide of Manganese on rocks. Wire-like or filiform masses are very characteristic of native silver.

Stalactites are irregularly shaped crystalline masses found in caverns hanging from the roof, and stalagmites are similar masses accumulated above the floor. These terms are rather geological (as indicating methods of origin) than structural.

When no indications of crystalline structure are apparent in a mineral aggregate it is said to be *massive*. Sometimes such masses are spoken of as *amorphous*. This, however, is an improper use of the term, which should only be applied to substances which fail to show crystalline structure when tested by optical and other methods, many perfectly well crystallised bodies often appearing structureless until so examined.

CHAPTER X.

MEASUREMENT AND REPRESENTATION OF CRYSTALS.

THE angles between two faces in a crystal may be measured in two different ways, namely, directly from the inclination of a pair of jointed blades striding over the edge, and indirectly by determining the angle through which the crystal must be turned in order to obtain the reflected image of an object successively from both faces. The instrument used in the first method is the hand or contact goniometer, which has not been materially altered from the form in which it was originally made by Carangeot for Romé de l'Isle and Haüy at the end of the last century, and is shown in fig. 314. It consists of a semicircular arc, divided into single or half degrees, according to size, having attached to it two steel blades, one of which, *k m*, is fixed, or rather has only a sliding movement in a straight line upon the pins *c d*, while the other has an angular as well as a sliding motion about *c*. The zero point of the graduation is on a line parallel to the direction of the fixed rule, so that when the latter is laid upon one of the faces containing the angle to be measured, and the movable one turned until it bears similarly upon the other face, the edge *g i* will indicate the value of the angle upon the divided arc. The object of the slots is to allow the length of the measuring arms to be varied to suit the size of the crystal. For the same purpose the arc is divided, and can be folded back upon a hinge at *b*, the supporting

FIG. 314.

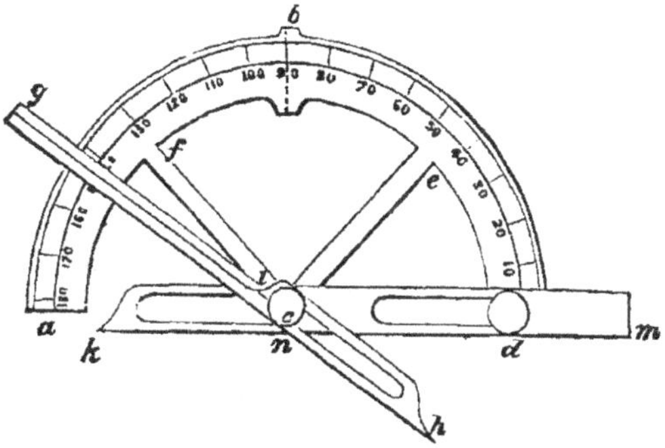

arm *f* being attached by a screw at the back, which can be taken out when greater freedom of manipulation is required, as, for instance, when working upon crystals in implanted groups. A more convenient arrangement is to have the arms separate from the measuring arc, in which case a complete circle may be advantageously used.

Under the most favourable conditions angles can be measured by the contact goniometer to within half a degree, when the crystals are of a certain size. It is therefore best adapted for trial measurements upon large and imperfectly lustrous faces, or for use in making models. When greater precision is required, the more exact method depending upon reflection from the faces must be used. Fig. 315 represents Wollaston's reflecting goniometer in one of its simplest forms. It consists of a circular disc E E, with a divided rim reading to single minutes of arc by a vernier at R, movable about a horizontal axis passing through the bearing C at the top of the pillar B, with a milled head for turning it at G. This axis is hollow, and an inner one, moved by the milled head I, is connected with the arrangement carrying the crystal at K. This consists of two bent arms K M, movable about a pin L, and the carrier proper O, which has a rotatory as well as a sliding movement in the collar N,

FIG. 315.

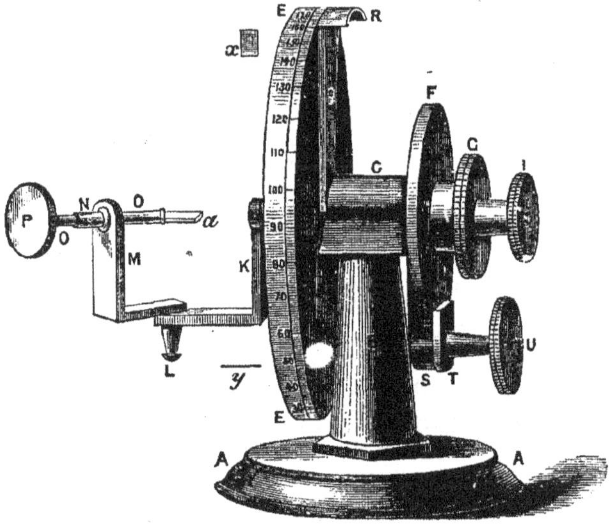

the crystal *a* being attached by a ball of wax to its outer end. The parts S T and U form a clamp upon the disc F for maintaining the circle in any particular position. The foot A is a block of wood; in the larger instruments it is of metal, with levelling screws.

In observing angles it is necessary to bring the edge exactly into the line of the axis of the instrument, which is done by the various movements of the carrier, which together form a kind of universal joint. Two signals are required, one of which, *x*, is seen by reflection, and the other, *y*, directly. These may be any small, well-defined objects, such as a bright spot seen through a screen placed before a window or a lighted lamp, and a window-bar, chalk-line or line of light, as far off as they may be conveniently seen. The edge is in adjustment when the image of a horizontal or vertical line reflected from either face coincides with that viewed directly from the same point, the eye being as close to the crystal as possible. The circle being clamped, the crystal is then turned by I until the reflected image of *x* is brought into coincidence with *y*, seen directly, when it is unclamped, and the whole circle is turned by G until the same thing is seen from the second face. The angle through which the circle is rotated will be the supplement of the dihedral angle required, if it was originally set to zero. As a rule, however, it is better not to start from this point, but to read the vernier after each adjustment, and take the difference of the readings, repeating them for several different positions on the circle to eliminate the errors of eccentricity as much as possible. In all cases the observation must be often repeated to obtain results of any value.

Numerous modifications and improvements of the reflecting goniometer have been introduced for the purpose of facilitating the operation of centering and adjusting the crystal, and obtaining greater precision in observation. For the latter purpose a sighting telescope of low magnifying power is used, the observed object being the image of the cross wires in a second telescope, or collimator; and for the former the carrier is made with a system of rectilinear and cylindrical, or sliding and rocking, movements analogous to those of the 'universal' machine tools. The more exact modern instruments of this kind are usually constructed upon the pattern introduced by Mitscherlich in 1843; another construction, that of Babinet, in which the divided circle is horizontal, being also considerably used. The latter has the advantage of placing the crystal upright, and not as in the preceding form overhanging the carrier; but as it requires signals that are not in the same vertical plane, it is not so easily used when without a telescope. The large instruments made by Fuess, of Berlin, on this pattern are remarkable for the extreme precision of their centring apparatus, which has two straight line movements at right angles to each other, and two cylindrical ones also at right angles, the former being used for centring or bringing the edge into view, and the latter for adjusting or making it vertical. This operation is facilitated by the addition of a lens of short focus to the objective end of the telescope, which for the time converts it into a microscope of low magnifying power, and allows the edge to be brought into coincidence with a vertical wire in the eyepiece.

When the observations are made with the unaided eye or a single telescope, the signals observed must not be too near, or there may be an appreciable error of parallax. A convenient object is a small gas-flame about half-an-inch high, from 15 to 30 feet distant, the observations being made in a darkened room. When the instrument has two telescopes the error of parallax may be eliminated, but the loss of light is so considerable that crystals with very perfect faces are required to give distinct images of the signal.

In all cases observations must be repeated several times, and, if possible, upon different crystals, the quality of the reflection obtained being specially noted. As a rule, the smallest crystals have the most brilliant faces, and therefore give the best results.

A goniometer for measuring the angles of minute crystals with dull faces has recently been described by Hirschwald. It is of the Wollaston pattern, but the position of the faces is ascertained by bringing them into the focus of a compound microscope with an objective of very short focal distance, which, when adjusted, can be traversed horizontally by sliding movements, the adjustment of the face being complete when its whole surface is equally well defined in the field of the microscope as the latter is moved over it.

The calculation of the crystallographic constants from the observed angles of crystals is the work of determinative crystallography, and the subject is beyond the scope of the present treatise. The student is referred to the special works on the subject, more particularly to Miller's tract on Crystallography, and the works of Klein and Mallard.

Methods of representing crystals. The larger number of figures in this volume, following the usual practice of works on Mineralogy, are drawn in so-called parallel perspective, which supposes the point of sight to be at an infinite distance from the object represented, so that all rays proceeding from the latter to the eye are parallel, or all lines and surfaces parallel to each other remain so in the drawing; and in order to bring in a sufficient number of faces the plane of projection is so chosen that all lines actually at right angles will be projected as acute or obtuse angles, and consequently all lines, except those parallel to the plane of projection, and all faces, will be shown in other than their true dimensions. This is done by adopting reduced lengths for the axes in the triaxial systems, which may be either diminished by the same amount, as in the so-called isometric projection, or two may be reduced in one proportion and the third in some other, as in the monodimetric projection; or, finally, each axis may be reduced in a different proportion, giving the anisometric projection. Of these methods the first is not suited for representing crystals, as the distortion of the faces is too great, but either of the other two may be used, according to circumstances, the anisometric projection, as a rule, giving the most natural figures, which, however, is accompanied with the disadvantage of an excessive foreshortening of the basal section.

The following table, by Weisbach, gives the elements of the projections most commonly used in this class of drawing: the notation refers to three equal rectangular axes taken in Weiss's order; the first group of figures gives the proportional lengths assumed for the axes, the second the corresponding reduced lengths compared with their true lengths, as given in their orthographic projections, and the last the angles between the axes at the origin o. The second of the monodimetric and anisometric series respectively will be found to be most generally suited for representing crystals.

Projections	Proportional lengths of axes			Reduced lengths of axes for real length = 1,000			Angles between axes		
	a	*b*	*c*	*a*	*b*	*c*	AOB	AOC	BOC
Isometric . .	1 :	1 :	1	0·817 :	0·817 :	0·817	120°	120°	120°
Monodimetric	1 :	2 :	2	0·471 :	0·943 :	0·943	131°.25′	131°.25′	97°.11′
,,	1 :	3 :	3	0·324 :	0·973 :	0·973	133°.24′	133°.24′	93°.11′
,,	1 :	4 :	4	0·246 :	0·985 :	0·985	134°.06′	134°.06′	91°.47′
Anisometric .	5 :	9 :	10	0·493 :	0·887 :	0·985	157°	107°.49′	95°.11′
,,	6 :	17 :	18	0·333 :	0·944 :	0·998	172°.50′	96°.23′	90°.47′
,,	8 :	31 :	32	0·250 :	0·968 :	0·999	174°.46′	94°.55′	90°.20′
,,	10 :	49 :	50	0·200 :	0·980 :	1·000	175°.52′	93°.58′	90°.10′

In drawing the projection, the axis c is made vertical, and the other axes are then set out from the point o, at their proper angles, with a protractor. The proportional lengths are then laid off upon each side of the centre, and if the drawing is sufficiently large, these lengths may be subdivided into equal parts. This gives the projection, or 'axial cross,' for three axes of the same length, or that of the unit form of the cubic system; and therefore, by applying to the vertical axis the proper value of c as given in works on Descriptive Mineralogy, the unit axis of any tetragonal species may be found, and similarly, by altering a and c, those of a rhombic species corresponding to $b = 1$.

In the hexagonal system the projections of the lateral axes h and l are found from those of the cubic system by making the front axis of the latter $a = \sqrt{3} = 1·732$, and drawing lines to the extremities of the right and left axis b, which in this case corresponds to k. This gives the projection of a right-angled triangle, or half the base of a rhombic prism, whose obtuse angles are 120°; a line parallel to $k\bar{k}$, bisecting a, truncates the acute angle or gives a third side of a regular hexagon, and lines drawn from these points of intersection through the centre, o, will be the projections of the axes required; the vertical axis is then changed to the characteristic value of *c*, as in the tetragonal system.

In the oblique system the position of the clinodiagonal axis is found by laying off its projections upon the rectangular unit axes *a* and *c*—that is, upon the former *a* sin. β, and upon the latter *c* cos. β—these being the sides of a parallelogram whose diagonal corresponds to the unit-length of an axis having the characteristic inclination required, the proper lengths corresponding to the fundamental parameters of the species being then substituted as in the rhombic system.

In the triclinic system a similar principle is followed. Starting from three rectangular axes of the same length, and calling the angle between the pinakoids 0 1 0 and 1 0 0, *c*, the diagonal of a parallelogram whose sides are *a* cos. *c* and *b* sin. *c*, will be the projection of a horizontal line, *b'*, whose length = 1 in the plane of 0 1 0. Setting off upon this, the length *b'* sin. α, and upon the vertical *c* cos. α,[1] the sides of another parallelogram are obtained, whose diagonal will be the projection of the axis *b*, its acute angle to the vertical being to the right or left of the centre according as the first quantity is taken on the positive or negative side of *b'*, the second being always positive. The projection of the front axis is found from the sin. and cos. of the angle β, in the manner given for the oblique system; the third axis, *c*, retains the vertical position unchanged. These oblique axes will all be of the same length, so that to obtain those proper to the species, *a* and *c* must be multiplied by their characteristic parameters, *b* being regarded as unity. From the projection of the axes the unit form of a species is obtained by joining their extremities by straight lines, which, in the cubic system, will represent the edges of the octahedron in the tetragonal, those of the unit-pyramid, and so on for the other systems. The cube, or, generally, any diagonal prismatic

[1] α being the angle between the axes *b* and *c*.

form, is found by ruling through the extremities of an axis lines parallel to the other axis in the same plane, which meet in the section on the zone plane, and lines through the corners of this plane parallel to the third axis will be the projection of edges of combination of faces belonging to the zone.

The construction of forms with other parameters is effected in a similar manner by lengthening the unit axes by the amounts indicated by the co-efficients in Weiss's method, or cutting off the reciprocal quantities, when working by the indices, and connecting the points so obtained by straight lines. This will determine a series of planes having the properties of the faces, and their convex portions will give the projection required. This method of construction is shown for the principal forms of the cubic system in figs. 15, 18, 25, and 28.

In drawing combinations it is to be remembered that the octahedron, or unit pyramid, is always the largest, and the cube rectangular prism, or pinakoid, the smallest of the constituent forms. It is therefore necessary, in the first instance, to determine approximately the shape of the latter faces from a freehand sketch or measurement of the crystal it is intended to represent. The parallelepiped formed by the pinakoids or analogous faces is then completed, and the other forms are put in by truncating its edges and angles according to their geometrical properties, as laid down in works on Descriptive Crystallography. Illustrations of the methods to be followed, with examples, will be found in the larger text-books, more particularly in those of Groth and Dana. Considerable practice, and, above all, a knowledge of the zonal relation of the faces, is required, in order to obtain the required results with the minimum number of construction lines. As a rule it is best to work from a carefully prepared projection of a cube, with inscribed octahedron and rhombic dodecahedron, to the same axes. Upon these should be shown the positions of the rhombic and trigonal interaxes and the points where they intersect the faces. If of a sufficient size, the derived forms may be added, but it is perhaps better as a rule not to put more than two or three forms into one figure, in order to prevent confusion by the multiplication of lines. A useful size is an octahedron of about six inches length of vertical axis. This should be drawn upon strong paper or cardboard, and used as a protractor for projecting smaller figures by means of parallel rulers or set squares upon other sheets. In the other systems similar projections for certain type species should be constructed. Among the most useful will probably be found Tinstone, Zircon, Idocrase, and Apophyllite in the tetragonal, Quartz and Calcite in the hexagonal, Topaz, Barytes, and Sulphur in the rhombic, Orthoclase and Augite in the oblique, and Albite and Axinite in the triclinic system. A hard lead pencil, with a fine tapered point, should be used in drawing, the lines being kept as fine as possible.

In all except the cubic system the method of representing crystals by projections upon the plane of the base, or that of some other prismatic zone, is attended with considerable advantage, as all similar faces are shown of similar geometrical form and edges parallel to the plane of projection of their true lengths. Such projections, however, being only horizontal or ground plans, do not indicate the shape of the prismatic faces of the zone, which only appear as lines. The best examples will be found in the excellent illustrations to Brooke and Miller's Mineralogy. The mode of construction is that of ordinary orthographic projection.

Generalised representations of crystals. The representation of an individual crystal by either of the methods of projection previously noticed involves a knowledge of its material shape, and is special to itself; and therefore, in those species whose series of forms is extensive, a great number of figures will be required, in order to obtain an idea of their crystallographic characters. This knowledge may be conveniently summarised by the use of other projections, in which the faces are represented by lines or points instead of surfaces, and which have the advantage of allowing the whole of the observed faces in any species to be combined into a single figure.

In the first of these methods, the linear projection of Quenstedt, the faces are represented by the straight lines, in which they intersect the plane of projection, supposing them all to pass through a point in the normal to that plane

FIG. 316.

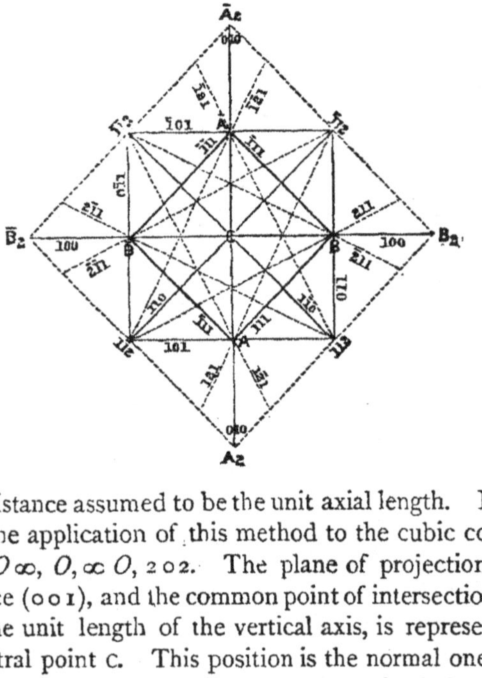

at the distance assumed to be the unit axial length. Fig. 316 shows the application of this method to the cubic combination $\infty O \infty$, O, ∞O, $2 O 2$. The plane of projection is the cube face (0 0 1), and the common point of intersection of the faces the unit length of the vertical axis, is represented by the central point C. This position is the normal one for the octahedron, for those faces of the rhombic dodecahedron whose indices have the order 0 1 1 and 1 0 1, and those of the icositetrahedron whose order is 1 1 2. The first form, therefore, will be represented by the projection of its horizontal edges, or the square whose diagonals are the unit lengths of the lateral axes A and B, the second by the square described upon these lengths, and the third by the square whose diagonals are 2 A and 2 B. The remaining faces of the cube and rhombic dodecahedron, being parallel to the vertical axis, will, when made to pass through it, be represented by straight lines intersecting at 45°, or the diagonals of the preceding squares. The eight remaining faces of the icositetrahedron having the order 1 2 1 and 2 1 1, intercept the vertical axis at 2 C; and therefore, when brought back to C, the projections of the first four will be the four lines enclosing a rhomb, whose diagonals are a and $\frac{1}{2}b$, and those of the second four the sides of a similar rhomb having the diagonals $\frac{1}{2}a$ and b. The zonal relations of the faces of the different forms entering into a combination are indicated in this projection by the crossing points of the lines, or the so-called zonal points. Thus (1 1 2) is the zonal point of a zone containing two adjacent faces of the rhombic dodecahedron, and the face of 2 *O* 2 truncating their edge, the latter being common to two octants of the crystal; B gives the zone formed by two adjacent faces of the octahedron, the face of the rhombic dodecahedron truncating their edge and two faces of 2 *O* 2, and so on for many others, which may be deduced from the figure. In the systems of lower symmetry the diagrams are less regular, and apparently less simple, but their construction is easily learnt when the leading principle is mastered. For the complete description of these the reader is referred to Quenstedt's works on Crystallography and Mineralogy. An ingenious application of this method in the construction of perspective figures of crystals is given in E. S. Dana's 'Text-book of Mineralogy,' p. 427.

Spherical projection. This method, originally introduced by Neumann, and brought into general use in its present form by Miller, supposes the crystal to be placed within a sphere, both having a common centre, and lines normal to the faces to be drawn through the centre to the surface of the sphere on either side. These lines will obviously be the diameters of the sphere, and the points of intersection their poles; and as the angle between the normals of two faces is equal to the supplement of their dihedral angle, the position of the poles will determine those of the faces by a system of points upon the spherical surface, which give rise to problems that can be solved by spherical trigonometry. If a sphere be supposed to be projected upon its equatorial plane, that being also parallel to the terminal pinakoid, or the equivalent face of the cube, the pole of that face will be the centre of the circle of projection, the poles of faces in the zone of the principal prism will lie in the circumference, their angular distance being given directly by the measured angles with the reflecting goniometer, and those of the

FIG. 317.

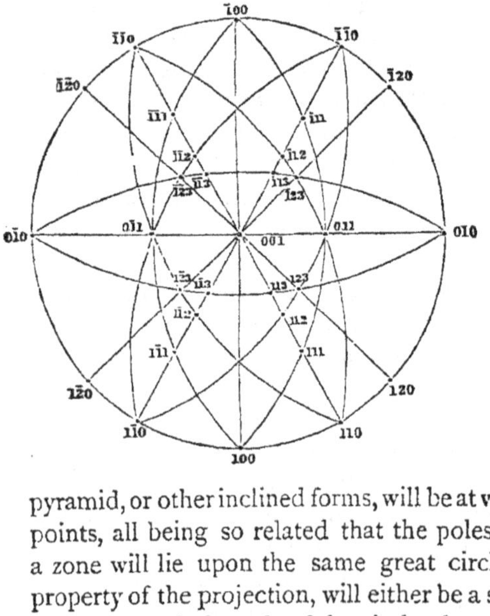

FIG. 318.

pyramid, or other inclined forms, will be at various intermediate points, all being so related that the poles of all the faces in a zone will lie upon the same great circle, which, from the property of the projection, will either be a straight line when it passes through the pole of the circle of projection, or an arc of a circle when in any other position. Fig. 317 is an example of the spherical projection of a crystal of Topaz, containing the forms ∞P, $\infty \breve{P} 2$, P, $\frac{1}{2} P$, $\frac{1}{3} P$, $\frac{2}{3} \breve{P} 2$, $\breve{P} \infty$, $\infty \breve{P} \infty$, $0 P$, as shown in horizontal projection in fig. 318. In it the principal zones, or those that include the terminal pinakoid, are represented by diameters, which are projections of meridians of the sphere, whose direction is found by laying off at the centre the supplements to the interfacial angles of the prism from the pole of 010; the position of the pole of any intermediate face in these zone circles being found by taking in them a distance equal to the tangent of half the supplement of the inclination of the face upon $0 P$, the radius being considered as unity. Zones that are not perpendicular to the plane of projection, or do not include $0 P$, are represented by arcs of circles, these being the stereographic projections of great circles oblique to the equatorial plane. This system has the advantage of representing the faces of crystals in the most general manner —namely, by points—and therefore there is no limit to the number of them that can be included in a single figure. Some very remarkable examples will be found in Descloizeaux's 'Text-book of Mineralogy.'

CHAPTER XI.

PHYSICAL PROPERTIES OF MINERALS:—CLEAVAGE, HARDNESS, SPECIFIC GRAVITY, ETC.

THE methods employed in the second great division of mineralogical research, or the investigation of the structural peculiarities of minerals, are essentially those of experimental mechanics and physics. All such investigations are based upon the assumed existence of physical molecules or indivisible particles of matter, all of the same kind and similarly arranged in the same substance. Without entering into the question of the actual nature of such molecules, or atoms, it appears to be certain that a great part of the results of observations may be explained by assuming that the molecular centres occupy the points of a reticular system,[1] and that the molecules may be considered as concentrated upon such points as attractive or repulsive centres of force. As a working hypothesis this involves the further supposition that if they are to be regarded as having definite size or dimensions in space, these dimensions must be very small as compared with the linear distance between any two. By the substitution of such centres of force for geometrical points in the reticular systems of the crystallographer, we arrive at the idea of the physical as distinguished from the morphological crystal, in which the consideration of molecular arrangement precedes that of form, the latter being a consequence of such arrangement. These investigations are especially applicable to those minerals whose crystals appear in forms common to two or more systems, such as the combination of three pairs of rectangular planes, or the regular six-sided prism, the true character of which can only be determined by a knowledge of their physical characteristics.

Without anticipating what will subsequently be treated in detail, it may be generally stated that the observations upon which conclusions as to molecular structure are founded are essentially those of elastic resistance to forces tending to

[1] There are two conjugate forms of the relation between the molecular centres and the points of a reticulation. They may coincide, or the molecular centres may occupy the centres of gravity of the (tridimensional) mesh. In a strictly cubical system this merely shifts the position in space of the whole reticulation. But it is quite different in the well-known case of round shot in pile. The centres of these lie on the knots of a fourfold reticulation, dividing space into octahedra and tetrahedra in the numerical ratio, 1 : 2, parallel to the faces of a regular tetrahedron; but their bounding planes form a network, each mesh of which is a right rhombic dodecahedron. Some writers use one, and some the other, hypothesis.

alter the equilibrium of the mass, including under this head the coarser mechanical agency necessary to produce fracture, as well as the more subtle evidence derived from the transmission of vibratory movements, as manifested in the action of light, heat, electricity, and magnetism. In some cases we find that the effects of these agents are similar in any direction, while in others there is decided dissimilarity in different directions, and these differences are intimately connected with crystalline symmetry. In addition to these there are two properties common to all solids alike, and therefore independent of structure—namely, density and hardness, or specific cohesive power—which are of value in Determinative Mineralogy, but it will be more convenient to consider first the structural property where relation to crystalline form is most apparent.

Cleavage. The resistance opposed by cohesion in homogeneous amorphous substances to forces tending to displace or separate their particles is similar in any direction, so that there is no reason why they should yield more readily in one way than another when strained beyond their elastic limits by forces of any kind, and therefore, when so treated they will break up into fragments of essentially irregular form. It is, however, different in the case of crystals, which, in a large number of instances, when similarly treated, show very decided tendencies to separate into fragments bounded by planes which are for the most part related in some simple manner to the unit form of the series proper to the substance; and when the direction of such surfaces is known, a comparatively slight cutting or wedging strain will be sufficient to produce a separation, while the resistance in other directions may be considerably greater. This property is known as crystalline *cleavage*, and the surfaces of separation are called *cleavage planes.* It is directly related to crystalline structure, but has no relation to specific cohesive power or hardness as measured by the resistance to abrasion, the hardest known mineral, Diamond, being one of the most commonly cleavable, as is also Gypsum, one of the softest; while among those of intermediate hardness, Quartz, Garnet, and Pyrites, are scarcely cleavable, but break like masses of glass or other amorphous bodies.

The readiest way of determining the cleavage of a crystal is to place the edge of a knife or small chisel upon a face parallel to that of some principal form, and strike a light blow with a hammer, when, if the direction is near that of a principal cleavage, a more or less flat-faced fragment will be removed. If, on the other hand, no cleavage is obtainable in the direction of the blow, the fractured surface will be uneven and irregular, or will show traces of step-shaped structure in the direction of the true cleavage plane. Thus Fluorspar, whose crystals are chiefly cubes, cannot be cleaved parallel to that form, but yields with the greatest ease in the direction of an octahedral face. Galena has an extremely perfect cleavage parallel to the faces of the cube; and Zinc-blende to those of the rhombic dodecahedron. Some minerals, such as Mica and Gypsum, are very easily cleavable, and may with slight effort be divided by the finger-nail, or the point of a knife, or needle, into laminæ of extreme thinness. In the case of Mica there seems to be no limit to the capacity for cleavage, as laminæ may be obtained thinner than the edge of any cutting tool that can be brought to bear upon them.

In some instances cleavages may be developed in imperfectly cleavable crystals by strongly heating and suddenly cooling them in water. Quartz crystals, when so treated, occasionally develop faces parallel to those of the unit rhombohedron; and under ordinary circumstances they break with a fracture like that of glass. Easily cleavable minerals, such as Salt, Galena, Fluorspar, and Calcite, usually decrepitate, or fly to pieces, when suddenly heated, the fragments obtained being regular cleavage forms.

In the cubic system, where all three axes are physically

equivalent, a cleavage parallel to any face of a form requires the existence of a similarly perfect one parallel to the remaining faces, and therefore the cleavage form will be a closed one, as are also the forms produced by pyramidal and rhombohedral cleavage in the tetragonal and hexagonal systems; but the prismatic and basal cleavage in these systems, as well as all those in the remaining systems, can only give open forms, and therefore, to obtain regular fragments, a cleavage in two or more directions is required in the systems of lower symmetry. As, however, these are not always found, or, when present, are of very unequal value, the general cleavage tendency of crystals belonging to these systems is to produce parallel plates from the extreme development of a principal cleavage as compared with others. This is very strongly marked in the Mica group, whose crystals are cleavable without limit parallel to the base, but are exceedingly tough and strong in other directions. In describing minerals with two or more cleavages, it is necessary, therefore, to indicate their quality as well as their direction. The terms used for this purpose are *highly perfect*, as in Mica; *very perfect*, as in Fluorspar, Barytes, and Hornblende; *perfect*, as in Augite and Chrysolite; *imperfect*, as in Garnet and Quartz; and *very imperfect*, when only traces of cleavage can be obtained.

The following are the principal directions of cleavage observed in the different systems, with a few examples of each, from which it will be seen that only the simpler forms, or those with low indices, are possible as cleavage forms:—

1. *Cubic system.*

Octahedral: Fluorspar, Sal-ammoniac, Diamond.
Cubical: Common Salt, Galena.
Rhombic dodecahedral: Zincblende.

2. *Hexagonal system.*

Pyramidal of either order . .	$P, P2$	Pyromorphite.
Prismatic of either order . .	$\infty P, \infty P2$	Apatite, Red-zinc Ore, Cinnabar.
Basal . . .	$0P$	Beryl, Red-zinc Ore.
Rhombohedral .	R	The Carbonates of the Calcite group, Quartz.

3. *Tetragonal system.*

Pyramidal of either order . .	$P, 2P\infty$	Scheelite, Copper Pyrites.
Prismatic of either order . .	$\infty P, \infty P\infty$	Rutile, Tinstone.
Basal . . .	$0P$	Apophyllite.

4. *Rhombic system.*

Prismatic . .	∞P	White-lead Ore, Natrolite.
Macrodomatic .	$\bar{P}\infty$	Barytes.
Brachydomatic .	$\breve{P}\infty$	Barytes.
Basal . . .	$0P$	Topaz, Prehnite.
Macropinakoidal .	$\infty\bar{P}\infty$	Anhydrite.
Brachypinakoidal .	$\infty\breve{P}\infty$	Barytes, Antimony Glance.

5. *Oblique system.*

Hemipyramidal .	$\pm P$	Gypsum.
Prismatic . .	∞P	Hornblende, Augite.
Clinodomatic .	$\not P\infty$	Azurite.
Hemiorthodomatic	$\pm\bar{P}\infty$	Euclase.
Basal . . .	$0P$	Orthoclase, Epidote, Mica.
Orthopinakoidal .	$\infty P\infty$	Epidote.
Clinopinakoidal .	$\infty\not P\infty$	Orthoclase, Gypsum.

6. *Triclinic system.*

Hemiprismatic .	$\infty P' \infty' P$	Labradorite.
Hemidomatic	$'\bar{P}' \infty$	Cryolite.
Basal	$0 P$	Triclinic felspar group.
Macropinakoidal .	$\infty \bar{P} \infty$	Cyanite.
Brachypinakoidal .	$\infty \breve{P} \infty$	Axinite.

Besides the cleavage surfaces, other divisional planes indicating structure may be obtained in some minerals by special treatment. Thus, by filing away two opposite edges of a cube of Salt parallel to the face of a rhombic dodecahedron, and screwing it up carefully in a vice, an internal fracture parallel to a face of the latter form will be produced. A similar face parallel to $-\frac{1}{2}R$ may also be readily produced in Calcite. Another method consists in the use of a conical steel point such as a centre punch, which, when placed on the face of a crystal and struck, often develops a system of cracks parallel to some principal crystallographic directions.

Fracture. The characteristic appearances of the surfaces of minerals, broken in directions that are not cleavage planes, are described by the following terms :—

1. *Conchoidal.* This is the characteristic fracture of homogeneous amorphous substances, the surfaces presenting an alternation of rounded ridges and hollows ; it is best seen in glass and imperfectly cleavable minerals, such as Quartz, Garnet, and Ice. According to the nature of the undulations of the surface, it is further characterised as flat- or deep-, coarsely or finely, conchoidal.

2. *Smooth*, when the surface, without being absolutely plane, presents no marked irregularities.

3. *Splintery*, when the surface is covered with partially separated splinters in irregular fibres, as in fibrous Hematite.

4. *Hackly*, when the surface is covered with ragged

points and depressions. This is especially characteristic of native metals.

In easily cleavable minerals it is, as a rule, difficult to develop any special fracture, but it may sometimes be done by striking a fragment a sharp blow with a blunt point, as that of a rounded hammer or pestle, when traces of characteristic fractures may occasionally be obtained, springing across from one cleavage surface to another.

Hardness. By this term is meant the resistance of the surface of a mineral to abrasion when a pointed fragment of another substance is drawn rapidly across it without sufficient pressure to develop cleavage separation. When the latter is the harder substance it will scratch the former, but when it is the softer the point will be blunted without the surface passed over being affected. As there are very considerable differences between minerals in regard to this test, it is one of the most important of their physical constants; but as there is no means of expressing the results in absolute measure, recourse must be had to an indirect method, in which comparative hardness is measured by a scale of typical minerals. This, known as Moh's scale of hardness, is as follows :—

1. Talc.
2. Gypsum or Rock Salt.
3. Calcite.
4. Fluorspar.
5. Apatite.
6. Orthoclase.
7. Quartz.
8. Topaz.
9. Corundum.
10. Diamond.

Breithaupt, while preserving these numbers, proposed to interpolate Mica as 2.5 and Scapolite as 5.5, but they have not been generally adopted.

The softer numbers of the scale, Talc and Gypsum, may be scratched by the finger nail, and those up to 6 by a file or the point of a knife, while Quartz and the higher numbers are all harder than steel.

In testing a mineral for hardness, it is applied successively to different numbers of the scale until one is found that can be scratched by it. The positions are then reversed, and if it is found that the scale mineral will scratch that under examination, both are said to be of the same hardness; but if it is not scratched, its hardness is said to be intermediate between that of the particular number and the next harder one. Thus, the hardness of Barytes, which scratches but is not scratched by Calcite, and is easily scratched by Fluorspar, is said to be 3.5, or between 3 and 4. Small differences in the hardness of substances may also be appreciated by drawing fragments of about the same size over a flat file, when the harder substance will give a sharper sound than the softer one; and if they are both of lower hardness than quartz, the softer one will leave the largest amount of powder or streak on the file.

The hardness of crystals that are easily cleavable often shows very decided differences in different directions. This subject has been investigated by Exner, who found that the minimum load upon a steel point necessary to produce abrasion when moving over a cube of rock salt parallel to the diagonal of the face was one-third greater than that required when the line of motion was parallel to an edge. Similarly the face of a rhombohedron of Calcite may be more easily scratched parallel to the principal cleavage than in any other direction. It is better, therefore, in making up a scale of hardness, to use irregularly broken fragments, when they can be obtained of sufficient purity, rather than perfect crystals, but when the latter are used the trial should be repeated with different faces.

For cabinet or indoor use a complete scale made up of a few pieces of each number, with the possible exception of the diamond, should be arranged in a divided box, together with a flat file not too coarsely cut; but for travelling use the numbers 3 to 8 will be found sufficient in most cases. The nail-trimming blade of a penknife is also extremely useful in trying the hardness of minerals.

In testing cut gems, care is required, especially with those of doubtful authenticity, not to disfigure them, and it is therefore best in the first instance to apply the test of the file cautiously to the border of the specimen, where a scratch may be hidden by the setting when the stone is mounted.

Tenacity. This term is rather loosely applied in mineralogy to the behaviour of minerals when subjected to the action of cutting or pulverising tools, the different degrees being distinguished as follows :—

1. *Brittle*: The substance breaks up into fragments without extending. By far the larger number of minerals are of this kind and no special examples need be given.

2. *Ductile*: The substance can be cut with a knife, but crushes to powder under a hammer. The property is seen in Copper Glance and Copper Pyrites, and in a higher degree in Silver Glance and Chloride of Silver; the latter cuts into shavings like horn, and can scarcely be powdered.

3. *Malleable*: The substance can be cut into shavings and beaten out under the hammer. The soft native metals, Copper, Silver, and Gold, are examples.

4. *Flexible*: The substance when divided into thin plates can be bent, and remain so without breaking, as Talc.

5. *Elastic*: A thin plate of the substance when bent springs back to its original form when the strain is removed. This happens with Mica plates that are not too thin. Another remarkable example is the elastic Bitumen, or Elaterite of Derbyshire.

Although not a common property, a few instances of extreme toughness are known in minerals, such as certain varieties of Serpentine, the massive Felspars, Jade, and Saussurite. These are, as a rule, uncrystalline substances, and their toughness is in no relation to their hardness. Malleable native Copper, especially when intimately mixed with siliceous vein stuff and some varieties of Hematite and Iron Pyrites, has the same property in a high degree. Probably the uncrystalline diamond or *carbonado* may be considered as combining toughness with extreme hardness.

Density and Specific Gravity. The density of a substance is the mass of its unit volume expressed in units of weight. Specific gravity is the ratio of the density of a substance to that of another assumed as a standard, or of the weights of equal volumes of both. Water at the ordinary temperature of air, or at that of its maximum density, is usually adopted as the standard substance. In the metrical system the same number expresses the weight of a cubic centimetre in grammes, or both density and specific gravity.

The term 'density' is generally used by French mineralogists, and 'specific gravity' by those of England and Germany, in describing minerals. Rammelsberg, however, employs the equivalent expression 'volume-weight.'

Specific gravity is one of the most important factors in determinative mineralogy, as it is found to be constant or variable within small limits in different varieties of the same species, while the differences between different species is often very considerable, the observed range being from 0.75—0.90 in some liquid hydrocarbons, to 21 or 22 in the metals of the platinum group. It will, however, be seen in examining a classified list of the specific gravities of minerals, such as those published by Websky and the Bureau de Longitude,[1] that species of analogous composition and constitution are generally near together, and that upon this characteristic a certain rough grouping is possible, as in the following examples :—

0·5—1.5. Most fossil resins, Petroleum and Bitumen.

1.0—2.0. Coal Lignite and carbonaceous minerals generally, many hydrated Alkaline Sulphates and Borates, Nitre.

2.0—2.5. Sulphur, Graphite, Nitrate of Sodium, Salt, Gypsum, and most Zeolites.

2.5—2.75. Quartz, the Felspars, Talc, Serpentine, Calcite Emerald.

[1] These, with other admirable tables of the physical constants of minerals, should be in the possession of every student. They are contained in the *Annuaire* for 1876 and later years.

2.8—3.0. Aragonite, Magnesite, Dolomite, Tremolite, Wollastonite, Mica.

3.0—3.5. Apatite, Fluorspar, Epidote, the Pyroxene and Hornblende groups, Tourmaline, Olivine, Axinite, Diamond, Topaz, the Sulphides of Arsenic.

3.5—4.0. Spinel, Corundum, Siderite, Limonite, Strontianite, Celestine.

4.0—4.5. Rutile, Zircon, Barytes, Witherite, Zincblende, Copper Pyrites, Magnetic Pyrites.

4.5—5.5. Hematite, Magnetite, Iron Pyrites, the Ruby Silver Ores, and many compound sulphides not containing Lead.

5.6—6.6. Arsenic, Arsenical Pyrites, Oxides of Manganese, and many compound sulphides containing Lead and Silver.

6.7—7.9. Antimony, Sulphide and Carbonate of Lead, Sulphide of Silver, Tinstone, and Pitchblende.

8.0—9.0. Tellurides of Gold and Silver, Sulphide of Mercury, and Copper.

9.0—11.0. Bismuth, Silver, and Palladium.

12.0—15.0. Mercury, and Amalgams.

15.0—20.0. Gold, Platinum, and several of its associated metals.

Above 20. The native alloys of Osmium and Iridum.

Metallic Iridum, when perfectly purified from the metals with which it is usually associated, is the densest substance known, having a specific gravity of 22.4.

The determination of specific gravity is in principle very simple, the substance being first weighed in air, and then in water, the difference between the two weights gives the weight of an equivalent volume of water, and the quotient of the original weight by the difference will be the specific gravity. An exact determination is, however, a matter of considerable nicety, and involves the use of delicate balances, such as are only found in laboratories. The general details of manipulation will be found in the larger treatises on practical chemistry. When the substance contains cavities, it is necessary to powder it before taking the specific gravity, and in such cases the determination is made by estimating the amount of water displaced by a known weight of the powder from a bottle that has been exactly filled with water and weighed at a standard temperature.

As most minerals enclose more or less of hollow or air-filled spaces, it is not remarkable that they should, as a rule, be denser when powdered than in the solid state. In irregular aggregates the difference is often very considerable. Thus, the spongy substances known as float-quartz and pumice in masses are apparently lighter than water from the large amount of air enclosed, although when pulverised their true density is found to be between two and three times as much.

In the determination of the specific gravity of mineral masses of large size and known weight, the method of gauging the volume of water displaced may be conveniently used. One of the best is that given by Mohr, which is susceptible of considerable accuracy. The gauging vessel is a glass cylinder, which is filled with water to a standard point formed by a needle projecting from a slip of wood across the top, the exact level being attained when the point of the needle and its reflected image in the water coincide. The weighed substance is then carefully lowered into the cistern, when it displaces its own volume of water, with a corresponding rise of the surface level. The amount of displacement is measured by drawing the water into a graduated tube or burette until the original level is restored. A convenient size of graduated tube is the ordinary alkalimeter used in volumetric analysis containing 1,000 grains, and divided into 5-grain spaces, or an equivalent one with metrical divisions. The level of the water may be adjusted with great nicety by a simple valve formed of a piece of glass rod inserted in the indiarubber delivery tube, the aperture of which can be varied by slight pressure of the finger upon the tube. This method, which has the advantage of not requiring a correction for temperature, is well adapted for taking the specific gravities of Coal, Limestone, and similar substances ranging from 2 to 3, which can be used in fragments of about half a pound weight.

Jolly's spring balance, another contrivance for the approximate determination of specific gravity, is recommended as being expeditious in use, fairly accurate, and dispensing entirely with the determination of absolute weight. It consists essentially of a pair of scale pans suspended one above the other; the upper one is attached to the end of a coiled steel spring, and the lower one is immersed in a cistern of water standing on a bracket, whose position can be adjusted by a sliding movement worked by a rack and pinion. The face of the upright bar to which the arm carrying the spring is attached has a silvered mirror with a scale of equal parts engraved upon it fixed to it in front, and the upper scale pan carries a pointer. When in use the level of the water vessel is so adjusted that the lower scale pan may be freely immersed when a reading is taken by bringing the pointer into coincidence with its reflected image in the glass scale. The mineral is then placed in the upper scale, whereby the spring is distended to a certain point, which is determined by a second reading, which when deducted from the first measures the weight in air. By removing it to the immersed pan, the strain is diminished, and the pointer consequently rises to an amount determined by a third reading, which when subtracted from the second measures the loss of weight in water. It is of course essential to accuracy that the spring should deflect equally for equal weights applied, or it must not be strained to anywhere near its elastic limits, while at the same time it must be sufficiently free to work to move through a measurable distance by a moderate change of weight. The instrument, which is very highly spoken of by Von Kobell, is made at Munich at a cost of 24*s*.

The density of large masses of an approximately regular figure may be roughly determined by weighing them and calculating their cubic volume from their measured dimensions. The specific gravity is found by dividing the weight by the contents in cubic feet multiplied by 62.4 lbs., or the weight of a cubic foot of water. The reverse operation of calculating the weight of a measured mass from the known specific gravities of its components is often useful on the large scale in estimating the probable yield of mineral deposits, and the student may therefore be recommended to become familiar with it by practice.

Specific gravity is in another class of approximate methods determined by immersion in a fluid of known density, when the substance, if heavier, will sink, but if lighter, it will float. This is specially useful in the discrimination of cut stones; the fluid used for this purpose, called after the inventor, Sonstedt's solution, is made by saturating a solution of iodide of potassium in water with iodide of mercury, which gives a liquid having a maximum density of 2.77, and which, when diluted with water, mixes without sensible change of volume, and therefore the specific gravity is proportional to the amount of the two fluids in the mixture. This is also useful as a means of separating mixed minerals for analysis, when they are so intimately associated as to be incapable of separation by hand; as, for instance, the Felspars and lighter silicates in a rock may be roughly divided from the denser ferrous and magnesian ones by a solution of a specific gravity of about 2.75, when the first mineral will just float, while the latter will sink readily. The chief drawback to the use of this substance is in its extremely poisonous character, and it can therefore be scarcely recommended except for laboratory use.

An extension of the same principle adapted for the separation of the denser constituents of mineral sands has been proposed by Bréon, who uses molten chloride of zinc or lead, or mixtures of both. These give fluids whose density ranges from 3.0 to 5.0. As these salts can be melted in glass tubes, and are alike soluble in water, the separated substances can be obtained in the pure state without much trouble. They have been applied in the separation of Tinstone, Rutile, Magnetite, and other heavy minerals from Quartz, and silicates in the microscopic investigation of sands.

The same method is applied on the large scale in the separation of Gold from Galena and Iron Pyrites by a fluid of intermediate density, namely Mercury, in the Hungarian Gold-mill, although in this case the result is not quite so simple, the metal being to some extent, at any rate, dissolved in the separating fluid.

CHAPTER XII.

OPTICAL PROPERTIES OF MINERALS.

UNDER this head are included the general and more apparent properties of colour, lustre, and translucency or opacity, which are common to all minerals alike, as well as others, requiring special methods of investigation applicable only to such minerals as are transparent. These include the determination of the quality, and numerical values, of the refraction constants in particular directions, the so-called axes of optical elasticity, and the relation of these axes to the principal crystallographic lines, and the phenomena of pleochroism. Now, although the practical application of these methods is, as a rule, beyond the power of beginners, as they involve the use of exact and somewhat expensive apparatus and carefully prepared crystallographic material, the results obtained are in many cases of such great interest and importance, especially in the determination of crystallographic characters from fragments of minerals in forms not otherwise definable, that a brief indication of their character and the reasoning upon which they are founded may not be out of place here. The reader is referred to the treatise on 'Physical Optics,' by Mr. Glazebrook, for the more complete discussion of the subject.

In order to understand how light may be used as an indicator of the structure of minerals, it will be necessary to consider as shortly as possible certain common optical terms, and first of all the nature of light itself. This, according to the undulatory theory of Huyghens, which forms the basis of all modern optics, is a consequence of the vibratory movement of molecules in an intangible fluid, or *ether*, which is assumed to fill the intermolecular space in all material substances—solid, liquid, or gaseous alike; and by its continuity forms a medium for the transmission of such movements from their point of origin, the luminous source, to the optic nerve, where they become apparent as light the character of the movements being essentially similar to those of waves traversing air, water, or any other substance. If a system of molecules in stable equilibrium be supposed to be arranged in line at regular distances apart, as in fig. 319, a straight line passing through them will include all their positions of repose. If, however, one of them be by some force displaced sideways, as for example *a* to *a'*, the tendency of the attractive force of the adjacent particle *b* will be to bring it back to the original position, and it will after a time return, arriving at *a* with sufficient impetus to carry it on to a point at the same distance as *a'* on the other side of the line of equilibrium, and so on backwards and forwards, a periodical vibratory motion, analogous to that of a pendulum, being set up in a plane determined by the direction of the moving force. When the particle *a* is displaced it will exert a disturbing influence over the adjacent one *b*, tending to draw it to itself as indicated by the diagonal arrow, which influence will be opposed by the attraction of *c* acting along

FIG. 319.

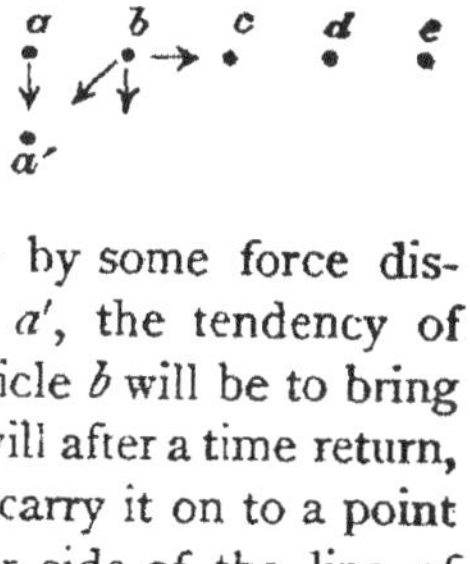

the horizontal arrow. The movement of *b* will, therefore, be in some intermediate direction, such as that of the vertical arrow, which it will swing parallel to *a*, and so on for any molecule in the train; the commencement of the movements being progressive, each molecule starting when the preceding one has traversed a certain dis-

FIG. 320.

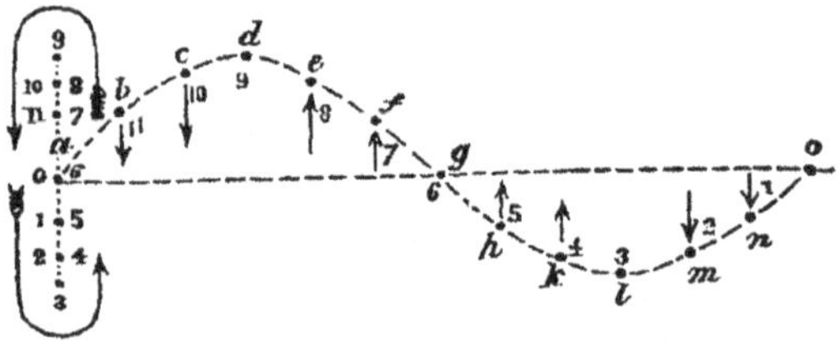

tance on its journey. If the particle *a* in fig. 320 be supposed to have just completed one vibration, having successively occupied all the positions indicated by the points 0 to 11 in the order shown by the arrows, the molecule *b*, which started when *a* was at 1, will at the same moment be at 11, moving downwards towards the centre line; *c* will be a little farther off at 10, but moving in the same way; *d* will be at the greatest distance above the line at 9; the points next following, *e* and *f*, will be travelling upwards, *g* will be passing through the central position, and so on to *o*, which will be just about to commence its first movement. The points *a* to *o* therefore represent the positions of a contiguous series of vibrating molecules, supposing their movement to be stopped simultaneously, and together they represent the successive positions passed through by each one or the phases in its movements: the line joining them will be a wave line, having a crest at *d*, a hollow at *l*, and nodal points at *a*, *g*, and *o*, the horizontal line passing through them being the line of propagation of the wave. The total distance from *a* to *o* is called a wave-length; *a g* and *g o* are half, and *a d*, *d g*, *g l*, and *l o* quarter wave lengths, quantities that are usually represented by the symbols λ, $\frac{\lambda}{2}$, and $\frac{\lambda}{4}$. Points in the series which are half a wave-length or whose difference of phase $= \frac{\lambda}{2}$ are vibrating in opposite directions, as shown by the arrows in fig. 320.

The surface containing the paths of the molecules is termed the plane of vibration, and the direction in which the wave movement progresses, the direction of propagation. In the simple case represented these are at right angles to each other, or the wave is a transverse rectilinear wave.

The distance *a*3–*a*9, termed the amplitude of vibration, determines the intensity or brilliancy of the light, and, as in the case of the pendulum, the time of vibration is constant, whether this distance be long or short, the movement of the particle when at its maximum speed being quicker in the former than in the latter case, but the total duration of one vibration is the same in either. The wave-length, on the other hand, being the distance travelled during the time of a single oscillation, will depend upon the period of the latter being longer or shorter, according as the motion is slower or faster. Upon these differences depends the colours of the light produced; that having the slowest period or greatest length of wave is called *red*, and that of the most rapid movement and shortest wave-length is *violet*, those of the colours orange, yellow, green and blue being of intermediate velocities and wave-lengths. When a luminous source produces waves of only one length, the light is said to be homogeneous or *monochromatic*, but when waves of different lengths are originated simultaneously the light is heterogeneous, mixed, or white. The latter is the character of ordinary sunlight, monochromatic light being characteristic of the vapours of elementary bodies when intensely heated —sodium producing yellow, thallium green, lithium red, &c. The light of candle, oil, or gas flames is also mixed, but differs from sunlight in the relative proportions of the constituent colours, containing less blue and more yellow, some portion of the latter coloured light being of a kind never seen in sunlight.

When a line of molecules vibrating in any direction is subjected to a new impulse acting in some other direction, its movements will be resolved according to the parallelogram of forces, and a new wave will be set up, compounded of both. In such cases the two waves are said to interfere, and the new one to be an interference wave. The effect produced by such interference may vary very considerably in character, according to the relation of the constituent waves. In the simplest case, where both are moving in the same direction, their planes of vibration being also the same, if one differs from the other by one or more complete wave-lengths, the amplitude of the new wave will be augmented, it being equal to the sum of those of the original ones. When the difference of phase is one half or any uneven number of half wave-lengths, the positions in the fresh wave will correspond to similar ones in the second, but their directions of motion will be dissimilar. If therefore the two waves be of different amplitudes, the phases of that produced by their interference will correspond to those of the larger one, but its amplitude will only be equal to their difference—that is, the intensity of the resultant wave will be considerably reduced. In the special case where both waves are of the same amplitude, that of the resultant = o, or the wave motion is entirely destroyed. For differences of phase other than the above, the interference waves produced are of different phases and amplitude. Thus, for waves of similar amplitude and quarter wave-lengths apart, the augmented wave of interference has a difference of phase $= \frac{1}{8} \lambda$ in advance of the first, while for three-quarter wave-length distance it is retarded by $\frac{1}{8} \lambda$.

The wave motions originated by a luminous point are propagated uniformly in every direction from it as a centre, so that at any particular instant a point in any one wave will be in exactly the same condition of vibration as one in any other at the same distance from the centre, or, in other words, points of similar phase on the wave paths will all be equidistant from the centre, and a surface including the whole of them will be a sphere whose radius is the common distance. This surface is known as the wave surface or wave front, and the radii representing the wave paths as light rays; any small portion of the surface and the rays determining it constitute a beam of light.

As the points included in a spherical wave surface all commence their vibrations simultaneously, they will affect

FIG. 321.

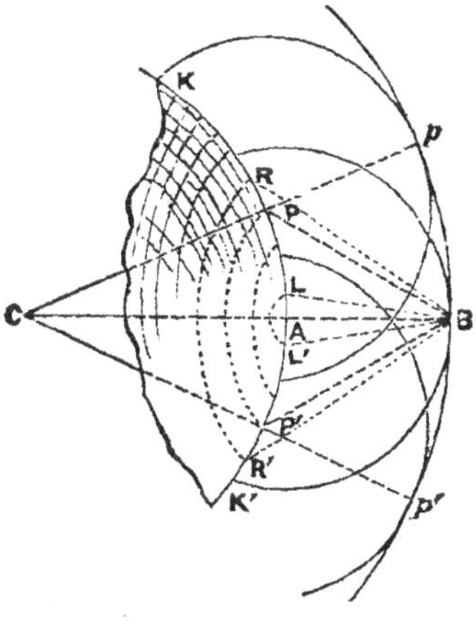

the equilibrium of the particles beyond them in a similar manner, so that each may be considered as originating a particular wave. In fig. 321 K K′ represents a portion of the surface of a wave originating at C, and P P′ a similar portion at a later instant. The point B in the second will be reached by movements coming not only from A but from every other point of the surface K K′. To determine the effect of such movements, suppose a series of circles L L′, P P′, R R′, &c., to be described upon the surface K K′ like the parallels of latitude on a globe about C A B as a polar axis ; all points in the circumference of any one of these, such as P and P′, will be equidistant from B, but at a different distance from those in any other circle. The waves proceeding from different circles will therefore have different distances to travel, and as their period and rate of progression are the same, they must arrive at B in different conditions of vibration. If further we suppose for every one of the circles another to be constructed of exactly half a wave-length greater distance from B, the undulations proceeding from each of them will by their interference be extinguished, and therefore have no effect at that point. Such circles can be described for every portion on the surface except A, and therefore the whole of the waves proceeding towards B will be extinguished except those on the line A B, and similarly only those on the prolongation of the line C P and C P′ will reach *p* and *p*′ respectively. From this it follows that although every point on the wave surface is the centre of a new wave, the motion of the latter will only be apparent at the point where it touches the surface forming a common envelope to the whole of the waves of the same class. This is, in the case in question, the spherical surface *p p*′, or the wave front corresponding to the later period. Upon this principle is founded Huyghens' wave construction, whereby the wave front at any given moment may be found if the originating surface and the rate of propagation be given. This is done by describing upon points of origin, such as A P and P′ semicircles whose radii are proportional to the velocity of wave transmission at those points, and drawing through the points *p* B *p*′ the common enveloping surface, which is the second wave front as before.

If the point of origin of the wave motion be situated at a great distance, the rays contained in a beam of light may be considered as sensibly parallel, and the included portion of the wave front as a plane surface. This is termed a plane wave. Ordinary solar or daylight is of this character, the distance of the luminous centre, the sun, being so great that a beam of light must have a diameter of nearly a thousand miles for the radii limiting it to diverge by an angle of one second from parallelism, so that in most instances of natural illumination the beam may be considered as a cylinder made up of parallel rays.

The power of transmitting wave movements uniformly in one or more directions supposes the medium possessing it to be homogeneous or uniform in molecular structure. If the structure is such that the rate of transmission is exactly the same in any direction, the medium is said to be *isotropic*. This represents the most complete and symmetrical kind of homogeneity.

In the second kind of homogeneous media, known as *anisotropic* or *heterotropic*, the rate of progression of wave movements varies with their direction, but is constant for any similar direction. In heterogeneous or non-homogeneous media, on the other hand, there is no relation between direction and velocity of wave propagation, the latter being subject to variation even when the former is constant. The movements in such a medium can only be reduced to order by supposing them to be made up of portions of homogeneous substances of different properties, and treating each one separately.

The velocity of propagation of wave movements changes when they pass from one medium into another, the different velocities being related to each other according to a fundamental proposition in mechanics, directly as the square roots of their coefficients of elasticity e, and inversely as the square roots of their densities; or if c and c' represent the different velocities, their ratio will be expressed by

$$c : c' = \sqrt{\frac{e}{d}} : \sqrt{\frac{e'}{d'}}.$$

When e is nearly of the same value in the two media the velocity will diminish as d increases, or generally the velocity of propagation is less in the denser medium.

In passing from one medium to another of different density, the entire wave movement is never completely transferred. If the second medium has the lesser density, the energy of a vibrating molecule at the limiting surface of the first will be more than sufficient to establish an equality of movement in a similar molecule in the second, and therefore part of the original motion will be returned in the form of a backward wave, whose phase at any point will be exactly the same as that of the original wave would have been at a point similarly situated in front.

On the other hand, if the second medium be the denser, not only will the original movement be entirely absorbed, but a return movement will be set up in the molecules of the first in the form of a wave of exactly opposite phase to the original one. In either case the change is accompanied with the same result—a portion of the original movement only enters or is transmitted by the second medium, the remainder being diverted or reflected at the surface of contact. This proposition holds good in all cases. Whenever light passes from one medium to another, a portion is invariably lost by reflection, and the intensity of the transmitted beam is diminished, the loss increasing with the difference in density of the two media. Where the difference is so great that none or very little of the light enters, the denser substance is said to be opaque; but if all, or nearly all, is transmitted, it is transparent or translucent. No substance is known to be either completely opaque or perfectly transparent, as the light reflected by the most characteristic examples of the former, properly polished metal surfaces, is always modified to some extent, showing that penetration has taken place, though only in a very small degree; and those that can be reduced to a sufficiently thin laminæ such as gold and silver, are actually found to be translucent, and the most limpid translucent substances in sufficiently thick masses sensibly diminish the light passing through them. The light so lost is said to be absorbed. When the absorption affects waves of all lengths similarly, the result is only a diminution in the brilliancy of the light; but selective or unequal absorption of particular kinds of waves alter the character of the transmitted light, giving rise to the phenomena of colour.

The nature of the wave surface produced by reflection may be found by the construction given in fig. 322. If

FIG. 322.

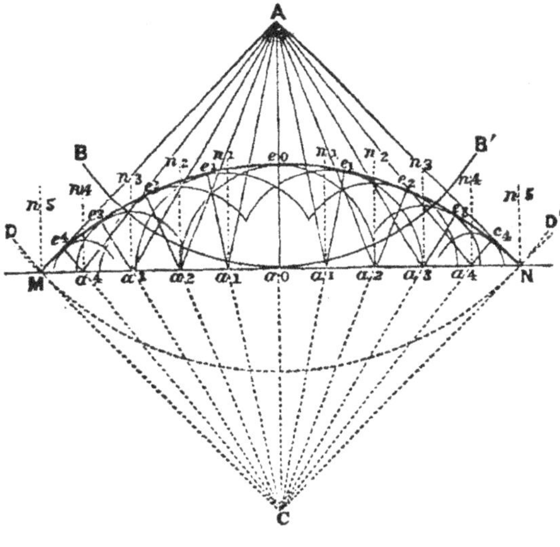

M N in fig. 322 represent the surface of contact of two dissimilar isotropic media, and A the point of origin of wave-motion in the first, the arc B B′ will be the section of the spherical wave-front at the moment when the ray A a_0 meets M N; the more divergent rays A a_1 to A a_4 on either side arriving at the same surface progressively at later periods. At each of these points a reflex wave-movement originates, that from a_0, being the earliest, arrives at the point e_0 at the same moment that those from a_1 a_2 a_3 and a_4 reach e_1 e_2 e_3 and e_4 respectively, and the outer rays A M and A N have just arrived at the surface M N. If, therefore, semicircles be described from a_0 with the radius a_0 e_0, from a_1 with a_1 e_1, from a_2 with a_2 e_2, etc., the curve forming their common tangent, the circular arc M e_0 N, whose radius is C a_0 + a_0 e_0, will be the wave-front required. If an arc be struck with the same radius from A passing through M N, it will represent the position that the original wave-front would have reached had the motion been continued in the first medium. Hence we see that the direct and reflected wave-surfaces are exactly similar in dimension, but reversed in position.

The direction of reflection in the central ray A a_0 is the same as that of its arrival or incidence—i.e., normal to the reflecting surface; but in the rays on either side these directions differ, making longer angles with each other as the obliquity of incidence increases. If through each of the reflecting points, a_1 to a_4, a line be drawn normal to M N, it will be symmetrical to the two rays at that point, or they will make equal angles with it and consequently with the reflecting surface. This is expressed in the following proposition: The angle of incidence is equal to the angle of reflection.

FIG. 323.

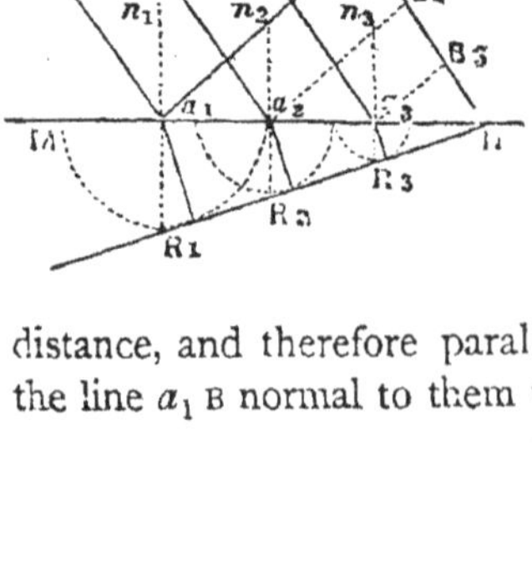

The direction of rays passing into the second medium is found by the construction shown in fig. 323. If A_1 A_2 A_3 A_4 represent rays coming from a luminous source at a great distance, and therefore parallel, the surface represented by the line a_1 B normal to them will be the plane wave-front at the moment of incidence of the ray $A_1\ a_1$ with the surface limiting the two media M N. Supposing the density of the second medium to be such that the velocity of wave propagation in it is only one-half of that in the first, the semicircles described about $a_1\ a_2$ and a_3, with radii corresponding to half the distances B_1 N, B_2 N, and B_3 N respectively will represent the positions of waves set up at these points at the moment when the most distant ray, A_4, arrives at the point M, and, according to the principles previously laid down, the plane represented by the line N R_1, which is tangent to all of them, will be the wave-front of the whole beam. As this is also a plane wave, the parallelism of the rays is not altered, but their direction, indicated by the lines normal to R_1 N passing through $a_1\ a_2$ and a_3 is changed, the individual rays being bent or refracted towards a normal to the plane of incidence passing through the same points. If, on the other hand, the first be the denser medium, the velocity of propagation will be doubled in the second, and the rays, as shown in fig. 324, will be bent away from the normals to the plane of incidence, or the refractive angles will be increased. The amount of the refraction is measured in fig. 325 for the rays A_1 O and O R_1 by the lines $A_1\ n_1$ and $R_1\ n_3$, which are to each other the sines of their respective angles with the vertical line nop, or calling the larger angle i and the smaller one r,

FIG. 324.

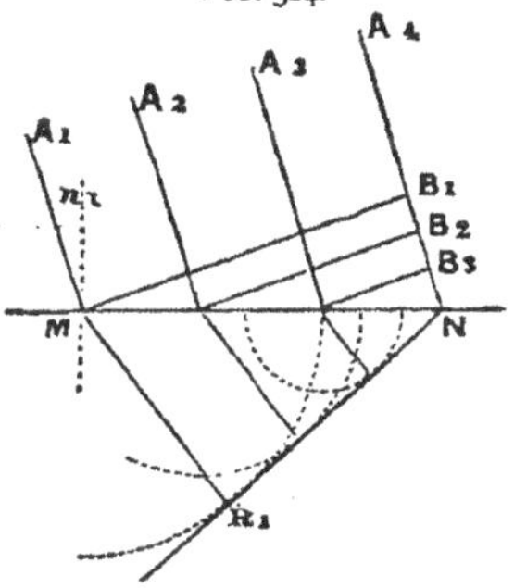

$$\frac{\sin.\ i}{\sin.\ r} = n.$$

This quantity, which is constant in an isotropic substance for any angle of incidence, and corresponds to the ratio of the respective velocities of wave transmission or $\frac{v}{v_1}$, is termed the *quotient*, *exponent*, or *index of refraction*, the last or the equivalent term, *refractive index*, being most generally used. The value of n varies for the greater number of transparent isotropic minerals between 1·3 and 1·7, that of air being taken as unity, the maximum of 2·4 being attained in Diamond. A few anisotropic substances such as the Ruby, Silver Ores, and Cinnabar, afford examples of still higher refractive indices, which reach or even slightly exceed 3·0.

A more oblique ray, such as $A_2 o$ in fig. 325, whose angle of incidence is i', is refracted in the denser medium in the direction $o R_2$, or has the larger angle of refraction r', but the ratio of the quotient of their sines is not altered—i.e.,

FIG. 325.

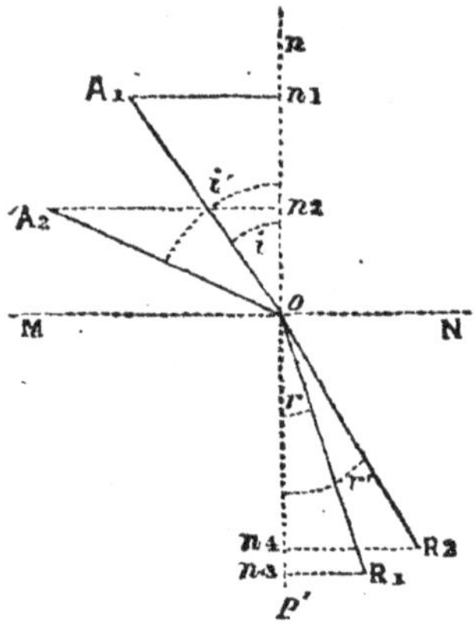

$$\frac{\sin.\ i'}{\sin.\ r'} = \frac{\sin.\ i}{\sin.\ r} = n$$

as before; or the deviation of the refracted ray varies with the angle of incidence. The maximum obliquity of incidence is in the direction M o, or parallel to the dividing surface, in which it is clear there can be no refraction, as the path of the ray is entirely within the first medium. Also, if the ray meet the plane of incidence at right angles, or $i = 0°$, the deviation of the refracted ray r also $= 0°$, or there is no deviation, and the ray follows the same direction in both media without refraction. For any other direction sin. i is always greater than sin. r, as sin. $r = \frac{\sin.\ i}{n}$. If the first medium be denser than the second, the velocity of transmission will be greater in the latter than the former, the wave-front of the transmitted beam R_1 N, fig. 326, will make a longer angle with M N than that of the incident one M B_1, the angles of refraction of the individual rays will be larger than their angles of incidence, and will increase with the deviation of the latter from the vertical, but more rapidly, as will be apparent by turning fig. 325 upside down and considering the positions of the refracted and incident rays to be transposed. In this case sin. $r = m$ sin. i, and as sin. $90° = 1$, when sin. $i = \frac{1}{m}$, $r = 90°$, or the refracted ray will coincide with the plane of separation. The construction for the special case $n = 2$ and $r = 90°$ is seen in fig. 326, where the vertical line through N is the tangent plane or wave-front of the refracted beam, and M n the direction of the refracted ray. The radius of the circle M K $= 2$ B_1 N, and $A_1 n =$ sin. $l = \frac{1}{2}$ or $l = 30°$. For this particular value of the refractive index, therefore, all rays whose inclination to the normal to the plane of incidence is thirty degrees or above are incapable of passing into the less dense medium, but are reflected at the surface of separation. This property is known as total reflection, and the minimum angle at which it takes place is the limiting angle of refraction, or the angle of total reflection.

FIG. 326.

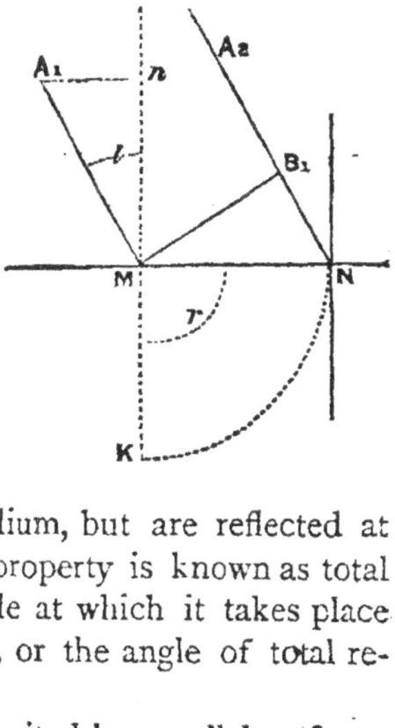

When the denser medium is limited by parallel surfaces, the angle of incidence of the refracted ray at the second surface will be the same as that of its first refraction, or $i' = r$ (fig. 327), and therefore its angle of refraction, r', on emergence into the first medium, will be the same as that of original incidence, i, or, in other words, the ray will resume its original direction. A ray of light is not therefore altered in direction by passing through any transparent substance bounded by parallel surfaces. If, however, the surfaces are not parallel, the ray at emergence will be more or less changed from its original direction, supposing the inclination of the surfaces not to be sufficient to produce total reflection. The amount of this deviation in direction may be calculated when the inclination of the limiting surfaces and the refractive index of the medium are known,

FIG. 327.

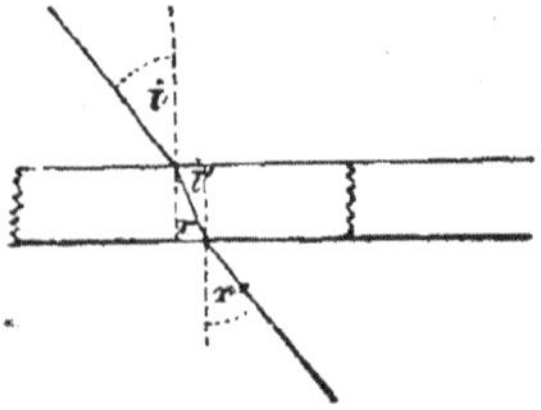

FIG. 328.

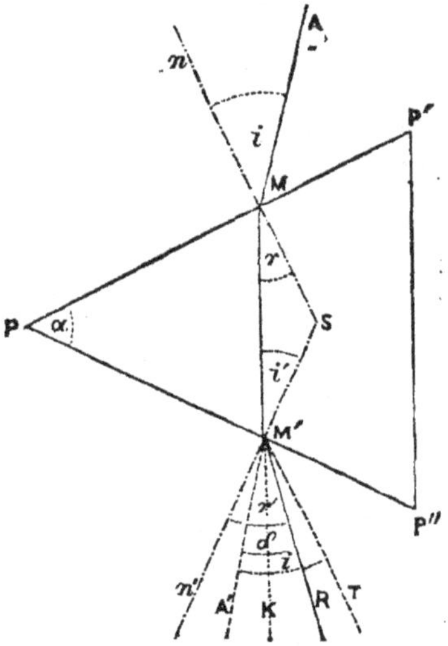

and conversely from the measured deviation produced by a refracting body or prism of known angle, the refractive index of the substance composing it may be found. The principle of this method is shown in fig. 328, where P P′ P″ is the section of a prism of the refractive angle α in a plane perpendicular to its refracting edge P. The ray A M incident at the angle i on P P′ is refracted at the angle r towards M′, where it becomes incident at the angle i' upon P P″, is refracted at the angle r', and follows the direction M′ R after emergence. If the line M′ A′ be drawn parallel to A M, the angle A′M′R $= \delta$ will measure the total deviation of the ray from its original course effected by the prism. Next, from the point M′ draw the lines M′ A′ and M′ T, the first parallel to the original direction of the ray A M, and the second to the normal n M, produce the common path of the first refracted and second incident ray M M′ towards K, and continue the normals n M n' M′ to their point of intersection at S. The angles formed at M′ will have the following signification:

$$n'\mathrm{M'K} = i' = \mathrm{KM'T} = r,\ \mathrm{A'M'T} = n\mathrm{MA} = i,\ n'\mathrm{M'R} = r',$$

and A′ M′ R $= \delta$, or the deflection of the ray from its original course by the prism. But

$$\mathrm{A'M'R} = \delta = \mathrm{A'M'K} + \mathrm{KM'R} = (\mathrm{TM'A'} - \mathrm{TM'K}) + (\mathrm{RM'}n' - n'\mathrm{M'K})$$

or $$\delta = (i - r) + (r' - i') = i + r' - (r + i').$$

In the triangle P M M′ $\alpha + (90° - r) + (90° - i') = 180°$, consequently, $\alpha = r + i'$ and $\delta = i + r' - \alpha$.

Fig. 328 is so constructed that $i = r'$ and $r = i'$, for which case $\delta = 2\,i - \alpha = 2\,r' - \alpha$, and has a minimum value.

If one angle is greater than the other, e.g. if $i = r' + \beta$, or $r' = i + \beta$, the deviation will be in the first case $\delta = 2\,r' + \beta - \alpha$, and in the second $\delta = 2\,i + \beta - \alpha$, either of which is obviously greater than $2\,i - \alpha$. This latter, therefore, is the minimum angle of deviation, and is produced when the ray makes equal angles with the faces of the prism both at incidence and emergence. Further, as

$$\delta = 2\,i - \alpha,\ i = \frac{\alpha + \delta}{2};$$

and as

$$r = i',\ \alpha = r + i' = 2\,r,\ r = \frac{\alpha}{2},$$

and

$$n = \frac{\sin. i}{\sin. r} = \frac{\sin. \frac{\alpha + \delta}{2}}{\sin. \frac{\alpha}{2}};$$

or the refractive index of the substance, if isotropic, is equal to the sine of half the sum of the refractive angle of the prism and the angle of minimum deviation divided by the sine of half the angle of minimum deviation. This is the simplest and most direct method of determining refractive indices, as two faces of a crystal may be used as a prism if their angle is not too large; from 40° to 70° are the most convenient angles when the refractive indices are moderately high (1·5 to 1·7). In all cases the determination must be made for particular parts of the spectrum, either by using monochromatic light or by observing the deviation of the principal dark lines in the solar spectrum. For the former purpose the flame of a Bunsen's burner coloured red by a bead of sulphate of lithium ignited at the end of a platinum wire, yellow by chloride of sodium, or green by sulphate of thallium, is the most convenient kind of illumination. The angle α is measured with a reflecting goniometer and the deviation by the same instrument arranged to allow the prism and telescope to move independently.

Another method of determining refractive indices originally proposed by the Duc de Chaulnes in 1767, and recently systematised and extended by Sorby and Stokes,[1] consists in measuring the displacement of the focal point of a lens or compound microscope adjusted to a distinct vision of an object in air when a parallel plate of known thickness of a denser substance is interposed between the object and the lens. The effect of this interposition is to bring the

[1] *Proc. Royal Soc.* xxvi. p. 384. *Journal of the Royal Microscopical Society*, 1878.

apparent place of the image nearer to the lens, so that the latter has to draw back through a small distance, *d*, whose amount depends upon the thickness of the plate *t*, and its refractive index *n*, if isotropic, the relation of these quantities being expressed by the formula, $n = \frac{t}{t - d}$.

This method has the advantage of being applicable to very thin and minute crystals or parallel plates, and is therefore useful in the microscopic study of rocks and minerals, but the measuring apparatus, which may be either a vernier scale or micrometer screw, but preferably the latter, must be made with great accuracy, as the quantities to be measured are usually very small.

In an isotropic substance the velocity of propagation of wave motion being alike in all directions, the refractive index will also be constant in any direction for light of any particular colour, being lowest for the red and highest for the blue end of the spectrum. Such substances are therefore said to be single-refracting or monorefringent. They include all transparent homogeneous gases and liquids, the latter with a few exceptions; and among solids, both those that are amorphous and those crystallising in the cubic system, all other crystallised substances being anisotropic.

The properties of a ray of ordinary light are similar in any radial direction about its line of transmission considered as an axis; it appears to vibrate simultaneously in these directions. There is, however, good reason to suppose that this is not actually the case, and that the directions or azimuths of vibration change continuously, but so very rapidly that the changes are not perceptible to the eye. Upon this supposition the plane of vibration of a molecule, *o*, fig. 329, would be changed by a small angle at each pulsation, or if the limit of its first vibration is *n*, on the line *o* N, that of the second will be somewhere to the right of N (if the change is in that direction), that of the third, somewhat further on the same side, and so on, until the direction of vibration is on the line *o e*. The motion in any intermediate plane, such as *o c*, may be considered as the resultant of two forces acting simultaneously, *o n* towards N, and *o e* towards E; and if the velocity be the same in any direction the amplitude of the vibration in *o c* will be exactly the same as those in the N S or E W planes, and consequently the intensity of the light will be unchanged, whatever may be the azimuth of its plane of vibration. These relations, however only hold good for isotropic media, as in anisotropic ones the velocity of wave propagation varies with the direction, the differences being greatest between directions at right angles to each other. Supposing N S to be such a direction of maximum velocity, E W will be that of the corresponding minimum, and the molecule *o* will vibrate in either of these with its full proper intensity; but when in the direction *o c* the intensity will not, as in the previous instance, be that due to the components *o n*, *o e*, as these impulses having different velocities cannot arrive simultaneously at *c*, but each will act independently, and consequently the ray will be resolved into two, whose intensities will vary with their azimuths, one of them having its greatest intensity in the direction *o* N, diminishing to *o n* in the direction *o c*, and to nothing in *o* E, or in proportion to the cosines of the azimuths, and the other having its maximum in *o* E and diminishing to nothing in *o* N or *o* S, or as the sines of the azimuth; but as these changes take place too rapidly to be followed by the eye, the resulting effect is the production of two rays, each of half the intensity of the original ray of ordinary light, and vibrating in one plane which is at right angles to the corresponding plane of the other ray. Such rays are said

FIG. 329.

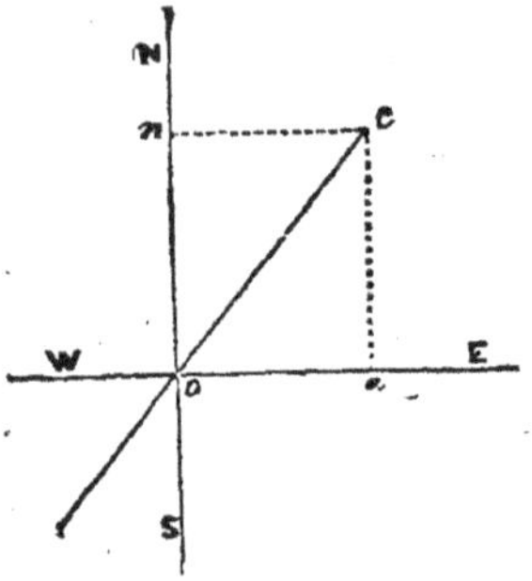

to be polarised in planes at right angles to their planes of vibration, and are no longer freely transmissible in any direction like those of ordinary light, and as a consequence of their differences in velocity they will have different refractive indices, or the substances producing them are doubly refracting or birefringent.

The general phenomena observed in a doubly refracting crystal are best seen in Calcite, in which the property is very strongly developed.[1] If *a b c d* (fig. 330) be the section of a cleavage rhombohedron so placed that *a b* and *c d* are the shorter diagonals of a pair of faces, *a c* and *b d* polar edges, and *a d* the principal axis, a ray of light, *r*, incident at *n* in a direction normal to *a b*, will be divided into two parts; one following the law of ordinary refraction will pass through unchanged towards *o*, while the other will be refracted, and assuming the original direction on emergence at *p* will travel in a direction parallel to the first, but at some distance from it, towards *e*, so that by looking through the crystal in the direction *n o* towards a brightly illuminated object on a dark ground, two images of it will be seen, one at *o* and the other at *e*. The first of these rays is called the ordinary and the second the extraordinary ray. If the direction of incidence be oblique, both rays will be refracted, but in different degrees: the ordinary ray following the law of sines and having a constant index, $\omega = 1{\cdot}658$ for yellow light, whatever may be the angle of incidence; while that of the extraordinary one varies with its direction in the

FIG. 330.

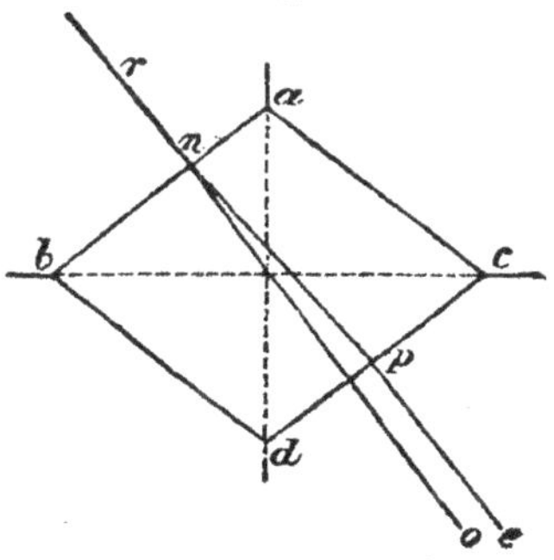

[1] By strong double refraction is meant that the indices of the rays differ considerably, not that their absolute values are very high.

crystal, being at a minimum of $\epsilon = 1{\cdot}486$ parallel to *cb*, and at a maximum which is the same as that of the ordinary ray when parallel to the principal axis *ad*. As the same relation holds good for any of the three lateral interaxial planes, which in common with all planes passing through the vertical axis are called *principal optical sections*, it follows that both rays will have the same index, or there will be only single refraction in the direction of the vertical axis. If, therefore, an object be viewed in this direction, which can be done if the polar solid angles are ground and polished into faces parallel to the horizontal plane, or a transparent natural crystal terminated by the basal pinakoid is used, only one image will be apparent, but in every other direction there will be two, their distance apart being greatest when seen at right angles to the principal axis. This direction of single refraction is termed an optic axis, and crystals in which one such direction is apparent are said to be optically uniaxial or uniaxal. These include all those belonging to systems in which a single principal axis can be distinguished—i.e., the hexagonal and tetragonal systems.

The velocity of propagation of light waves being inversely proportional to the refractive indices of the medium, that of the ordinary ray will be the same in any direction, or its wave surface will be a sphere; but in the extraordinary ray it differs according to the angle made by the ray with the optic axis, being the same as that of the ordinary ray when parallel, and at a maximum difference from it when at right angles to that axis. For any intermediate direction, such as *op*, fig. 331, it will be equal to the radius vector, corresponding to the angle ζ, of the ellipse, whose semi-axes are the greatest and least velocities, *oc*, *oa*. But, as these conditions hold good for any plane containing the optic axis, it follows that the wave surface of the extraordinary ray will be the spheroid or ellipsoid of revolution generated by the rotation about *ao*, of the complete ellipse of which *apc* is a quadrant. In Calcite, where the direction of minimum velocity is parallel to the optic axes, the spheroid is oblate, or the extraordinary encloses the ordinary wave surface; but in the other class of uniaxial species, of which Quartz may be taken as the type, the optic axis is the direction of maximum velocity, and therefore the extraordinary wave spheroid is prolate, and enclosed by the sphere representing the ordinary wave surface. These conditions are represented in fig. 331, if *oc* be turned into the vertical position, when *ca'* will be the section of the ordinary wave, while *cpa* remains the extraordinary one. In the former class, for any doubly refracted ray in the principal optic section of the crystal, the extraordinary has a lower refractive index than the ordinary ray, and will therefore be deviated from the original direction less than the latter, or make a larger angle with the optic axis than the ordinary ray. These are called negative crystals. In the other or positive class of uniaxial crystals, in any direction involving double refraction, the higher index is always that of the extraordinary ray, which will therefore make a smaller angle with the optic axes than the ordinary ray.

FIG. 331.

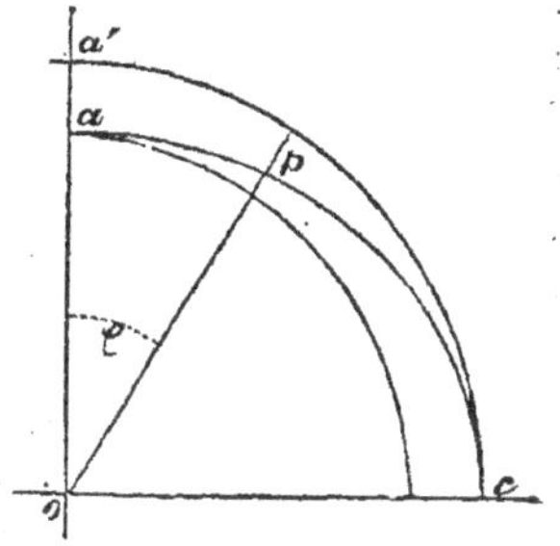

For the determination of the refractive indices of a uniaxial crystal by the method of minimum deviation, a prism is required having a refracting edge either parallel to the optic axis, or perpendicular to it; but in the latter case the faces of the prism must make equal angles with the optic axis. The deviation of the two rays is observed successively in one operation, each one being alternately extinguished by a Nicol's prism interposed in the path of the beam in the manner subsequently described.

In the second great division of anisotropic crystals,

those belonging to the rhombic, oblique, and triclinic systems, the optical phenomena are more complex, the differences in the velocities of wave transmission and refractive indices in different directions not being related to the principal crystallographic lines in any simple manner. In any such crystals there will be two directions in which the molecular elasticity, and, consequently, the velocity of wave transmission, will be respectively the maximum and minimum of all the possible values; and these are at right angles to each other in the same plane, while in a third direction, at right angles to both, the value will be an intermediate one. These directions are known as the axes of maximum, minimum, and mean[1] elasticity, respectively. In such a system, all three axes being dissimilar, direction of equivalent refractive value cannot be symmetrical to any one axis, as in the uniaxial class, and therefore the wave surface cannot be a spheroid.

If O X and O Z, fig. 332, be the axes of maximum and minimum elasticity, and the position of O Y, the axis of mean elasticity, be imagined as perpendicular to the plane of the page, a ray of ordinary light, following the direction O X, would vibrate successively in every possible azimuth in the normal plane containing O Y and O Z, and, consequently, on entering the crystal, would be resolved into two rays, each vibrating parallel to one of these axes and moving with its characteristic velocity. If, therefore, the motion originate at O, the faster moving ray, whose velocity will be that corresponding to the mean elasticity, will arrive at B, when the slower one has only reached C, which, with corresponding positions on the opposite side, will be points on the wave section. Similarly, a ray following the direction of minimum elasticity, O Z, is resolved into two vibrating parallel to O X (maximum) and O Y (mean), whose proportional paths are O A and O B. For any intermediate direc-

[1] The word *mean* is here used in the sense of intermediate, and does not refer to any particular mean.

tion, such as O L, the paths of the resolved rays are O B, and as the velocity of the slower ray is greater than the minimum, but less than the mean, the points B and L′ are nearer together than B and C; while for O S, which is nearer the maximum than the minimum axis, the corresponding paths are O B (mean) and O T′ (greater than the mean and less than

FIG. 332.

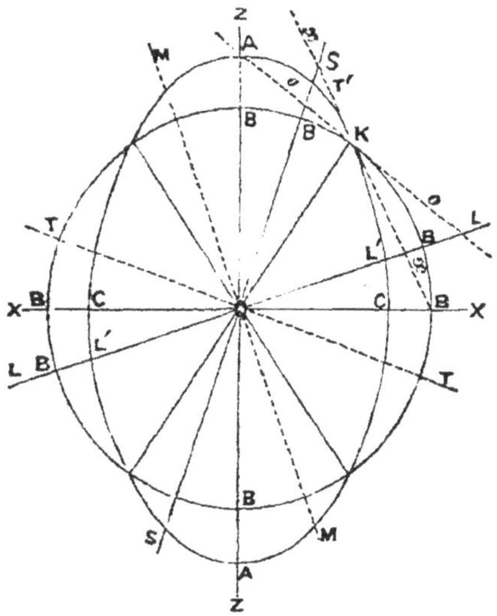

the maximum), and consequently T′ and B are nearer together than A and B are. The points B being equidistant from O, the curve passing through them all is a circle whose radius is proportional to the mean velocity; while O C, O L′, O T′, and O A, are radii of the ellipse having the semi-diameters O C and O A, whose lengths represent the corresponding minimum and maximum velocities. The wave is therefore bounded by two surfaces, of different curvatures, which cross each other at K. This point being common to both curves, the resolved rays following O K would have the same velocity in the crystal; but as the tangent planes *o o*, *x x*, representing their plane wave surfaces are differently inclined to O K, owing to the difference in the curvature of the two surfaces, they will be dissimilarly refracted at emergence. The two surfaces have, however, a common tangent plane, whose contact line—a small circle of the diameter U U′ (fig. 333)—is the base of a cone whose apex is at O. The rays, therefore, whose paths in the crystal are on the surface of the cone will be only singly refracted, and emerge as a cylinder, whose direction, the normal to U U′, is called an *optic axis.*[1] As, however, the wave surface is symmetrical to the two axes in the plane O X and O Z, a line, O V, to the left of O, and equally inclined to O Z, will have similar properties, or will also be an optic axis, and the crystal will be *biaxial* or *biaxal.*

FIG. 333.

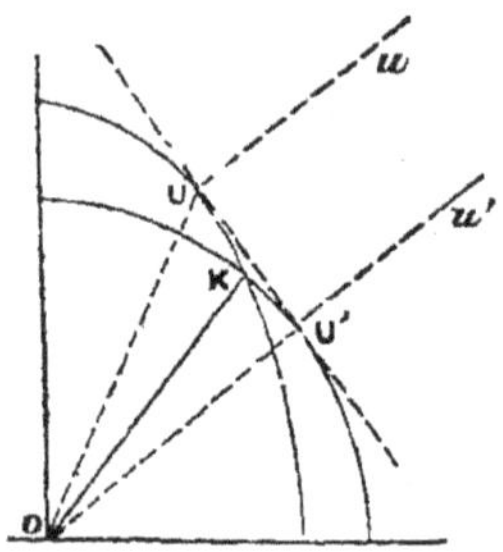

The section of the crystal in which all these axes lie is called the plane of the optic axes, and the axes of elasticity which bisect the angles between the latter are their *median*, or *mean lines*, or *bisectrices*. These are further characterised as first, or principal, and second, or supplementary, mean lines; the former bisecting the acute, and the latter the obtuse or supplementary angle of the optic axes.

The position of the optic axes, with reference to the median lines, depends upon the differences of velocity of wave transmission in the three axes of elasticity. If the mean value is nearer to the maximum than to the minimum, or the radius O B approaches closer to the major than to the minor semi-axis of the ellipse, as in fig. 334, the acute angle of the

[1] In this figure the divergence of the rays at the apex of the cone is far greater than any known example. The divergence is very small in all substances but those with very strong double refraction, such as Aragonite, whose maximum and minimum indices differ by about 10 per cent., and in which the angle U O U′ is 1° 55′.

optic axes 2 V will have the axis of minimum elasticity for their first median line ; but if it is nearer to the minimum, or if O B is but very little greater than the semi-axis O C (fig. 335), the axis of maximum elasticity is the first median line. Crystals of the first class are called positive, and those of the second negative. The value of the angle 2 V may therefore vary within wide limits, but must be always greater than 0° or less than 90°. It is tolerably constant, and often a most

FIG. 334.

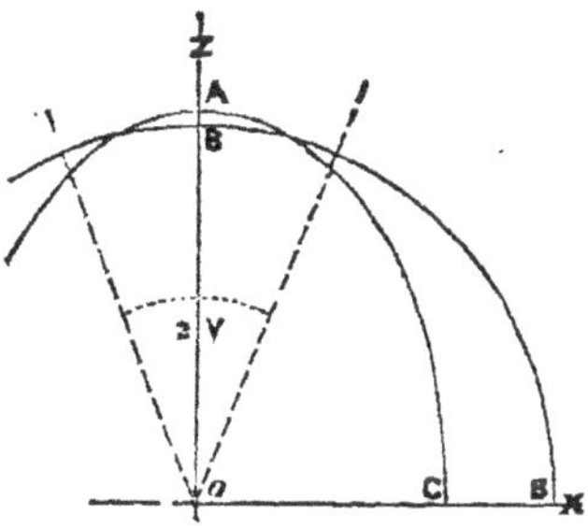

FIG. 335.

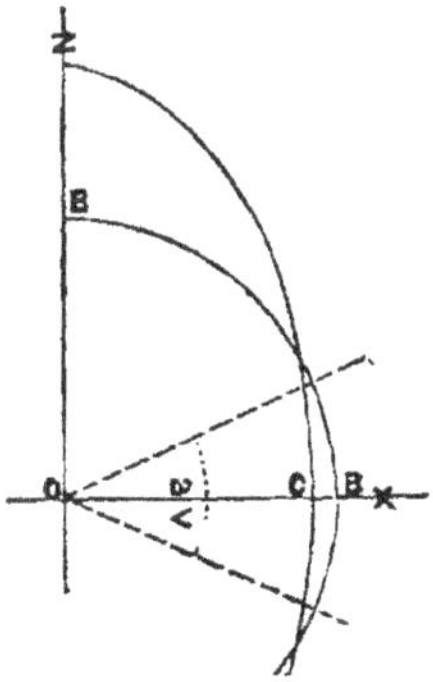

characteristic element in such minerals as are of comparatively simple constitution ; but in those containing numerous elements, and giving complex molecular formulæ, such as mica felspar and topaz, it may vary very considerably in different specimens of the same mineral, or even in different parts of the same crystal.

The curves formed by the intersection of the wave surfaces with the other two principal optic sections will be, in the plane containing O X and O Y (fig. 336), an ellipse whose semi-axes are proportional to the maximum and mean velocities enclosing a circle whose radius is proportional to the minimum velocity ; and, in that containing O Y and O Z (fig. 337), a circle whose radius is proportional to the maximum velocity enclosing an ellipse having its semi-diameters proportional to the minimum and mean velocities. In each of the three sections, therefore, one ray will follow the law of ordinary refraction, and if the indices corresponding to these be determined by prisms whose refracting edges are respectively parallel to the three axes of elasticity, three values will be obtained, which are the three principal refractive indices. These are usually distinguished by the symbols α, β, and γ; the first referring to the minimum, the second to the mean, and the last to the

FIG. 336.

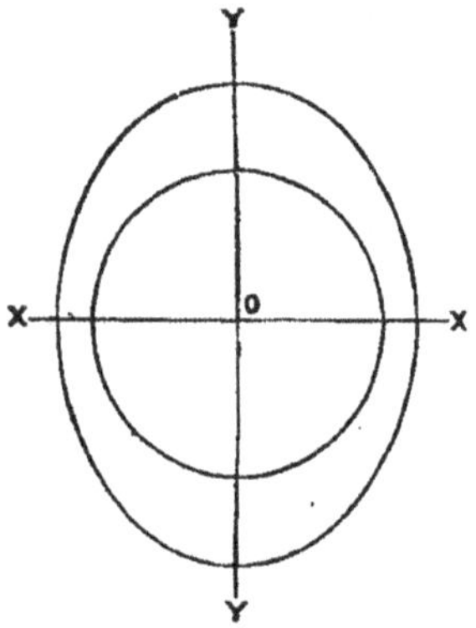

FIG. 337.

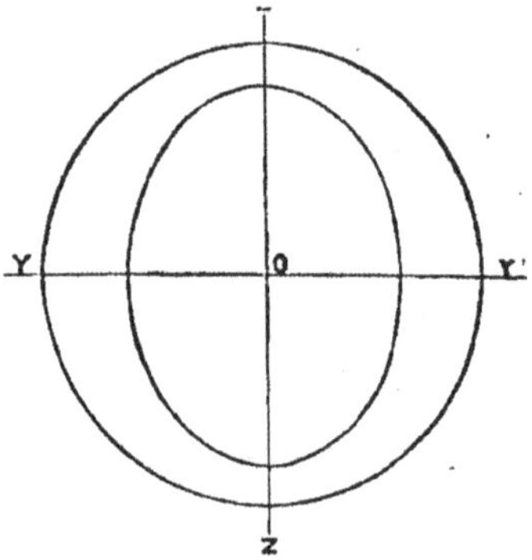

maximum value. For their d termination three prisms are not absolutely necessary, as two will suffice if they are so cut that the refracting angle is either bisected symmetrically by, or one face is parallel to, a principal optic section. The three refractive indices and the corresponding axes of elasticity are the optical constants of a biaxial crystal from which the actual angles of the optic axes are calculated.

The method described in page 234 may also be used, but the phenomena are more complicated than with isotropic media, the distance through which the objective must be shifted being different, for ordinary and extraordinary rays, from the production of what are called by Sorby 'bifocal' images. For the details of the special appliances required the reader is referred to the original memoirs.

The statements contained in the preceding pages of this chapter may be conveniently summarised as follows :—

Supposing light to be a succession of spherical waves, the wave generated at each point being spherical, it is the line of greatest intensity which represents the visual ray, and this is ordinarily, in a homogeneous and isotropic medium, a straight line. But when, in consequence of definite arrangement, there is not isotropy, all the waves are not spherical, and the path of the visual ray is not a mere straight line, speaking generally. The chief cases are :—

1. Isotropy in general.

a. Absolute amorphism, such as a homogeneous glassy or colloid medium.

b. Cubic symmetry.

Any three pairs of equal forces at right angles are in equilibrium, as in fig. 338, the third being perpendicular to the page, and may be replaced by three other

FIG. 338.

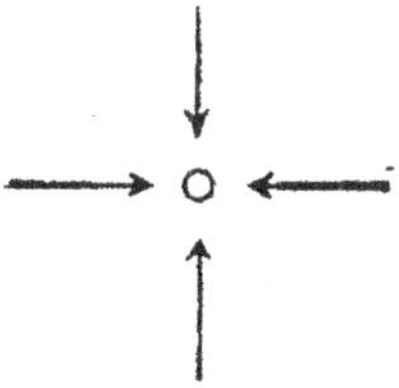

FIG. 339.

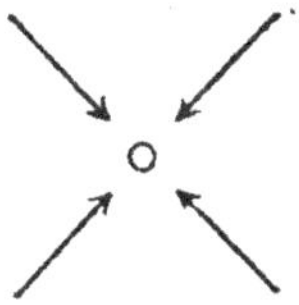

pairs of equal orthogonal forces (fig. 339), arbitrarily chosen as regards aspect (or direction in space): any equilibrium which can be represented by one such system is therefore isotropic as regards all forces of the same nature

2. Isotropy about an axis.

a. Amorphism polarised in one direction, as in an amorphous elastic and isotropic mass, subjected to definite pressure or tension.

b. Quadratic symmetry in planes at right angles to the axis. The same remark applies about two pairs of orthogonal forces in a plane, as about three pairs in solid in the case of cubico-symmetry.

c. Ternary plane symmetry, or senary (fig. 340). The plane forces may be replaced by two pairs of equal orthogonal forces, as far as mere equilibrium is concerned. This symmetry about an axis constitutes the isotropy of uniaxial systems, the isotropic axis being the optic axis.[1]

3. Symmetry. If any three pairs of forces, equal in pairs, keep a material point in equilibrium (fig. 341), the third pair lying out of the page, they may be replaced by

FIG. 340.

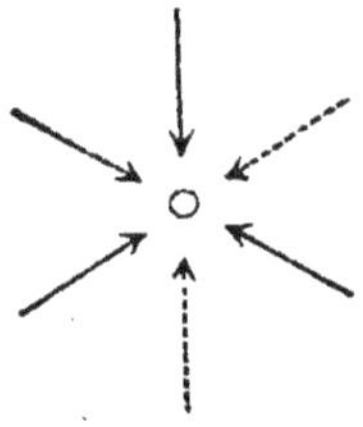

FIG. 341.

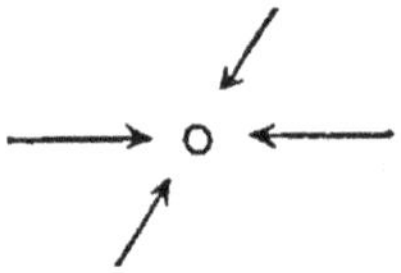

three pairs of orthogonal forces of definite magnitude, whose position and magnitude depend upon the position and magnitude of the original forces. These orthogonal forces are called the axes of optical elasticity. Taking them as the axes of an ellipsoid, the optic axes are lines at right angles to its circular sections.

The optical characters of minerals are most strikingly evidenced by the interference of phenomena observed when parallel sections cut in known directions are examined under polarised light. The observation depends upon the general principle that the two rays resulting from the passage of a plane polarised ray through a double refracting crystal are

[1] There is symmetry about a plane normal to the axis in all such crystals. It is instructive to compare this with the unsymmetrical, but isotropic, character of a vertical line in the earth's atmosphere, taken as a refracting medium. There is, however, no polarisation, as the air is not under tension in any one direction more than another.

capable of interference like ordinary light when reduced to the same plane of polarisation by passing through a second polarising medium. The instruments used for this purpose, known as polarising microscopes or polariscopes, differ somewhat in construction, and chiefly as regards the polarising agent employed. Those most commonly used are a bundle of parallel glass plates, and Nicol's prism; the latter is a cleavage rhombohedron of calcite, whose length is about twice its breadth, the smaller end faces being ground down until the angle with the longer edges of the other four faces is reduced from 71° to 68°, when the principal section will resemble A B C D (fig. 342). The crystal is then sawn in half, on a plane at right angles to the principal section, whose projection is the line B C, and the surfaces, when perfectly polished, are cemented together parallel to the original position by a layer of Canada balsam; the side faces are blackened, and enclosed in a metal tube, the ends being exposed. If a ray of light meets one of the end faces in a direction parallel to the length of the prism, it will be doubly refracted; the extraordinary ray, having the lower refractive index—nearly the same as that of Canada balsam, namely 1·536—passes through almost without deviation in the direction *r e*; the ordinary ray, on the other hand, having the higher refractive index—1·654—is deflected through a larger angle, and consequently meets the optically lighter layer of balsam at such an inclination that it is totally reflected, and passes into the blackened covering at *o'*, where it is absorbed. Of the two rays, therefore, produced by double re raction, only the extraordinary ray, whose planes of vibration and polarisation are respectively parallel and perpendicular to the principal section, is transmitted; and such a ray will pass freely through a second prism of the same kind, if the principal sections, represented by the longer diagonals of the end faces, of both are parallel; but

FIG. 342.

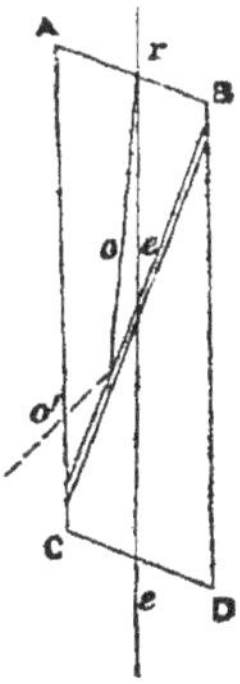

if their principal sections are crossed at right angles, the light is completely extinguished. For any intermediate azimuth of vibration, the light transmitted will be the component of the resolved ray that vibrates parallel to the principal section, and, therefore. will be less in pro ortion as the angle between the principal section of the Nicol and the plane of vibration of the ray increases; and the same relation holds good for two plane polarised rays, such as originate by the passage of ordinary light through a doubly-refracting medium, vibrating in different planes, and entering the prism simultaneously. In the latter case, the portions of the rays transmitted are said to be reduced to the same plane of polarisation, and are in condition to interfere in the same manner as rays of ordinary unpolarised light.

Fig. 343 is a generalised representation of the common form of polariscope. It consists of two Nicol prisms, A and B, attached to a vertical pillar, with a plane mirror for reflecting light in the lower prism, and a carrier D, for the crystal under examination, which may be, ei her fixed to the pillar, or, as is more generally the case, mounted in a divided ring which rotates on the top of the case of the lower prism. The latter, which is called the polariser, is fixed, and is usually of a larger size than the upper one, or analyser, which is movable, both about its own axis and in a vertical direction, by a rack-work fixed to the pillar. This arrangement is used for the examination of minerals under parallel light, the whole of the rays passing through the instrument being parallel to its axis and to each other; but to produce the characteristic phenomena which are observed when the crystal is traversed by rays of different degrees of obliquity, a combination of condensing lenses is placed above the polarising prism, and a corresponding combination in the reverse direction below the analyser, which are so adjusted that their common focal point is in the centre of the plate of mineral under investigation. These are represented in section in fig. 344; the lower series of four plano convex lenses, c_1, c_2, c_3, c_4, form the condenser, and the upper, o_1, o_2 o_3, o_4, the objective. These combinations are equivalent to single lenses of short focal length and low magnifying

FIG. 343.

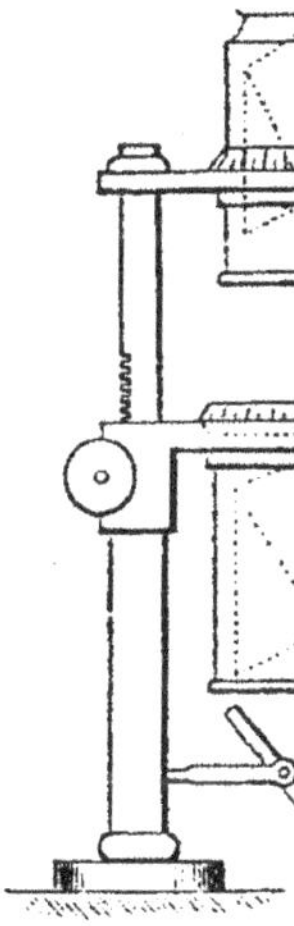

FIG. 344.

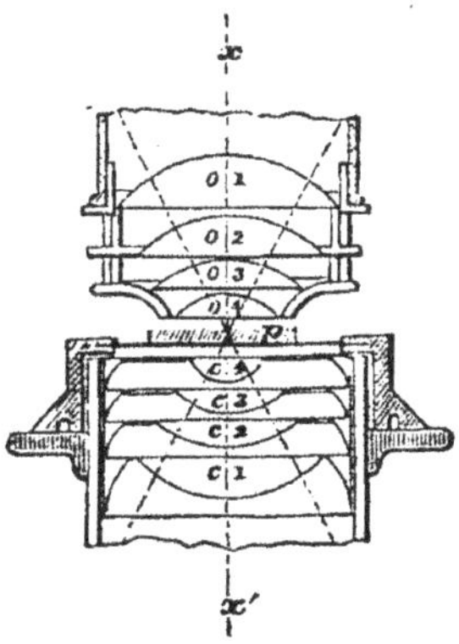

power, but giving a large field of view. The size of the image may also be amplified by a double convex lens, forming an eye piece, placed below the analysing prism.

When a parallel-sided plate of a transparent isotropic substance is placed upon the carrier D, it will have no effect upon the light passing through the instrument—that is, if the planes of vibration of the two Nicols be parallel, the field will be illuminated; if they be crossed at right angles, there will be complete, and at any other angle partial obscurity, exactly the same as there would be by the mere action of the prisms themselves; and further, no change in the light will be observed when the plate is turned round in its own plane. These conditions, which apply both to amorphous substances and those crystallising in the cubic system, hold good both for parallel and convergent light—such substances being said to be without *depolarising* effect.

If a plate of a uniaxial crystal, which may be either an artificial section cut perpendicularly to the optic axis, a sufficiently thin crystal with well-developed basal pinakoid faces, or a basal cleavage plate, be examined in parallel polarised monochromatic light, it will appear light or dark according to the position of the Nicol's[1] prisms, in the same way as an isotropic substance, the direction followed by the light being that of single refraction in the crystal. If, however, the section be cut parallel to the optic axis, it will when placed between parallel Nicols appear bright when its principal section corresponds to their planes of vibration, and if turned in its own plane the light will diminish with the angular deviation and be completely extinguished at 45°, from which point the light will again increase to a maximum at 90°, disappear at 135°, and so on through the complete rotation, the field of the instrument being four times light and dark alternately. When the Nicols are crossed, the light is extinguished, when the principal section of the plate is parallel to the plane of vibration of either Nicol, or at the azimuths 0°, 90°, 180°, and 270°, and is brightest at the intermediate or half quarter points of the circle, 45°, 135°, &c. There are, therefore, four alternations of light and darkness during the revolutions of the plate, as before, but the order of their occurrence is inverted. When white light is used and the plate is sufficiently thin to cause a retardation of one of the refracted rays corresponding to a half-wave length of any particular colour, that colour will be extinguished, and consequently, at the position of maximum illumination, the plate will show the residual or complementary tint; or if a spectroscope be used over the analyser, the spectrum will show an absorption band corresponding to the missing colour. These positions with crossed Nicols are at the half-quarter points, but with a plate whose retardation is a complete wave length, these points correspond to

[1] Or, as they are usually called, 'Nicols.'

those of extinction, the greatest illumination being at 0°, 90°, 180°, and 270°. When the plate is much thicker, so that the retardation amounts to $\frac{3}{2}$, $\frac{5}{2}$, $\frac{7}{2}$, or a greater number of half wave lengths. so much colour is extinguished by interference that the residual tint is scarcely distinguishable from white, forming the so-called white of the higher order, and therefore the plate shows four times light and dark in a revolution, nearly in the same way as in homogeneous light. In these cases the spectrum of the plate will be crossed by several dark bands; these are especially well seen in Quartz. When the Nicols are parallel, the colours seen at the points of maximum illumination are complementary to those obtained when they are crossed.

Under convergent homogeneous light, the appearances presented by a section perpendicular to the optic axis between parallel Nicols are seen in fig. 345. A central bright space corresponding to the direction of the optic axis, or that of single refraction, is surrounded by segments of concentric dark rings, which are broadest along lines 45° and 135° from the plane of vibration of the Nicols, and fade gradually towards the lines of 0° and 90°, producing the effect of a bright four-armed cross, traversing the rings symmetrically. The dark rings nearest to the centre correspond to the positions in which the rays produced by double refraction in the crystals emerge, with a difference of half a wave length, and therefore, when reduced to the same plane of polarisation by the analyser, interfere and produce complete extinction on the line of the principal section, *d d'*. In any other principal section making a smaller or larger angle with the line E W, light incident at the same angle will be similarly refracted and retarded, but the rays will not be equally resolved by the analyser; that one whose direction

FIG. 345.

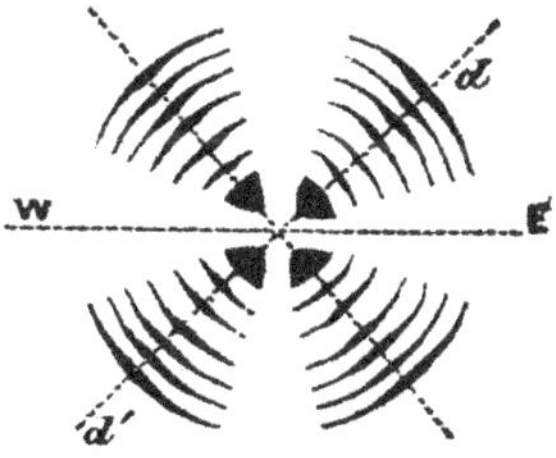

of vibration is nearest to E W being more completely transmitted than the other, which vibrates at right angles to it; and therefore, although there will still be interference, the extinction will not be complete, owing to the unequal amplitude of the two rays transmitted by the analyser. This difference is at a maximum in the principal sections that are respectively parallel and perpendicular to E W, as in these only one of the refracted rays is transmitted by the analyser, and therefore the field remains bright, forming the arms of the cross. With increased distance from the centre, as the obliquity of the incident light increases, the paths of the refracted rays in the crystal are lengthened, and their difference of phase at emergence becomes greater, so that at a point some distance beyond the first dark ring, where it amounts to two half-wave lengths, the light is augmented by their interference producing a bright ring. Beyond this it diminishes again, total obscurity being produced when the difference of phase is $\frac{3}{2}$, and so on, the field being bright and dark from the interference of waves differing in phase by even and uneven numbers of half-wave lengths respectively. With increased obliquity of the light, however, these interferences recur more rapidly, and therefore towards the border of the field of view, the rays are narrower and closer together than in the centre. If the plate be viewed between crossed Nicols, the complementary phenomena shown in fig. 346 are observed: a series of continuous dark rings taking the places of the bright ones in fig. 345, with a dark rectangular cross, one bar of which corresponds to the projection of the plane of vibration of the polariser, and the other to that of the analyser, crossing them symmetrically.

FIG. 346.

In white light, the rings, instead of being dark, are coloured, the tint of any one corresponding to the residual colour after extinction of the particular light which is retarded by a half-wave length in one of the refracted rays. As these are constant for the same angle of incidence, the colour will be the same at any similar distance from the centre of the plate, from which circumstance the rays are called *isochromatic* curves. In uniaxial crystals these are circles if the section be cut perfectly, or very nearly, square to the optic axis; but if it is sensibly oblique, they will be somewhat elliptical, and the cross, instead of changing from black to white by the rotation of the analyser, will be resolved into a looped figure like that of a biaxial crystal.

The colours of the rings recur in the order of the Newtonian interference scale, those of higher refrangibility are generally seen only in the rings nearest the centre, alternations of red and green of diminishing intensity appearing towards the margin of the field. These latter, the so-called colours of higher orders, can only be seen within much narrower limits than the black rings produced in monochromatic light, and therefore the figures produced with the latter are much sharper and better defined than with white light. The distance between the rings varies with the strength of the double refraction, or the difference between the refractive indices, of the ordinary or extraordinary rays, the thickness of the plate, and the refrangibility of the light used. For plates of equal thickness, of different substances, they will appear closest in those crystals that have the stronger double refraction and higher refractive power. For the same substance the distance between the rings will be increased by diminishing the thickness of the plate, and to a less degree by illuminating with less refrangible light. Figures 345, 346, represent the rings seen in a section of Calcite $\frac{1}{8}$ inch thick, in nearly monochromatic red light, produced by passing solar light through a plate of ruby glass; but if a blue light be used in the same way, the rings will be crowded much closer together, but will become invisible at about two-thirds of the former breadth of field. When the section is very thin, $\frac{1}{50}$th of an inch or less, no rings, but only a black cross or four black spots are seen, according to the position of the analyser. With minerals of weak double refraction, such as Idocrase or Apophyllite, thicker sections are required in order to see the rings.

The method of determining the sign of the double refraction uniaxial crystals will be described subsequently.

Circular Polarisation. A section of a quartz crystal perpendicular to the optic axis viewed by monochromatic light between crossed Nicols, unlike other uniaxial crystals, allows some light to pass through, and it is only by turning the analyser through a certain angle that the light can be extinguished; but when this position has been determined the plate will remain dark when turned in its own plane in the same manner as an ordinary uniaxial crystal, showing that the light enters the analyser in the condition of plane polarisation, but that the direction of its plane of vibration has been changed. This condition of the light, called circular polarisation, is produced by all substances, whether isotropic or uniaxial, that crystallise in plagihedral forms—that is, the tetartohedral forms of the cubic, the trapezohedral tetartohedra of the hexagonal, and the plagihedral hemihedra of the tetragonal systems. Examples of the first and last of these are only known in artificially crystallised salts, but the property is most strikingly developed in the hexagonal system, in the minerals Quartz and Cinnabar. The angular deviation of the plane of polarisation produced by a plate of Quartz increases with its thickness, as well as with the refrangibility of the light used for illuminating, being greatest for the blue and least for the red end of the spectrum, or for any particular thickness it varies inversely as the square of the wave length. With a quartz plate of $\frac{1}{25}$th of an inch thick, the rotations are:—

Red 19·00°, Yellow 24·00°, Blue 32°, extreme Violet 44°.

The specific rotatory power of Cinnabar is very much

greater, being, according to Descloizeaux, fifteen times that of Quartz. As a consequence of the difference in rotation for different colours, a circularly polarising crystal can never appear dark in parallel polarised white light, whatever may be the relative positions of the polariser and analyser. When the latter are crossed, the rays whose plane of vibration most nearly coincide with that of the analyser will be most completely transmitted, while those making larger angles with it will for the most part be extinguished. The prevailing colour of the plate, therefore, will be mainly that of the freely transmitted rays, and this will be constant for any plate of the same thickness. By turning the analyser its plane of vibration becomes successively parallel to those of the different coloured rays, and the colour of the field changes with each coincidence. The order in which the colours meet each other is not the same in all cases, and to produce them in the order of their refrangibility from red to violet it is sometimes necessary to turn the analyser from right to left, and sometimes the reverse. These differences correspond to differences in the position of the plagihedral faces in the crystals from which the plates are cut, the former indicating right-handed and the latter left-handed forms. In some instances, however, the sections show more or less irregular patches, some having right- and others left-handed rotation, proving that the crystals from which they are derived are complex twin structures.

In convergent polarised white light, Quartz gives a series of isochromatic rings, barred by black or bright arms, according to the position of the analyser, in the outer part of the field; but in the centre where the light is either parallel or but slightly inclined to the optic axis, the conditions are similar to those described for parallel light, and therefore, instead of a complete cross, the central space within the first dark ring will have the colour due to the thickness of the plate, when the Nicols are crossed, and will change when the analyser is turned as in parallel light. With a right-handed plate the rotation of the analyser to the right causes an apparent expansion of the rings, and the change from a light to a dark field, and to the left a corresponding contraction; with a left-handed plate the reverse conditions prevail.

When two plates of Quartz, perpendicular to the optic axis of the same thickness, but of opposite rotation, are superposed in parallel polarised light, the rotatory power of one will neutralise that of the other, or the field will remain dark as with an ordinary uniaxial crystal. When the two sections are of unequal thickness, such as the parallel-sided plate obtained by the superposition of two similar wedges inclined in opposite directions, there will be a dark line or neutral band where the thicknesses are exactly alike, on either side of which the characteristic colour due to the preponderance of the thickest plate at that point will appear. In convergent light, the interference figures known as Airy's spirals are produced, which, in addition to the concentric rings of an ordinary uniaxial crystal, show a series of four spiral arms crossing in the centre of the field, the direction in which the arms are coiled being determined by that of the rotation of the lower plate. Thus, in fig. 347, the left-handed plate is below the right-handed one; but if it were above, the arms would coil in the opposite direction, the figure being otherwise unaltered.

FIG. 347.

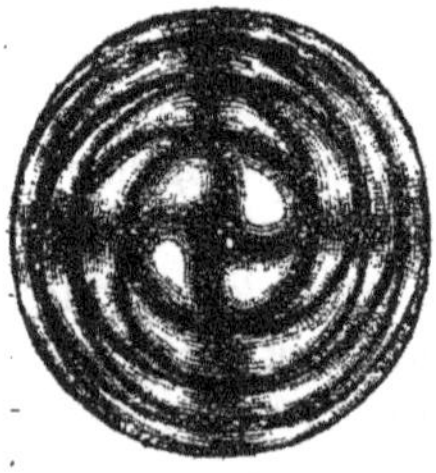

These phenomena are sometimes seen in plates cut from apparently simple Quartz crystals, proving them to be parallel alternations of right- and left-handed ones.

The property of circular polarisation in Quartz crystals is confined to the direction of the optic axis; in every other direction, therefore, they behave like ordinary plane polarising crystals. Even in sections normal to the axis a certain thickness is required to produce a sensible rotation, a very thin one shows merely a black or dark grey cross, between crossed, and four dark blue spots with parallel, Nicols.

In the few known cases of cubic tetartohedral crystals the circular polarisation is independent of direction, and any section of the crystal shows the rotation of the plane of polarisation equally well. The principal salts in which this has been observed are the Bromate of Sodium, Chlorate of Sodium and Nitrate of Barium. The last two have been shown to be geometrically tetartohedral, but in the first the development is inferred from the circular polarisation.

Optical examination of biaxial crystals. In parallel polarised light, a section of a biaxial crystal, when rotated between crossed Nicols, will appear four times dark and light (or coloured if sufficiently thin) in each revolution, except it be perpendicular to an optic axis, when it will give a dark field in all positions. This latter case, however, rarely arises in practice, as such sections are not easily prepared. In the rhombic system the crystallographic axes coincide with those of optic elasticity, and therefore when a section of a rhombic crystal parallel to either a pinakoid or a prismatic face, whether prism or dome, is placed between crossed Nicols, the field will appear dark whenever a crystallographic axis in the section is parallel to the plane of vibration of either Nicol, as well in white as in homogeneous light of any colour. In such cases the *directions of extinction* are said to be parallel to the crystallographic axes.

In the oblique system only one axis of optic elasticity coincides with an axis of form—namely, with the orthodiagonal, the other two lying in the plane of symmetry at right angles to each other, but oblique to the crystallographic axes in that plane. If therefore a thin tabular crystal or a section parallel to either the base or the orthopinakoid be placed between crossed Nicols, it will extinguish the light in the same manner as a rhombic one—i.e., parallel to the crystallographic axes; but if the section be parallel to the clinopinakoid, it will only appear dark when the vertical axis of the crystal is inclined to the vibration plane of a Nicol at some angle which will differ with the colour of the light employed, a property known as the dispersion of the axes of elasticity. If the section be parallel to a face of a prism, the same kind of dispersion will be observed, but in a different degree, the angle between the vertical axis and the direction of extinction being less for the same colour than in the plane of symmetry, and these directions approach nearer to parallelism, as the angle of the plane of the section with the orthopinakoid diminishes—that is, as the front or obtuse angle of the prism increases. In the triclinic system, there being no coincidence in direction between the axes of form and optic elasticity, the directions of extinction in any plane will be oblique to the axes of the crystal.

The directions of extinction, therefore, afford a means of determining the character of the symmetry of a crystal, or its crystallographic system, and as such their observation is one of the most important operations in physical crystallography. An exact determination, however, is a matter of some nicety, and cannot practically be made directly, as the light fades so gradually near the point of maximum-darkness that the difference corresponding to a change of azimuth of 2 to 3 degrees is not easily appreciable. An indirect class of observation is therefore adopted, in which, instead of the extinction of light, changes in the appearance of the interference cross of a plate of Calcite normal to the optic axis when viewed between crossed Nicols, with a plate of the mineral under examination interposed, are used as a test of the change of position of a known crystallographic line in the latter. The instrument for this purpose, called, from the nature of the observation, a stauroscope,[1] in the original form given to it by Dr. Von Kobell in 1855, is essentially a polariscope, arranged for parallel light with a Calcite plate fixed below the analyser, the latter being crossed to the polariser so as to give a perfect dark cross. Above the polariser is a perforated metal plate,

[1] Figured and described at length in Rutley's treatise on rocks in this series.

forming an object carrier, connected with a rotating divided ring; lines are ruled on the plate parallel to the vibration plane of the polariser, and the section of the crystal is cemented to it with a known edge, preferably one parallel to the vertical axis, as nearly parallel to one of the lines as possible. It is then placed on the instrument, and adjusted to the zero point of the divided ring, when, if the edge is parallel to an axis of elasticity, there will be no change in the form of the cross; but if these lines are appreciably inclined to each other, the figure will appear to be distorted, and it will be necessary to turn the divided plate through a certain angle in order to bring it back to its proper form.

The results obtained by this may be regarded as accurate to within one degree or thereabouts. A more sensitive test has been proposed by Brezina, who uses a compound plate formed of two sections of Calcite whose planes are not perfectly at right angles to the principal axis, so joined that the optic axes of both lie in the same plane, but crossing each other in direction. This gives an interference figure made up of peculiar curves, the essential element being an elliptical ring lying with its longer axis right and left, and divided symmetrically by a vertical bar which extends to the limit of the field of view in a continuous line, when the Nicols are exactly crossed; but a very slight disturbance produced by a crystal whose planes of vibration are not parallel to those of the instrument produces a break in the line, the part within the ring becoming inclined to those above and below it, the change being said to be appreciable with an angular deviation of a very few minutes.

In all cases stauroscopic observations must be made with monochromatic light, the dispersion of the axes varying with light of different colours.

In convergent polarised light a section of a biaxial crystal perpendicular to an optic axis between crossed Nicols shows a series of nearly circular rings whose intervals, as in the cases already considered, depend upon the thickness of the plate and the strength of the double refraction, crossed by a single dark line instead of the four-armed cross seen in uniaxial crystals, and which revolves with the rings when the plate is turned on its own plane. When the section is taken perpendicular to the first median line, the Nicols being crossed and the line corresponding to the projection of the optic axial plane parallel to the vibration plane of one of them, the system of rings, represented in fig. 348, will be seen. These are curves known as lemniscates, the points where the optic axes meet the surface of the plate or their poles are the centres of independent series of rings, the innermost, or polar rings, being nearly circular, while the succeeding ones are lengthened progressively towards the opposite axis, until about third or fourth they meet, forming the cross-looped or figure of 8, the crossing points of the loops making the pole of the median line. Beyond this they are continuous round both axes, with a slight compression in the centre, which is less marked in each succeeding ring, so that those at the outside of the visible field are very similar in appearance to ellipses. The lines corresponding to the vibration planes of the Nicols are, as in the case of uniaxial crystals, marked by a dark rectangular cross, but the bars of unequal intensity, that joining the optic axes being as a rule darker and better defined than the vertical one.

FIG. 348.

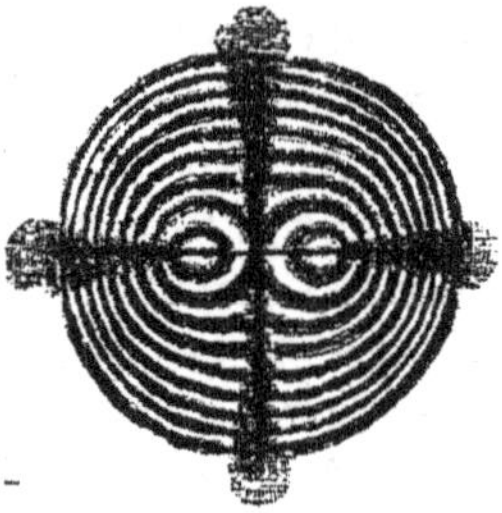

When the plate is turned through 45° in its own plane, the position of the Nicols being unchanged, the rings keep their relative positions, though revolving with the plate, but the cross, as seen in fig. 349, is resolved in two dark bands, usually sharply defined where they cross the central rings, but becoming wider and indistinct towards the extremities of the field. These bands, or, as they are usually called, 'brushes,' are portions of the two branches of an hyperbola, and the points where they cross the line of the optic axial plane, mark the extremities of the optic axes. The distance between the summits of the curves, when at their widest separation, is an indication of the amount of the inclination of the optic axes, and if the field of the instrument be provided with a micrometer scale of equal parts, the angle between them may be approximately measured.

FIG. 349.

By reducing the thickness of the plate, the distance traversed by the light is so much shortened that only the rays of more oblique incidence have sufficient length of path in the crystal to emerge with the difference of phase ($\frac{1}{2}\lambda$) necessary to produce extinction of light by interference, and

FIG. 350.

FIG. 351.

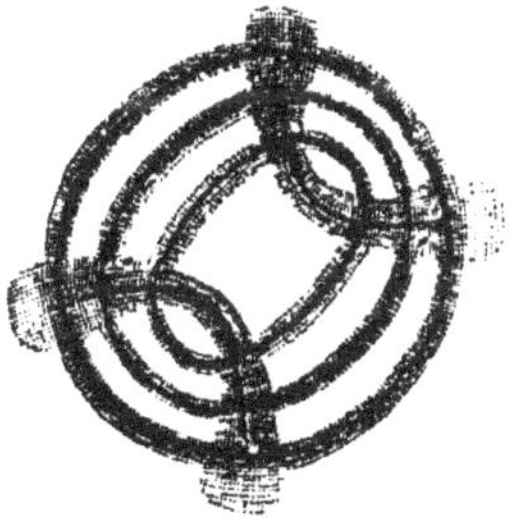

therefore the rings in such a plate will be further apart, and the innermost one will be at a greater distance from the centre than in a thicker plate, the polar rings and the looped figure generally disappearing, but the cross and brushes will not be altered. The character of these figures is seen in figs. 350 and 351, the first corresponding to the position in fig. 348, and the second to that in fig. 349. The same kind of figure is seen with thick plates when the double refraction of the mineral is weak.

Figs. 352–354 represent the successive disappearance of the inner rings in films of mica, when the thickness is

FIG. 352.

FIG. 353.

FIG. 354.

reduced. Fig. 352 is the figure seen in a plate producing a retardation of phase of $\frac{5}{4}\lambda$, fig. 353 of $\frac{3}{4}\lambda$, fig. 354 of $\frac{3}{8}\lambda$,

FIG. 355.

FIG. 356.

fig. 355 of $\frac{1}{4}\lambda$, and fig. 356 of $\frac{1}{8}\lambda$. In the last, scarcely any portion of the rings is seen, but the positions of the hyperbolic arms or brushes are equally well marked in all.

When the angle of the axes is very small, the figure is often scarcely distinguishable from that of a uniaxial crystal, the rings being nearly circular, and the four arms of the cross similar in intensity. The difference, however, becomes apparent on rotating the plate, when the arms divide into hyperbolic branches, instead of remaining unchanged, as they should do if the substance were actually uniaxial.

The exact measurement of the angle of the optic axes is effected by a polarising instrument with a goniometer attached to it, a rod in the prolongation of the axis of the latter forms the object-carrier, and allows the plate to be turned about an axis in its own plane. The instrument is placed horizontally, with the planes of vibration of the prisms crossed, and the plane of the optic axes so arranged as to show the maximum separation of the brushes. The plate is then turned about its axis of suspension by the rod so as to bring each of the optic axes into the centre of the field by intersecting the cross lines of the micrometer successively, the vernier arm of the goniometer being read for each position ; the difference between these readings is the observed angle between the optic axes, or twice that between either one and the first median line.

This angle so obtained, called the apparent angle of the optic axis, and indicated by the symbol $2\,E$, would be the same as the true angle $2\,V$, found by calculation from the three principal refractive indices, by the formulæ on p. 276, if the crystal plate were shaped either to a sphere having the first median line for a polar axis, or to a cylinder with its axis parallel to that of its mean elasticity, in either of which cases the rays emerge perpendicularly to the surface of separation, and therefore preserve their direction in passing from the crystal to the air. In all other cases there will be sensible deviation in the rays passing from the denser to the lighter medium, and therefore the apparent, will always be greater than the true angle of the optic axis, the amount of difference increasing with the absolute value of the latter, and the specific refractive energy of the crystal. We have already seen that of two rays originating by double refraction, whose paths are in the optic axial plane, one, vibrating parallel to the axis of mean elasticity, is in the condition of an ordinary ray having the constant refractive index β in any direction, while the other is an extraordinary ray in all directions but those of the optic axes, where its index also $= \beta$. It is only requisite, therefore, to know the value of the mean refractive index of the substance in order to convert the apparent into the real angle of the optic axes, or the reverse ; the relation of the angles V and E, or those of an optic axis to the median line, being expressed by

$$\sin. E = \beta \sin. V, \text{ and } \sin. V = \frac{\sin. E}{\beta}.$$

As the real angle of the optic axes may have any value up to 90°, it will often be found that the apparent angle exceeds a right angle, so that it is not always possible to see the rings or brushes about both axes at once : the field of the polarising instrument, as ordinarily constructed, taking in about 125° or 135° at the most. It also frequently happens with crystals whose optic axes are very divergent, especially when their refractive power is high, that the rings cannot be seen in the instruments, the rays following the optic axes being totally reflected in air. In such cases the angle can be measured by the deviation of the refracted rays, diminished by observing their emergence in a denser medium than air. This is done by placing a parallel-sided glass cistern, filled with a colourless oil whose refractive index is known, in the space between the condenser and objective of the instrument, into which the crystal plate is immersed ; the angle between the axes is then determined by turning the plate about its centre, and reading the goniometer as before. This gives the apparent angle in oil, $2H$, and from it, the refractive index of the oil $= n$, and β, the true angle is found by the formula :

$$\sin. V = \frac{n}{\beta} \sin. H.$$

The angle of the optic axes may also be calculated without knowing the refractive indices β and n, if a second plate can be obtained normal to their obtuse bisectrix or second median line, with which the obtuse angle is measured in oil. Calling this latter $2H'$, and the corresponding true angle $2V'$, their relation will be expressed, as in the preceding case, by—

$$\sin. V' = \frac{n}{\beta} \sin. H',$$

which requires a knowledge of the refractive indices. But as the sum of the acute and obtuse angles for the same coloured light is always 180°, $V + V' = 90°$ and $\sin. V' = \cos. V$. Substituting this value in the preceding equation, it becomes:

$$\cos. V = \frac{n}{\beta} \sin. H',$$

which, divided into the formula for the acute angle,

$$\sin. V = \frac{n}{\beta} \sin. H,$$

gives

$$\tan. V = \frac{\sin. H}{\sin. H'},$$

which expression determines the true angle of the optic axis without the use of any refractive indices; and if the acute angle can be measured in air, as the true angle is known, the mean refractive index can also be found, because,

$$\beta = \frac{\sin. V}{\sin. E}.$$

This method is one of considerable practical value, as it often allows the determination of the principal optical constants of a mineral to be made when the crystals are too small to furnish more than one prism of a sufficiently accurate form for direct observation, a very minute fragment of a parallel-sided plate, which need not be cut so exactly true for direction as the prism, being sufficient to show the interference figures in convergent polarised light.

The angle of the optic axes found by the above method is only true for the particular colour of the light used, and it will have some other value, either greater or less, for the other colours of the spectrum. The amount of the difference, which is known as the *dispersion of the optic axes*, is specific for any particular mineral, being sometimes only a few minutes, while at others it may be from thirty to forty degrees or more. Furthermore, the dispersion may take place in several different ways, each being characterised by a particular effect on the interference figures in white light.

In the rhombic system the median line is common for light of all colours, the axes being dispersed in the same plane symmetrically to it, and therefore the colours of the interference rings are arranged symmetrically to both arms of the cross when the plane of the optic axis is parallel to that of one of the crossed Nicols, but the colours of the individual rings vary with the dispersion. If the angle for red light be less than that for violet, which is indicated by the symbol $\rho < v$, the field enclosed by the polar rings about either axis will show a red margin on the inner and a blue one on the outer side, as in fig. 357, the colour being exactly similarly distributed on both sides of the horizontal line, and when the plate is turned through 45 degrees, the brushes will

FIG. 357.

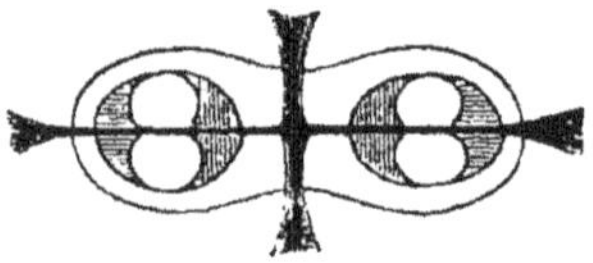

FIG. 358.

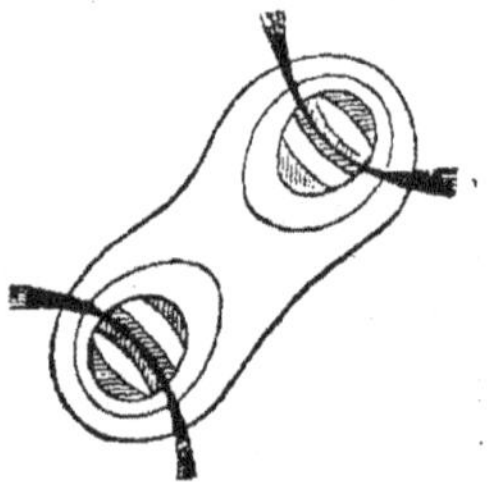

be bordered with blue on their inner or convex sides, and red on the outside, the colour within the rays remaining unaltered as in fig. 358.[1] This class of dispersion is characteristic of Nitre; while the opposite condition, where $\rho > v$ and the hyperbolas are bordered with red within and blue without, is seen in the allied minerals Aragonite and White

[1] The reader is recommended to colour these and the succeeding diagrams, when the differences will be more readily appreciated.

Lead Ore. In the latter, owing to the high specific refractive and dispersive power, the phenomenon is very strikingly shown, the brushes appearing as broad red and blue stripes, without an intervening dark space.

In a small number of crystallised substances belonging to the rhombic system, of which Brookite is the principal natural example, the axes of different colours, though having the same median line, are not only widely dispersed, but lie in different planes, those for red being in the plane indicated by one bar of the cross, while those for green are in the other bar at right angles to it. In white light these crystals give very peculiar interference figures, like fig. 359, the rings being replaced by a series of curves symmetrical to the cross upon a particoloured ground, the maximum of red being about the horizontal bar of the cross, and that of green near the vertical one. If, however, this is viewed in homogeneous red light, it is resolved into a series of lengthened oval rings, whose poles are at *r*, *r'*, while with green light another series, more nearly circular in form, are seen, having their poles at *r*, *r'*. The same phenomena are perhaps more strikingly seen in the triple Tartrate of Potassium, Sodium, and Ammonium, known as *Sel de Seignette*, which is not only more readily obtained than a crystal of Brookite, but, being colourless, shows the colours of the field more vividly. With a section of this salt the red and blue ring-systems may be seen superposed, when the middle part of the spectrum is extinguished by a tolerably thick cobalt blue glass.

FIG. 359.

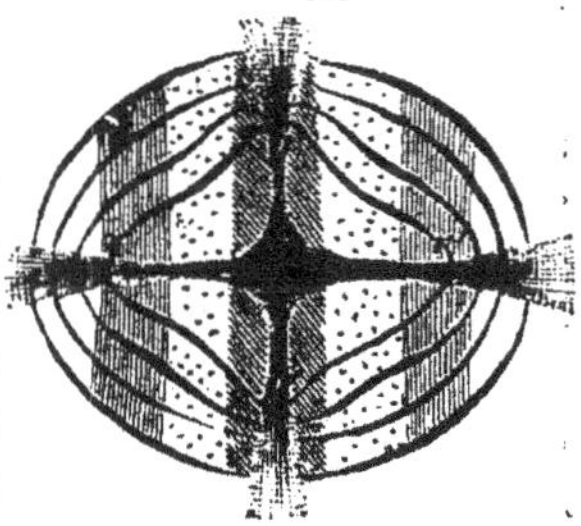

In the oblique system, the optic axes corresponding to different colours have not necessarily a common median line, and therefore a new element is presented for consideration, namely, the ***dispersion of the median lines.*** This may take place in three different ways, each having a more or less characteristic effect upon the interference figures, and to these the names *inclined*, *horizontal*, and *crossed dispersion* have been applied by Descloizeaux, who first systematically investigated them.

The first case, that of inclined dispersion, arises when the optic axes lie in the plane of symmetry dispersed at inclinations which are usually small, their median lines making angles with each other varying from a few minutes to one or two degrees. Of the latter, therefore, only one, usually that for the middle of the spectrum, can coincide with the normal to the plate, and consequently those for the other colours will be unsymmetrically placed right and left of the centre. When such a crystal is placed with its axial plane parallel to that of one of the Nicols, there will be often seen a marked difference in the shape and size of the polar rings, one being nearly circular, while the other is a lengthened

FIG. 360.

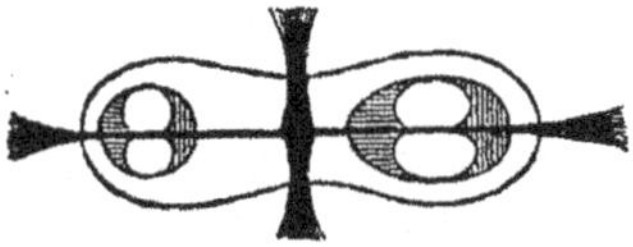

FIG. 361.

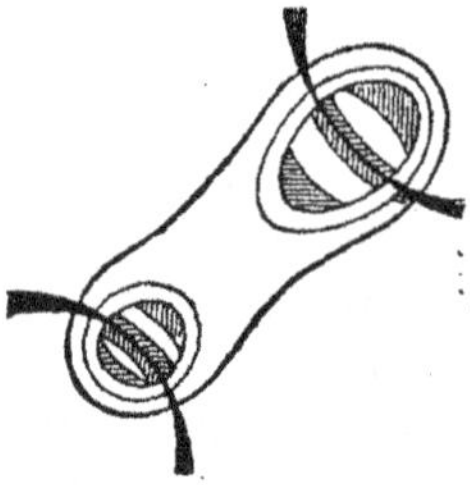

ellipse, as in fig. 360, the vertical bar of the cross is nearer to the latter than the former, and the colours are generally brighter about one pole than the other, the same order being observed when the plate is turned through 45°, as in fig. 361; but the colours of the brushes in the latter position may be either opposed, one being blue inside and red outside, and the other blue outside and red inside, or *vice versâ*, or similar, the phenomena being complicated by differences in the dispersion of the optic axes, as well as by that of their median lines. This kind of dispersion is best seen in Diopside, Gypsum and other minerals having large angles between their optic axes; the contrast between the form of the polar rings is seen in the artificial salt, Platino-Cyanide of Barium. When no polar rings are apparent, as in the small angled Felspars, there is only a slight difference of brilliancy of the colours of the brushes, which is not clearly appreciable without practice in observing.

When the plane of the optic axis is perpendicular to the plane of symmetry and parallel to the orthodiagonal, the latter may be either perpendicular to the first meridian line, or parallel to it. The former position corresponds to the second case, or that of horizontal dispersion, the axes corresponding to one colour only, being situated in a plane parallel to the orthodiagonal, while those for other colours are in planes making a slightly different angle with the clino-diagonal axis, but all their meridian lines lie in the plane of symmetry, so that the horizontal bars of the cross, in the interference figures for the extreme colours will not lie in the same line, but will be parallel to each other with more or less horizontal displacement. If therefore the plate is cut true for a mean colour, as yellow or green, the horizontal bar in the position of greatest obscuration will be bordered

FIG. 362.

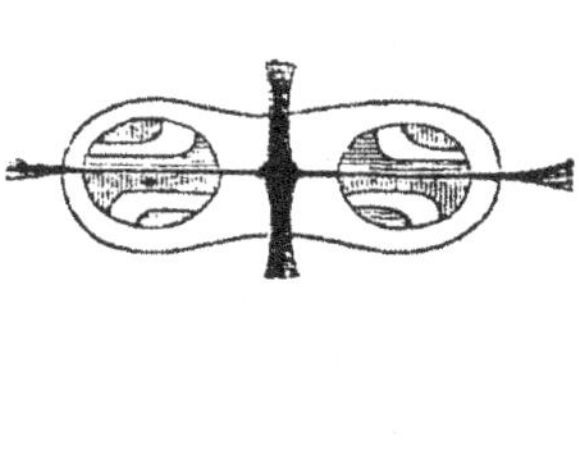

FIG. 363.

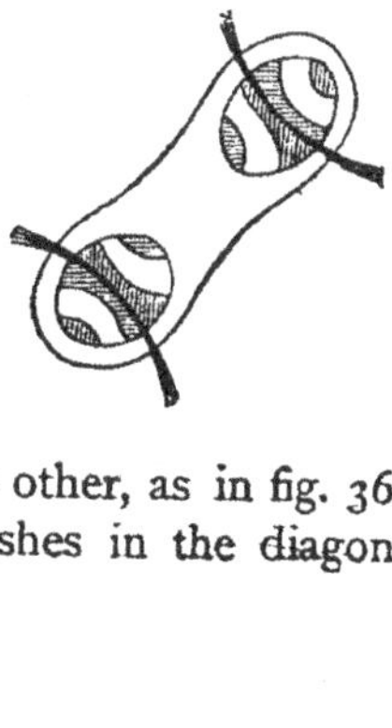

with red on one side and blue on the other, as in fig. 362, the corresponding colours of the brushes in the diagonal position being seen in fig. 363.

In the third case, that of crossed dispersion, the median lines for all colours coincide in direction with the orthodiagonal, but the planes of their optic axes cross at small angles. The colours of the rings and brushes are therefore disposed chequerwise, as in fig. 364, the maximum of red in the left

FIG. 364.

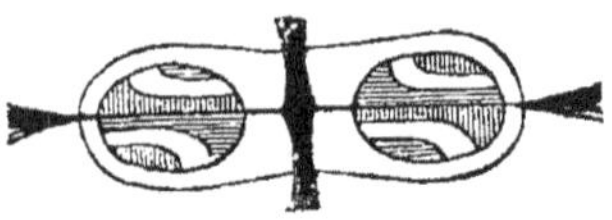

FIG. 365.

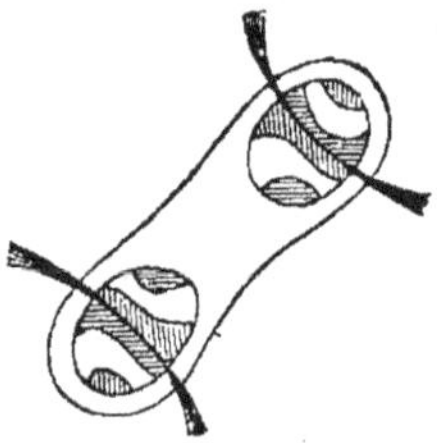

polar field being above the horizontal bar and to the left, and that of blue, below and to the right; while in the left one the positions are reversed, a similar contrasted arrangement is also apparent in the diagonal position, fig. 365. This kind of dispersion is well seen in Borax crystals, and to a lesser degree in Gay Lussite.

In the triclinic system no direct relation is apparent between the crystalline form and the interference figures, two or more kinds of dispersion being sometimes seen in the same crystal.

Determination of the sign of the double refraction. The positive or negative character of a doubly refracting mineral may be determined in most cases from a section showing its interference rings, by placing below the analyser a plate of another mineral whose sign is known, and observing the effect upon the figure. With uniaxial crystals the simplest method that can be used is to place above the section under examination a plate of some other uniaxial crystal, when, if both have the same sign, they will act together, producing the effect of an apparent thickening of the lower plate, and the interference rings will appear closer together; but if they are of dissimilar signs, one will partially neutralise the effect of the other, and the rings of the first will be expanded as though its thickness had been reduced. This method is, however, but of very limited application.

When a section of a uniaxial crystal, parallel or but slightly inclined to the optic axis, is examined in convergent polarised light between crossed Nicols, no coloured rings are observed except it be very thin, when a series of hyperbolic bands symmetrical to a central cross appear. If, therefore, a plate of sufficient thickness be placed so as to give a field of maximum brightness, in which position its planes of vibration are at 45° to those of the Nicols, and a tapered wedge of Quartz, whose length is parallel to its optic axes, be passed below the analyser, first in the direction of one vibration plane and then of the other, the hyperbolic bands will be seen in one or other position. The reason of this is, that Quartz being positive, its optic axis coincides with that of its minimum elasticity, and the particular direction in which it is without effect or appears to augment the thickness of the plate is the axis of like elasticity in the latter, while the other direction at right angles to the first is that of maximum elasticity in the plate, and opposed to that of the Quartz, which therefore acts as though the plate were thinned. If, therefore, the latter direction is that of the optic axis, the double refraction of the crystal is opposed to that of the Quartz, or is negative, while in the other case it is similar, or positive. The particular thickness required for this so-called compensating action varies with that of the plate, and therefore a certain length of Quartz is required to give the necessary range. As ordinarily made, the wedge is about $1\frac{3}{4}$ inches long and $\frac{1}{10}$ inch at the thickest end.

The same method is applied with sections of biaxial crystals, showing the interference rings and brushes by inserting the Quartz wedge, first in the direction of the line joining the optic axis, and then at right angles to it. If the rings appear to expand in the first position the crystal is positive; but if they are unchanged or completely effaced, it is negative, and the expansion or production of the hyperbolic bands will take place at right angles to the axial plane.

A third method which is specially applicable to uniaxial crystals and to biaxial ones in sections that are too thin to show the polar rings, depends upon the use of circularly polarised light, either in the polariser, or the analyser. If a plate of biaxial Mica sufficiently thin to produce a retardation of phase of exactly a quarter wave length between the refracted rays be placed in the path of a plane polarised beam, with the plane of its optic axes that of 45 degrees to the polariser, the light will emerge in a condition of circular polarisation, and the dark line marking the plane of vibration of the polariser will no longer be apparent.[1] A section of a uniaxial crystal viewed under these conditions no longer presents the concentric rings and dark cross between curved Nicols, but that shown in figs. 366–7, the rings being broken

FIG. 366.

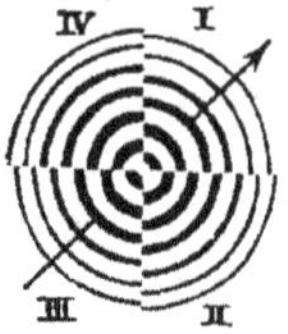

FIG. 367.

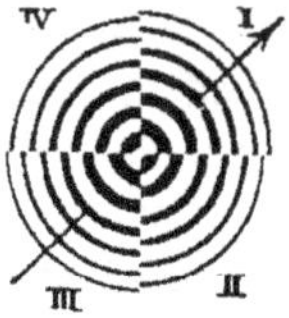

into four parts, which, when compared with their original forms, are expanded and contracted in alternate quadrants. The contracted parts having the greatest intensity of obscuration, those of the innermost ring will appear as two dark or coloured spots, and a line joining them will be either parallel or perpendicular to the axial plane of the Mica plate. If, therefore, the latter is always used in one position, with its axial line diagonal to the first and third quadrants of the circle if the spots lie upon it, as in fig. 366, the crystal is negative ; but when they are in the second and fourth quadrants, or at right angles to it, as in fig. 366, the crystal is positive. The same thing is observed with biaxial

[1] An explanation of this change of polarisation will be found in Spottiswoode's treatise on polarised light, p. 50.

crystals of small angle, but in those of wide angles the rings are expanded above and below the horizontal line, so that if the field be supposed to be divided into four quadrants as before, a negative crystal will show a spot above the line in the first, and below it in the third quadrant (fig. 368) ; while

FIG. 368.

FIG. 369.

in a positive one the spots will be below the line in the second, and above it in the fourth quadrant, as in fig. 369.

If the light be circularly analysed as well as polarised, by the use of a second quarter undulation plate below the analyser, the interference figures will appear as complete rings without any dark cross, and will behave in the same way as those observed in a circularly polarising crystal, the rings expanding or contracting as the analyser is turned one way or the other. If the polariser and analyser be placed parallel with the axial plane of the plate under examination in the vertical line of the field, and the two quarter undulation plates with their axial planes crossed at right angles or at 45° to that of the plate, the latter will, if negative, behave as a right-handed crystal, or the rings will expand when the analyser is turned to the right ; but if positive, the rotation must be to the left to produce the same effect. This is one of the best methods for use with crystals whose interference rings are small and close together.

Irregular polarisation. In some cases crystals belonging to the cubic system, when examined under polarised light, exhibit traces of double refraction not compatible with the assumption of uniform structure required by the system. The most notable examples are Alum, Boracite, and Senarmontite. These have been variously explained ; as for example, in Alum, by the assumption of lamellar structure due to the successive layers of the crystal not being in absolute contact, and therefore capable of polarising light in the same way as a bundle of glass plates, or by the existence of strains in the interior of the crystal, producing a structure analogous to that of unannealed glass, and in Boracite by the symmetrical inclusion of doubly refracting substances in very minute crystals. The polarising character of such minerals is not constant, differing in different parts of the same section, and being quite independent of any particular direction of the crystal. Similar irregularities are often observed in uniaxial crystals, the interference rings and cross behaving like those of biaxial ones of small angle. Here again, however, the disturbance is commonly local, and a part of the plate may often be found giving the proper figure, and in the disturbed figure the innermost ring is not a continuous curve as that of a truly biaxial crystal should be, but is made up of disconnected portions of circles of different radii.

A general explanation of the anomalous optical behaviour of minerals applicable to all the known cases has been put forward at great length in an admirable memoir by Professor Mallard,[1] who supposes these phenomena to be indications of polysynthetic structure, simple crystals, either asymmetric or of systems of low symmetry, by many repetitions producing groups which simulate simple forms of more complex symmetry.

The presence of foreign substances in transparent minerals, as well as of twinned structure, is often very strikingly shown in parallel polarised light. Fig. 370 is an example of a section of an apparently homogeneous Quartz crystal perpendicular to the optic axis, in which all the light parts have one rotation and the shaded ones the opposite, thus proving it to be a really complex twin of many right- and left-handed individuals with irregular contact surfaces.

[1] *Annales des Mines*, 3 sér. vol. x. p. 60.

Fig. 371 is a section of a twinned group of Aragonite, in which the planes of the optic axes in the alternate individuals are inclined to each other as shown by the lines and rings. If parts of the two adjacent bands be seen together

FIG. 370.

FIG. 371.

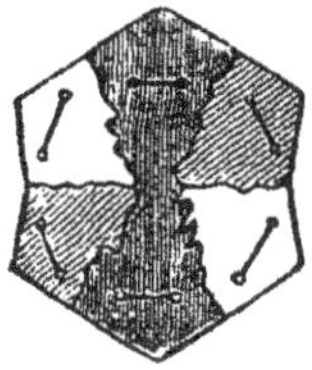

in convergent light, both systems of interference rings, crossed in direction, will often be apparent at the same time, but by shifting the plate slightly either to right or left one of them will disappear. The same thing is seen in Carbonate of Lead and Strontianite, minerals whose crystals are analogous in structure.

The subject of minute inclosures of foreign minerals, as well as that of the preparation of thin sections suitable for optical examination, has already been treated at length in the treatise on Rocks in this series, to which the student is referred. It may, however, be useful to remark that the sections required for showing interference rings are, as a rule, much thicker than those prepared for microscopic investigation under parallel light. In the greater number of the cases the most useful thicknesses will lie between $\frac{1}{8}$ and $\frac{1}{10}$ of an inch; but the actual size of the fragment, apart from its thickness, is immaterial; for example, the whole of the rings may be seen and the character of the double refraction determined on a plate of Mica of two or three hundredths of an inch on the side.

The preparation of sections for the polariscope is much facilitated by the existence of a perfect cleavage parallel to the optic axis in uniaxial, or to the first median line in biaxial, crystals, as with these cleavage plates will usually suffice without specially grinding or polishing. Among the former are the tetragonal variety of Sulphate of Nickel, which separates from a saturated solution at temperatures above 15° Cent.; the native Phosphate of Copper and Uranium also tetragonal. The transparent crystals of Molybdate of Lead (Wulfenite) from Utah, which are tabular to the basal pinakoid, may be used without any preparation; but as the specific refractive power is very high, only the thinnest will give a system of rings of any great width. Among biaxial substances the best examples are the species of the Mica group, which are susceptible of almost unlimited cleavage, and are therefore well suited for illustrating the alteration of the rings with the thickness of the plate. Topaz, Sugar and Bichromate of Potassium have perfect cleavages normal to one of the optic axes, so that from them plates may be easily obtained showing the rings about one axis only.

The following table shows the optical constants of the principal transparent minerals. It is for the most part compiled from that given in the 'Annuaire du Bureau des Longitudes,' and the works of Descloizeaux and Groth.

[*Note to page* 263.]—The optical elements of a biaxial crystal are related in the following manner:

Axes of elasticity	a :	b :	c
Refractive indices	(min.) α :	(mean) β :	(max.) γ
Velocities	$\frac{1}{\alpha}$:	$\frac{1}{\beta}$:	$\frac{1}{\gamma}$
Coefficients of elasticity	$\frac{1}{\alpha^2}$:	$\frac{1}{\beta^2}$:	$\frac{1}{\gamma^2}$

The true angle of the optic axes with the median line is found, when the three refractive indices are known, by the formula—

$$\text{Cos. } V = \sqrt{\frac{\frac{1}{\beta^2} - \frac{1}{\gamma^2}}{\frac{1}{\alpha^2} - \frac{1}{\gamma^2}}}$$

OPTICAL CONSTANTS OF THE PRINCIPAL TRANSPARENT MINERALS.

ISOTROPIC.

Amorphous and Cubic	Ray	Refractive index
Water	—	1·336
Hydrophane, dry	red	1·387
Do. saturated with water	,,	1·439
Hyalite	,,	1·437
Opal, iridescent variety	,,	1·446
Quartz, melted	,,	1·457
Fluorspar, green	,,	1·433
Alum	,,	1·458
Sylvine	yellow	1·482
Analcime	red	1·487
Plate glass (mean)	—	1·530
Agate, light coloured	,,	1·537
Rock Salt	yellow	1·543
Sal Ammoniac	,,	1·642
Boracite	,,	1·667
Spinal (rose colour)	red	1·712
Arsenious Acid	,,	1·748
Garnet Almandine	,,	1·772
,, Cinnamon Stone	,,	1·741
Senarmontite	,,	2·073
Zincblende (yellow)	,,	2·341
Diamond (colourless)	,,	2·414
,, (brown)	,,	2·487
Ruby Copper Ore	,,	2·849

ANISOTROPIC UNIAXIAL.

Tetragonal positive	Ray	Indices	
		ω	ϵ
Leucite	—	1·508	1·509
Apophyllite	red	1·5317	1·5331
Scheelite	,,	1·918	1·934
Zircon	,,	1·92	1·97
Phosgenite	orange	2·114	2·140
Anatase	—	2.554	2·493

Tetragonal negative	Ray	Indices	
		ε	ω
Mellite	yellow	1·525	1·550
Meionite	,,	1·560	1·595
Melinophane	red	1·592	1·611
Idocrase	yellow	1·717	1·719
Wulfenite	red	2·304	2·402

HEXAGONAL.

Positive	Ray	Indices	
		ω	ε
Ice (mean index) . . .	yellow	1·309	—
Quartz	,,	1·544	1·553
Parisite	red	1·569	1·670
Phenakite	,,	1·654	1·670
Dioptase	green	1·667	1·723
Greenockite	yellow	2·688	—
Cinnabar	red	2·816	3·142

Negative	Ray	Indices	
		ε	ω
Nitrate of Sodium. . . .	yellow	1·336	1·586
Hedyphane	red	1·463	1·467
Pyromorphite	,,	1·465	1·474
Calcite	yellow	1·486	1·659
Apatite.	—	—	1·657
Dolomite	,,	1·503	1·612
Nepheline	,,	1·537	1·542
Pennine	red	1·576	1·577
Emerald	green	1·578	1·584
Tourmaline (green) . . .	red	1·620	1·641
Corundum (Ruby) . . .	,,	1·759	1·767
Ruby Silver Ore (dark) . .	,,	2·881	3·084
Do. light (Proustite) . .	yellow	2·792	3·088

ANISOTROPIC BIAXIAL.

	Ray	Indices			Angle of optic axes		Dispersion
		Min.	Mean	Max.	Real 2 V	Apparent 2 E	
RHOMBIC-POSITIVE—					° ′	° ′	
Thenardite . .	red	—	1·470	—	—	—	
Natrolite . .	,,	1·477	1·480	1·489	59·29	94·27	
Struvite . .	,,	—	1·497	—			
Harmotome . .	,,	—	1·516	—			
Anhydrite . . .	,,	1·571	1·576	1·614	—	43·30	
Electric Calamine	yellow	1·614	1·617	1·636	46·09	78·39	$\rho > v$
Topaz (white) . .	,,	1·612	1·615	1·622	56·39	100·40	$\rho > v$
Celestine . . .	,,	—	1·625	—	—	89·36	
Barytes . . .	,,	1·636	1·637	1·648	—	63·12	$\rho < v$
Olivine (Peridote)	red	1·661	1.678	1·697	87·46	—	$\rho < v$
Zoisite . . .	,,	—	1·700	—		98·2	
Diaspore . .	yellow	—	1·722	—			
Chrysoberyl . .	,,	1·747	1·748	1.757	—	42·50	$\rho > v$
Staurolite . . .	red	—	1·753	—	—	—	$\rho > v$
Anglesite . . .	,,	1·874	1·880	1·892	66·40	89·49	$\rho < v$
Sulphur . .	yellow	1·958	2·038	2·240	69·40	70° to 75°	$\rho < v$
-NEGATIVE—							
Sulphate of Sodium .	,,	—	1·440	—			
Sulphate of Magnesium . . .	,,	1·4325	1·4554	1·4608	—	77·50	
Sulphate of Zinc .	,,	1·457	1·480	1·484	—	70·16	
Nitre	,,	1·333	1·5046	1·5052	7·12	—	
Aragonite . . .	,,	1·530	1·682	1·686	17·50	30·14	$\rho < v$
Andalusite . . .	red	1·632	1·638	1·643			
Autunite . . .	,,	—	1·572	—			
Cordierite . . .	orange	1·562	1·561	1·563			
Carbonate of Lead	yellow	1·804	2·076	2·078	8·07	16·54	
OBLIQUE-POSITIVE—							
Ferrous Sulphate .	,,	1·471	1·478	1·486	—	85·27	
Gypsum . . .	,,	1·521	1·527	1·530	—	61·24	inclined
Euclase . . .	,,	1·652	1·655	1·671	—	87·59	inclined
Anthophyllite . .	red	—	1·636	—	81·05	—	$\rho > v$
Diopside . . .	yellow	1.673	1·679	1·703	58·59	111·34	inclined
Sphene . . .	red	—	1·903	—	—	53·30	$\rho > v$
-NEGATIVE—							
Borax . . .	yellow	1·447	1·469	1·472	39·36	59·23	crossed
Adularia S. Gotthard.	,,	1·519	1·524	1·526	69·43	121·06	horizontal
,, Eifel . .	red	1·517	1·523	1·5240	13·34	20·45	inclined
Muscovite (ural)	yellow	—	1·541	1·574	40·21	64·14	$\rho > v$
Tremolite . . .	,,	—	1·622	—	87·31	—	inclined
Actinolite . . .	,,	—	1·629	—	80·04	—	inclined
Epidote . . .	red	1·731	1·754	1·761	73·36	—	inclined
Malachite . . .	—	—	1·88	—	43·54	89·18	inclined
TRICLINIC-POSITIVE—							
Sillimanite . . .	,,	—	1·66	—	44·0		
-NEGATIVE—							
Sulphate of Copper	yellow	1·516	1·539	1·546	—	96°	$\rho < v$
Axinite . . .	red	1·672	1·678	1·681	71·38	158·13	$\rho < v$
Amblygonite (Montebras) . . .	,,	—	1·592	—			
Cyanite . . .	,,	—	1·720	—			

CHAPTER XIII.

OPTICAL PROPERTIES OF MINERALS—*continued.*

Translucency. In systematic mineralogy, minerals are classified as transparent, semi-transparent, translucent in different degrees, and opaque, according to their power of transmitting ordinary light through their mass; these terms being used in the popular sense, without reference to the homogeneity or colour of the substance. The test of transparency is the power of discerning an object through a parallel-sided plate or crystal of a certain thickness. Rock-crystal, Calcite, Gypsum, and Barytes, and among the ores of the heavy metals, Zincblende in its lighter-coloured varieties, are among the most transparent substances known. When the object is only imperfectly seen, the substance is semi-transparent; when only a cloudy light like that seen through oiled paper or ground glass is transmitted, it is translucent; when no light is transmitted, it is opaque. These terms are to a certain extent relative, particularly in the lower degrees, where the thickness of the substance must be considered, especially when it is dark coloured. Flint and Obsidian, for example, are said to be translucent at the edges, or in thin splinters, while in thicker masses they are apparently opaque. Ferric oxide and its hydrates are also fairly translucent in minute microscopic crystals, but opaque when sufficiently large to be apparent without magnifying. Magnetite, on the other hand, does not appear to be susceptible of transmitting light under any condition, and is therefore opaque, as are also the native metals, and most of the heavy metallic sulphides.

Colour. When a transparent substance has the power of absorbing light-rays of different refrangibility unequally, it will, when viewed in ordinary light, appear to be of the colour of the light of greatest intensity transmitted. This may be a single colour or a mixture, its true nature being easily determined by examination with a simple spectroscope. In describing minerals, however, only the apparent colour, as seen by the unaided eye, is taken into account; such as are transparent and without selective absorption in white light being said to be *colourless*, while others are classified according to a scale laid down by Werner, which has been adopted by mineralogists in all countries, and is one of the few instances in which a uniform terminology has been obtained. It is founded upon the use of familiar coloured objects as standards of reference, distinction being made between metallic and non-metallic colours as follows :—

METALLIC COLOURS.

Copper red. As in metallic Copper and Red Nickel Ore.

Bronze red. Slightly tarnished Bronze and Magnetic Pyrites.

Bronze yellow. Perfectly fresh Bronze, and newly fractured Magnetic Pyrites.

Brass yellow. Freshly fractured Copper Pyrites.

Golden yellow. Pure unalloyed Gold.

Silver white and *Tin white.* These are used in a conventional sense for any brilliant opaque mineral without any strongly marked colour.

Lead grey. Galena, Antimony Glance.

Steel grey. Platinum, Fahlerz.

Iron black. Magnetite, Graphite.

NON-METALLIC COLOURS.

There are eight of these—namely, white, grey, black, blue, green, yellow, red, and brown, each being divided into numerous tints or varieties, in the following order, the purest or most characteristic tint being placed first.

Whites. Snow-white, yellowish-white, reddish-white, greenish-white, bluish- or milk-white, greyish-white.

Greys. Ash-grey, bluish-grey, greenish-grey, yellowish-grey, reddish-grey, smoke-grey, blackish-grey.

Blacks. Velvet-black, greyish-black, brownish- or pitch-black, reddish-black, greenish- or raven-black, bluish-black.

Blues. Prussian-blue, blackish-blue, azure, violet, lavender, smalt, indigo-blue, sky-blue.

Greens. Emerald-green, grass-green, verdegris, celadon, mountain-green, leek green, apple-green, pistachio-green, blackish-green, olive-green, asparagus-green, oil-green, siskin-green.

Yellows. Lemon yellow, sulphur, straw-yellow, wax-yellow, honey-yellow, ochre-yellow, wine-yellow, Isabella-yellow, orange-yellow.

Reds. Carmine, aurora or fire-red, hyacinth-red, brick-red, scarlet, blood-red, flesh-red, cochineal-red, rose-red, crimson, peach-blossom red, colombine-red, cherry-red, brownish-red.

Browns. Chestnut-brown, clove-brown, hair-brown, yellowish-brown, wood-brown or umber, liver-brown, blackish-brown.

According to intensity, colours are further qualified as light or dark, pale or deep.

As a distinguishing character of minerals, colour is of very unequal value; being constant, or showing but slight variation from a single tint in particular species, such as the native metals, the crystallised salts of Copper, and most natural metallic sulphides; while in the larger number of so-called non-metallic minerals a single species may, without any great variation of composition, be either colourless or show a considerable range of colours. In such cases, however, a particular tint may often be taken as an indication of partial replacement of one metallic constituent by another, as. for example, the Silicates of Magnesium and Calcium, which, when pure, are colourless, pass through various shades of green to nearly black, in proportion as Magnesium is replaced in part by the analogous dyad metal, Iron.

When Manganese is substituted in the same way, as in certain Micas and other silicates, they become red or purple, and so on in many other cases.

A mineral, ordinarily colourless, may also, if transparent, appear to be coloured, by reason of included foreign substances. Familiar examples of this are afforded by the numerous varieties of Quartz: the purest, or Rock Crystal, being colourless and transparent, while Amethyst, Chrysoprase, Cairngorm, and Eisenkiesel, show different colours, including purple, green brown, black, and red, due either to minute traces of metallic elements in combination, or to particles of ferric oxide, carbon, or other opaque substances, visibly included. Such minerals are said to be *allochromatic*, or adventitiously coloured, while those whose colour is uniform and due to their own proper absorption, are self-coloured, or *ideochromatic*. Strictly speaking, these latter are the only ones that can be properly said to be coloured minerals.

Streak. Useful aid in the determination of minerals is in many cases afforded by comparing the colour of the mass with that of the streak or powder produced by rubbing on a file or upon a piece of unglazed porcelain. With transparent minerals this is generally of a lighter colour, and with opaque ones darker than that of the mass. For instance, Gold, Copper Pyrites, and Iron Pyrites are often sensibly of the same brassy yellow tint, but the first gives a streak of its proper colour, while that of Iron Pyrites is black, and that of Copper Pyrites dark brown. Hydrated and Anhydrous Ferric Oxide, or, as they are commonly termed, brown and red Hematite, also differ sensibly in their streak, which is brown in the first and red in the second.

Pleochroism. The selective absorptive power producing colour is, in those substances that are isotropic, constant in any direction, so that they will appear of the same colour whether in ordinary or polarised light, but in anisotropic ones it may vary with the direction, so that a crystal may appear to be differently coloured by transmitted light, according to the direction in which it is viewed. This property, to which the general term of *pleochroism* is applied, depends upon the unequal absorptive capacity of the crystal for refracted rays vibrating in different planes in a manner analogous to the differences in optic elasticity. The most remarkable example is afforded by tourmaline, which absorbs the whole of the ordinary ray, whose plane of vibration is perpendicular to the optic axis, or the crystal is impermeable to light in the direction of that axis, while the extraordinary ray vibrating parallel to it passes freely, and in some cases almost without change of colour, in the direction of the lateral axes, the latter property being utilised as a method of obtaining a single plane polarised ray by double refraction and absorption in the polariscope known as the Tourmaline tongs. In less extreme cases, the absorption parallel to the axis will only be for particular colours, and the light transmitted will show the residual colour, while in a perpendicular direction the light will be of another colour, usually complementary, or nearly so, to the first. As the difference will be greatest in directions corresponding to maximum differences in velocity, a uniaxial mineral may have two distinct axial colours, or be *dichroic*, and a biaxial one *trichroic*, a different tint being apparent in the latter in the direction of each of the three principal axes of optic elasticity, when the light transmitted parallel to either of them is examined separately. This may be done by a Nicol's prism placed with its shorter diagonal parallel to the plane of vibration of each ray successively. For instance, a rhombic crystal whose axes of form and optic elasticity coincide, when viewed in the direction of the vertical axis, or through the base, shows the colours of the rays transmitted parallel to a lateral axis, when the principal section of the Nicol's prism is parallel to that axis, and those proper to the vertical and a lateral axis may be seen in the same way through either of the other pinakoids. Such differences of colour, though possible in all coloured doubly-refracting minerals, are not always apparent, the property being possessed in very unequal degrees by different minerals, and in many of them it is extremely feeble. In such cases it may often be rendered apparent when heightened by contrast, the colours proper to the two axes in the same section being seen side by side at the same time, which can be done by using a rhomb of Calcite or double-image prism instead of a Nicol's prism. The most convenient arrangement of this kind is Haidinger's Dichroiscope, fig. 372. It consists of a long cleavage rhombohedron of Iceland Spar, *a*, mounted in a tube, having a glass prism *b b′* of 18° refracting angle cemented to either end which parallelise the incident and emergent rays to the larger edges of the prism. A metal cap with a square hole, *c*, is fixed to one end, and a convex lens, *d*, is cemented to the glass prism at the opposite end, which brings the two images produced by double refraction in the Calcite exactly parallel to each other when viewed through the eye-piece cap *o*. In white light these images will be exactly alike except as regards intensity, but when a pleochroic mineral is placed in front of *c* they will be differently coloured, certain rays being extinguished in the ordinary beam, and their complementaries in the extraordinary one.

Fig. 372.

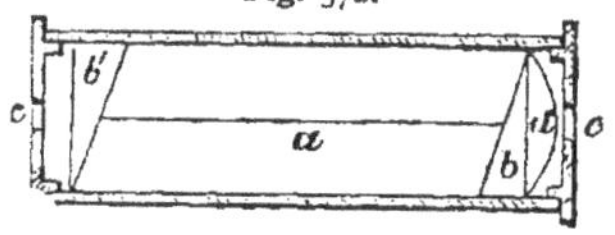

Some few minerals are so strongly pleochroic that the differences of tint are apparent to the unaided eye. Iolite or Dichroite, for instance, appears to be dark sapphire-blue in certain directions, and pale smoky grey or brown in others. These, however, are not pure axial colours, but are the approximate complementaries to those rays that are most completely absorbed, and for exact determination the use of polarised light is necessary. Zircon, Diaspore, Hornblende, the darker coloured Micas, certain varieties of Chlorite,

Andalusite, Axinite, and Epidote, are other examples of strongly pleochroic minerals.

An interesting application of the principle of selective absorption consequent on direction is afforded by Dove's test of the optical character of the Mica groups, many species of which give apparently uniaxial in erference figures when viewed in convergent polarised light, owing to the very small difference between their minimum and mean refractive indices. A polarising instrument is arranged for parallel light, and a plate of unannealed glass or calcite is viewed by a plate of mica used as an analyser. Supposing the latter to be truly uniaxial, its absorptive power will be equal for rays vibrating in any azimuth, and therefore no figure will be apparent, but if it be biaxial, however small the angle of its optic axis may be, every ray entering it will be divided into two, whose planes of vibration will be parallel to the axes of minimum and mean elasticity respectively, the one making the largest angle with the direction of the plane of the optic axes in the mica will be most completely absorbed, and the interference figure corresponding to the other will be rendered visible, though usually in very faint colours.

The success of the test depends on the power of absorption, and therefore it is only suited to those micas that are somewhat coloured and transparent in moderately thick plates, amongst which, however, the apparently uniaxial varieties are not generally found. The same remark also applies to the use of Tourmaline plates as polarisers, the colourless varieties, being almost without absorptive power, are comparatively valueless as compared with the brown or green ones.

Lustre. Light, when reflected from the faces of crystals or other surfaces, is partly returned in a regularly reflected beam, and partly irregularly reflected or dispersed, the joint effect of reflection and dispersion being to produce upon the surface the peculiar appearance known as *lustre*, *glance*, or *brilliancy.* In the definition of lustre, which is often very useful in the proper determination of minerals, two points are considered, namely, its quality or kind, which is specific and depends upon the refractive energy of the substance, and its intensity or degree, which varies in the same substance with the character of the reflecting surface.

The kinds of lustre, commencing with the highest, are:

1. *Metallic lustre.* This is the peculiar and brilliant appearance seen upon a perfectly polished metal surface. It is essentially characteristic of the native metals and heavy metallic sulphides, and of the few dark-coloured transparent ones, such as the Ruby Silver Ores and Cinnabar having refractive indices above 2·5.

2. *Adamantine lustre.* The typical example is the Diamond, but it is also characteristic of transparent minerals, whose refractive indices are from 1·8 to 2·5, which include the natural Sulphide, Carbonate, Chloride, Tungstate, and Molybdate of Lead. The heavy lead glass, known as flint glass or crystal, has also an adamantine lustre when polished.

3. *Vitreous lustre*, or that of a glass not containing Lead, is characteristic of most of the transparent crystals known as gems, whose refractive indices are below 1·8, Quartz or rock crystal being a most familiar example. Some of the minerals of this class, when dark-coloured or imperfectly transparent, show a resinous lustre, as that of boiled pine resin or *colophonium.*

4. *Fatty or greasy lustre* resembles that of a freshly oiled reflecting surface, and is characteristic of slightly transparent minerals, such as Serpentine, Nepheline, and Sulphur.

5. *Nacreous lustre*, or that of the Mother-of-Pearl shell, is a common characteristic of minerals having very perfect cleavages, and is best seen upon cleavage surfaces, such as those of Gypsum and Stilbite.

6. *Silky lustre* is essentially characteristic of imperfectly translucent and fibrous aggregates of crystals. The fibrous

variety of Gypsum known as Satin Spar, and the native Alums, are among the best examples.

There being no absolute standard of classification in regard to lustre, intermediate terms are often used in describing that of minerals which do not exactly correspond to a particular kind in the judgment of the observer. Thus anthracite is said to be semi-metallic in lustre, certain varieties of Carbonate of Lead metallic-adamantine, &c. The definitions based upon the refractive power of the substances given are those of Professor W. H. Miller.

It is not uncommon to find different faces of the same crystal different in lustre, and such differences are often of considerable value in fixing the position of the forms. Thus in Quartz, of the two rhombohedra making up the unit hexagonal pyramid, that considered as the positive one is generally brighter than the other; and the so-called rhombic faces, those of the acute pyramid of the second order $2P2$, are so much more brilliant than those of the associated forms that they may be often detected by the naked eye, even when extremely small.

As regards degree or intensity of lustre, minerals are said to be splendent, shining, glistening, or glimmering, as the proportion of dispersed to reflected light increases; when no distinct reflection is obtained, the substance is said to be dull. The terms are, however, even looser than those defining the quality of the lustre.

Iridescence. The appearance of a colour, either singly or in variegated bands and patches on their surfaces, when viewed under oblique incident light, is a well-marked character of many minerals, some of the most striking examples being furnished by the massive cleavable Felspar of Labrador, which very generally shows patches of deep blue alternating with green upon the brachydiagonal cleavage planes, and in rarer instances a much more extended range of colour, including rose-red and orange-yellow. Hypersthene also shows a bronze red tint upon the same surfaces. These appearances have been attributed by different investigators either to structures proper to the mineral, such as small pores or cavities regularly arranged, or to the interference phenomena produced by very minute crystals of Hornblende or other minerals, interspersed in the same regular manner. In other cases, minerals which are transparent and colourless, sparkle with a golden light by the reflection from interspersed opaque crystals, producing the so-called *avanturine* structure seen in Felspars and Quartz, the included substances being in the first case scales of Hematite or Goethite, and in the last flakes of a golden-coloured Mica. The large Calcite crystals from Lake Superior are sometimes coloured in the same way by small crystals of native Copper disseminated through them.

The brilliant iris of the Opal is attributed to the presence of minute faces crossing the mass in different directions, and in one variety known as Hydrophane, the opalescence disappears when the substance is soaked in water.

In minerals with very perfect cleavage coloured rings are frequently seen at different points in the interior. These are the ordinary colours of thin plates produced by minute films of air included between cleavage surfaces. They may be seen in almost any large clear specimen of Mica, Gypsum, or Calcite.

Surface iridescence or peacock colour is also due to the formation of very thin layers of one substance upon the surface of another, and therefore marks the commencement of alterations in the second. It is usually spoken of as iridescent tarnish, and is seen in many sulphides, such as Antimony Glance, Copper Pyrites, and Purple Copper Ore. The Hematite of Elba is also remarkable for the brilliant rainbow colouring often seen upon the crystals.

Asterism. Certain varieties of Corundum known as star sapphires, when ground to a spherical form and polished, show a pale blue six-rayed star when a strong light is reflected from the polished surface. This is due to repeated parallel twinning producing a finely lamellar structure, the laminæ of which act in the same way as the lines in a diffraction plate. The biaxial Mica of South Burgess, Canada, shows a similar but more sharply defined star when held in front of a candle flame or other luminous point. This is attributed to the inclusion of minute crystals of uniaxial Mica arranged along lines crossing at 60°. The same thing may be seen in thin sections of Labradorite, Aragonite, Proustite, Brookite, and generally in transparent minerals that either contain minute crystals of other substances symmetrically enclosed, or whose crystals show repeated parallel twinning. Thin plates of Aragonite and Strontianite, when held at a distance of about six or eight feet from a candle flame, often show an extended series of diffraction spectra on either side of the central image.

Fluorescence, or the property of rendering visible rays of higher refrangibility than are ordinarily apparent in white light, is not a very common property of natural minerals, but it is well marked in certain varieties of Fluorspar, especially the large transparent crystals from Weardale, which transmit an emerald or grass-green light, but reflect a deep sapphire-blue. More striking examples are furnished by the liquid hydrocarbons known as Petroleum, which are colourless, or transmit various tints of yellow to brown, but by reflection show the light blue rays above the violet of the ordinary spectrum.

Phosphorescence, or the power of emitting light in a dark place is characteristic of a certain small number of minerals, and may be variously developed by exposure to sunlight, friction, heating, or an electric discharge. The first method is completely effective with Diamond and some varieties of Zincblende, and less perfectly so with Strontianite and other earthy carbonates. The second method is applicable to Quartz, two pieces of rock crystal or any other variety of this mineral emitting a pale yellow light when rubbed together in the dark. This is still more markedly shown by loaf sugar. Heating is, however, the most generally effective method, as many minerals which are rendered phosphorescent when their temperature is raised are not so affected by sunlight. Fluorspar, when heated to 200 or 300 degrees Cent., becomes strongly luminous, the light being usually blue; but with the green variety known as Chlorophane it is of a brilliant emerald green. Phosphorite in the same way emits a yellow light. Topaz, Diamond, Calcite, and some silicates also phosphoresce when heated, but to a less degree than the typical examples, Fluor and Phosphorite. Phosphorescent minerals generally lose that property when strongly heated, but it may be more or less restored by subjecting them to a series of discharges from an electrical machine.

Crookes has shown that, when exposed to electric currents of high tension in an extremely rarified atmosphere, Ruby and Sapphire phosphoresce with intense red and blue, and Diamond with a vivid green light.

The best examples of phosphorescence are afforded, however, by the sulphides of the earthy metals—Barium, Strontium, and Calcium—which, though not natural minerals, are prepared by heating the native sulphates of these metals with carbon. Sulphide of Barium is the so-called Bologna phosphorus, and was the first substance in which the property of phosphorescence by sunlight was discovered. These substances are applied in the production of clock faces, which emit sufficient light to show the time in the dark.

CHAPTER XIV.

THERMAL AND ELECTRICAL PROPERTIES OF MINERALS.

Thermal relations of minerals. A crystallised substance placed in the path of a pencil of rays emitted from a heated body affects the latter in the same way as it would a beam of light—that is, it may either transmit them freely without sensible absorption, or it may absorb them entirely or in part. The former case, of which Rock Salt is the most complete example, corresponds to heat transparency, or *diathermancy*, and the latter, exemplified by Alum, to heat opacity, or *athermancy*. Sulphur and Fluorspar are also diathermanous, but less perfectly so than Rock Salt, while Tourmaline, Gypsum, and Amber are nearly as opaque to heat as Alum. Heat rays are also subject to the same laws of reflection, refraction, and polarisation, as those of light, being refracted singly by Rock Salt or other isotropic substances of sufficient heat-transparency, and doubly by those crystallising in the anisotropic systems. A parallel beam of heat rays from a Rock Salt lens falling upon two Mica plates is more completely transmitted by the latter when their planes of polarisation are parallel than in any other position, the amount being reduced to a minimum when they are crossed, showing an extinction exactly analogous to that of luminous rays under similar conditions.

The thermal conductivity of amorphous or cubic minerals is in like manner similar in any direction, while in those crystallising in the other systems it varies with the crystallographic symmetry, being more perfect in some directions than others, or, in other words, such minerals show axes of conductivity analogous to those of form and optic elasticity. The principal investigations upon this subject are those of De Sénarmont, who determined the conductivity in sections of crystals cut in known directions by coating one surface with wax,[1] and inserting into a hollow in the centre the point of a platinum wire heated by a lamp at some distance. When the heat received is transmitted equally in all directions, the wax will be melted in a circular patch around the wire; but if the rate is unequal, it will be an ellipse whose axes will correspond to those of maximum

[1] A modification of the apparatus in which the heating is effected by an electric current is shown in Tyndall's *Heat, a Mode of Motion*, p. 202, 4th edition.

and minimum rates of transmission in the particular section under investigation. In Quartz the conductivity is higher in the direction of the optic axis than at right angles to it, while in Calcite it is less, the two minerals in regard to this property being positive and negative in the same way as they are optically. In the rhombic system the three crystallographic axes are also axes of dissimilar thermal conductivity, while in the oblique and triclinic systems the latter, like the optic axes, are not directly related to the axes of form.

The dilatation of crystallised substances by heat is also, in all but those of cubic symmetry, attended by change of form, as the rate of expansion may be much greater in one principal crystallographic direction than another. This subject has been elaborately investigated by Fizeau, by a method of extraordinary delicacy, in which a plane surface of the crystal under investigation is covered by a very slightly concave glass plate, the curvature being so large as to allow the formation of Newton's rings when the surface is illuminated by a monochromatic yellow light. The crystal is supported upon a tripod of platinum, and heated in an air-bath, when if the expansion be uniform the surface will approach the glass regularly, and the rings change their positions symmetrically, but if it be unequal the surface will be distorted, and the distance from the glass will change more in some directions than others, so that the position of the rings will also be distorted, and by careful observation of these the alteration of form due to very small changes of temperature may be determined.

The general conclusions derived from these experiments are as follows:—

Cubic crystals have three axes of dilatation corresponding to those of form, the coefficient of expansion being similar for all, and consequently for any direction, in the crystal.

Uniaxial crystals have a principal axis of dilatation

corresponding to a particular coefficient of expansion, which may be either greater or less than that in directions perpendicular to it. In extreme instances the expansion may be positive in one direction and negative—that is, the substance may contract—at right angles to it; but, in any case, the arithmetical mean of the values obtained along three axes at right angles[1] to each other, will correspond to that observed along a line inclined at 54° 40′ to them. The relation of these lines is that of the trigonal interaxes, or the normals to the octahedral faces to the three principal axes of the cube.

In the rhombic system, the axes of dilatation, like those of optic elasticity, correspond to those of form; in the oblique system, one of the former is parallel to the axis of symmetry, the orthodiagonal, but the others make different angles, not only with the other crystallographic axes, but with those of optic elasticity and thermal conductivity. The relation of these different directions for Orthoclase is shown in fig. 373, taken from Fizeau's memoir, which represents the distribution of two of each of the three kinds of axes; E_2 and E_3 being two of the axes of optic elasticity, the first and second median lines of the optic axes; D_2 and D_3 two of the three axes of dilatation, C_2 and C_3 two of the axes of conductivity, and A $\bar{A}$ the clinodiagonal, and C $\bar{C}$ the vertical axis. In Gypsum the direction of the two axes of dilatation and the optic median lines very nearly coincide. In the triclinic system, no determinations of the axes of dilatation have been made, but from the analogy of those of optic elasticity they probably have no simple relation to the axes of form.

FIG. 373.

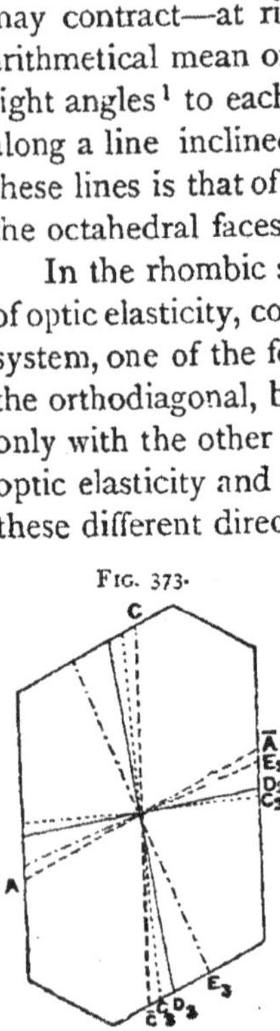

As a consequence of their unequal linear dilatation in

[1] These are in the tetragonal system three crystallographic axes, and in the hexagonal, the principal axis, one lateral axis, and the interaxis at right angles to the latter.

different directions, the geometrical characters of anisotropic crystals are subject to change with alterations of temperature. For example, Calcite elongates in the direction of the principal axis, and contracts at right angles to it, so that the angle of the polar edges of the rhombohedron becomes more acute by heating. The amount of this alteration was found by Mitscherlich to be 8′ for a range of 90°, the observed angle at 10° being 105° 4′, while at 100° it was only 104° 56′. Aragonite and Gypsum are examples of minerals in other systems whose rates of linear expansion in different directions are sufficiently different to show variations in the angles when heated. In the greater number of instances, however, the alterations are too small to be directly measurable, but the principle is important, as establishing the difference between substances crystallising in different systems, even when they may have forms of exactly similar geometrical characters. For example, the coincidence between a cube and a rhombohedron of 90° is only true for one particular temperature, as the latter will become either more acute or more obtuse when heated or cooled, according to the molecular arrangement of its substance, while a cube is rectangular at all temperatures.

The parameters of the unit or fundamental form of a series in like manner are altered by heating, but as the same change extends to all the forms of the series, these relations are not changed thereby. For instance, if the parameters of the form (1 1 1) be changed by heat from $a : b : c$ to $a' : b' : c'$, those of the form (2 2 1) will change at the same temperature from $a : b : 2c$ to $a' . b' : 2c'$, but the symbols will not be altered, as the new length of the vertical axis in the second will be twice the new lengths of the others. This is expressed in the statement that the principle of the rationality of the axes is independent of temperature.

The effect of heating substances is generally to increase their volume and diminish their density, and this is accompanied by a change in optical properties, more particularly in the refractive power, which is, as a rule, reduced, and in biaxial crystals, where the principal indices alter unequally, the change often affects the position of the optic axes. These changes are most apparent in the earthy and alkaline sulphates. In Barytes, Celestine, and Sulphate of Potassium, the inclination of the optic axis is increased, but in Felspar and Gypsum it is diminished, by heating. In the latter mineral the alteration is very considerable, even when the temperature is but slightly changed; the apparent angle of the optic axes which lie in the plane of symmetry is about 90° for red light at the ordinary temperature of the air, but at 116° it is 0, or the substance is apparently uniaxial. At higher temperatures the axes diverge again, but in a plane nearly at right angles to the original one. On cooling, the same changes take place in the reverse order. In some varieties of Felspar the change in angle of the optic axes is to some extent permanent—that is, it does not return exactly to the original value if the crystal has been exposed to a red heat.

Electrical properties of minerals. All minerals become electric by friction, but the positive or negative character of the electricity developed varies according to circumstances. For testing, an electroscope is used, consisting of a light metallic needle with a knob at either end, and suspended in the centre by a thread of raw silk, or balanced upon a steel point by an agate cap. This, when rendered positively or negatively electric by a rod of glass or sealing-wax, is either attracted or repelled by an excited mineral, according as the latter is charged with opposite or similar electricity. If the mineral under trial happens to be a conductor of electricity, it will be necessary to insulate it in order to obtain an effect upon the electroscope. Several minerals may be rendered electric by pressure, Calcite possessing this property in the highest degree; a cleavage fragment of Iceland Spar, when slightly pressed between the fingers, becoming positively electric. Topaz, Aragonite, Fluorspar, and Quartz are similarly affected, but in a less degree.

Minerals that become electric by change of temperature are said to be thermo- or pyro-electric. Axinite, Tourmaline, Calamine, Topaz, and Boracite are among those that show this property best. When the crystals show opposite electricities at different points, they are said to be polar, and the points at which the changes of electricity are observed are poles. These cannot, however, be distinguished as positive and negative, as both kinds of electricity are developed at either pole alternately, one during heating, and the opposite one in cooling. It is customary, therefore, to call the points which are rendered positively electric by heating or negative by cooling, 'analogue poles,' and those becoming negative by heat or positive by cold, 'antilogue poles.' The positions of these vary in different crystals. In the prominently hemimorphic species, Tourmaline and Electric Calamine, the property of thermo-electricity is most characteristically developed, and are the poles at opposite ends of the principal axis. Boracite has eight poles corresponding to the solid angles of the cube, while in Quartz the electric poles are situated at the ends of the three lateral axes.

Magnetic Properties of Minerals. Nearly all minerals containing iron are magnetic, but in very unequal degrees. Practically, only native Iron, Magnetite, and Pyrrhotine, or Magnetic Pyrites, are sufficiently magnetic to affect a compass needle upon a pivot, but with a more delicate astatic instrument many other minerals, including ferrous sulphates and silicates, show slight traces of magnetism. Native magnetic oxide of iron, or Lodestone, is always strongly magnetic, and is often found in masses showing distinct polarity, but natural magnets capable of supporting considerable weights are rare. Masses of Magnetite of a regular figure, not naturally polar, may, however, be rendered so by touch, in the same way as a steel bar is magnetised. The allied minerals, Chromic Iron and Franklinite, are generally magnetic, but it is not certainly known whether this is a special property or caused by finely interspersed Magnetite. Spathic Iron Ore, or Ferrous Carbonate, becomes strongly magnetic when heated to redness from the formation of magnetic oxide, as do most of the Sulphides of Copper and Iron when fused in an oxidising atmosphere.

For common purposes of testing, a magnetic needle with a brass or agate centre, to be used upon the same pivot as the electroscope, will generally be sufficient. When in use, it should be covered with a glass, to protect it from the disturbing action of currents of air.

CHAPTER XV.

CHEMICAL PROPERTIES OF MINERALS.

IN the determination of the nature of a mineral a knowledge of the kind of matter constituting it—that is, of the elementary substances it may contain, and the relative proportion of such substances to each other when more than one are present—is of primary importance. The technical methods whereby such knowledge is attained are known as qualitative and quantitative chemical analysis, and it will be assumed that the reader is familiar with the details of such methods, as laid down in the treatise on chemical analysis by Professor Thorpe in this series, or in some other standard work on the subject.

A certain knowledge of qualitative analysis is of great use in the identification of minerals of obscure habit, and for this purpose a course of simple testing by methods not requiring the resources of a complete analytical laboratory has been developed by mineralogical chemists, the results obtained by such tests being given in systematic descriptions of minerals under the head of chemical characteristics, or some equivalent term.

The methods employed in testing are divisible into two groups, the first involving the application of heat to minerals, either alone or in the presence of certain reagents, being known as analysis by the dry way, and the second, where liquid solvents or reagents are used, as the wet way. In the former, or dry way, the mineral under examination is subjected to the flame of a lamp, urged by a blast of air, which is usually produced by the mouth blowpipe, although the blast of a bellows or the flame of a Bunsen burner may also be used. The latter arrangements are, however, only to be found in laboratories, while the blowpipe is for the mineralogist essentially a portable instrument, and may, with the necessary apparatus and fluxes, be packed into a case of small volume, forming a portable laboratory, ready for use with very small preparation, and as such is invaluable to the travelling mineralogist.

The most convenient form of blowpipe is that of Gahn, as modified by Plattner. This consists of a brass or German silver tube, from 8 to 9 inches long, diminishing in diameter from one-third of an inch above to one-sixth or one-seventh below. A trumpet-shaped mouthpiece is fitted to the larger end; the smaller one fits air-tight in a cylindrical chamber about half an inch in diameter, from which a jet-pipe about an inch long projects at right angles; this is also slightly coned, from one-sixth to about one-tenth of an inch diameter, and is terminated by a conical nozzle of platinum perforated with a hole about $\frac{1}{50}$ inch diameter. For some purposes, a second nozzle, with a larger aperture, is desirable. All the parts are ground together so as to fit air-tight without screws. The trumpet-shaped mouthpiece is more convenient in use than the cylindrical form, from the support given to the lips when the blowing is carried on continuously for some time. Of the other forms of blowpipe, that known as Dr. Black's is the best, when made of an appropriate size, the great fault of most of the cheaper instruments being the disproportionately small diameter of their tubes.

The flame used with the blowpipe may be either that of a candle, a lamp, or gas jet, the best being that of an oil lamp with a flat wick, the top of the wick being cut with a forward slope of about 20°. When in use, the jet of the blowpipe is held parallel to the slope of the wick, and a current of air is forced into the flame by the action of the muscles of the cheeks, breathing being kept up through the nostrils in a manner which, though difficult of description, is easily learned with a little practice. When the jet is placed so that the air enters the flame at the higher side of the wick, a large proportion of the unconsumed gases of the dark interior part is carried forward, producing a pointed flame of a certain brilliancy, which is of a neutral or non-oxidising character, and is called the reducing flame. If, however, the point is laid above the middle of the wick, so that the air is brought into contact with the dark part of the flame, total combustion of the gases is produced, and the resulting flame is small, slightly luminous—having the characteristic blue colour of burning carbonic oxide, and much hotter than the reducing flame. This is called the oxidising flame, as any substance heated in it is subjected to the full effect of the unconsumed oxygen of the blast or of the adjacent atmosphere. The maximum oxidising effect is obtained immediately before the point of the flame. The power of producing a clean flame, or one that is entirely reducing or oxidising, is one of the first essentials to success in blowpipe work, and should therefore be practised by the learner, with the tests recommended by Plattner. These are borax beads, saturated with molybdic acid and oxide of manganese respectively; the former becomes colourless when melted for some time in an oxidising flame, but turns black in a reducing flame; while the latter is of a dark violet colour in the oxidising flame, which is entirely discharged by the reducing flame. To obtain these effects perfectly by the use of the blowpipe alone, considerable nicety of manipulation is required. It is also essential that the metallic oxides used should be pure, especially that of manganese, which must be free from iron, otherwise the action of the reducing flame will be obscured by the bead taking a more or less green tint.

The apparatus required (in addition to the blowpipe and lamp) is as follows :—

A pair of forceps with platinum points, closing by a spring. Pieces of platinum wire about three inches long, bent into a loop, about one-eighth of an inch diameter, at one end. A small platinum spoon, and a piece of platinum foil. A jeweller's hammer, small bright steel anvil, and an agate mortar. Pieces of glass tube, about a quarter of an inch bore, in lengths of three inches, open at both ends, and a few shorter pieces closed at one end. A few watch glasses; a small bar magnet, which may have a chisel point at one end. Charcoal for supports: the best are pieces of straight grained pine charcoal, about three inches long, and three-quarters of an inch to one inch square; these are, however, with difficulty obtainable, and should therefore be used carefully. For most purposes, artificially moulded blocks, formed of charcoal powder cemented with starch and subsequently carbonised, are sufficient; they are made in various sizes, and may be purchased of dealers in chemical apparatus. Hard-wood charcoal, made from brushwood, is to be avoided, as it usually decrepitates when heated, besides leaving a large amount of ash. The use of a plate of aluminium as a support for minerals giving coloured sublimates has been recommended by Colonel Ross.

The most essential fluxes, or reagents, are :—

Borax, calcined, but not fused into a glass; phosphorus salt (ammonic sodic phosphate); dried carbonate of soda, which must be free from sulphate; nitre; bisulphate of potassium; nitrate of cobalt; fluorspar; cupric oxide; and test papers, both of litmus and turmeric. In addition to these, a few liquid reagents, such as hydrochloric and nitric acids, and ammonia, are useful, although it is better to dispense with them as much as possible, as they cannot be easily carried when travelling.

The course of operations followed in a systematic examination of minerals by the dry way is the following :—

1. *Heating in tube closed at one end.* A fragment of the mineral (generally called the *assay*) is placed at the bottom of a tube closed at one end, and heated first over the flame of a lamp, and subsequently in the blowpipe flame. Hydrated oxides and salts by this means give off water, which condenses in visible drops on the cooler surface of the tube. Nitrates and the higher oxides of manganese give off oxygen, which can be recognised by the ignition of a glowing splinter of wood. A few minerals, such as Sulphur, Arsenic, Antimony, Mercury, and their oxides and sulphides, sublime without residue ; but are redeposited in the case of antimony and arsenic in the form of black metallic mirrors, a short distance above the assay, while the sulphides of the latter element are recognisable by the red or yellow colours of their sublimates. The higher sulphides, such as iron pyrites, give a sublimate of sulphur ; arsenical iron pyrites gives both sulphur and arsenic, recognisable by the red and yellow sublimates, corresponding to the minerals Realgar and Orpiment. These, however, are products of the decomposition, and do not exist as such in the mineral. Aragonite, Calcite, Magnesite, and Dolomite give off carbonic acid, with the formation of caustic lime and magnesia, when strongly heated, while Siderite leaves a residue of magnetic oxide of iron. The carbonates of zinc, copper, and lead are also decomposed, producing oxides of the metals. The sulphates of alumina and ferric oxide give off sulphuric and sulphurous acids, the acid character of the vapours being recognised by a slip of test paper inserted in the mouth of the tube.

2. *Heating in the open tube.* The fragment of mineral under examination is placed about half an inch above the lower end of the tube, which is held in a slightly inclined

position, and the blowpipe flame is directed upon it, so that the assay is heated in a full current of air, when sulphides give off sulphurous acid, which is easily recognised by its characteristic odour. Arsenides and selenides give the peculiar odours characteristic of Arsenic and of Selenium, and sublimates of their oxides, which deposit at a greater or less distance from the assay, according to their degrees of volatility. Antimony gives a similar sublimate of antimonious acid. Many sulphides which do not give an indication of sulphur in the closed tube are decomposed with the formation of sulphurous acid in the open tube.

3. *Fusibility.* For this test a fine splinter of the mineral, held in the platinum forceps, is exposed at the point of maximum heat in the oxidising flame. This supposes it to be very refractory; but in some instances exposure to the flame without blowing is sufficient to effect fusion. The range of the melting points of minerals being very great, even excluding those that are fluid at ordinary temperatures—water, mercury, &c.—it has been proposed by Von Kobell to express fusibility by a scale of typical minerals, analogous to that employed in describing hardness. This scale comprises the six following numbers; but, owing to the indefinite nature of the gradations between the different numbers, it is not much used :—

(1) Antimony-Glance. Fuses readily in the flame of a candle.

(2) Stilbite. Fuses without the help of the blowpipe.

(3) Almandine Garnet. A large or thick fragment can be melted before the blowpipe.

(4) Actinolite Hornblende of Zillerthal. Thin splinter melts in extreme point of the oxidising flame.

(5) Felspar (Adularia of S. Gothard). Similar to No. 4; but fuses with greater difficulty.

(6) Bronzite of Kupferberg. Thin splinters can only be rounded on the edges.

When the edges of a splinter cannot be rounded or

softened in the hottest part of the flame, the mineral is said to be infusible. This statement is of course only true in relation to the means employed: many substances that are infusible by the mouth blowpipe can be easily melted by the oxyhydrogen blowpipe or other powerful sources of heat.

The manner of fusion, as well as the character of the fused mass, is a point of considerable importance, and must be carefully observed, some minerals fusing quietly to a glass or enamel, while others intumesce or swell up, from loss of water or other volatile components, and become scoriaceous or slaggy masses; others, again, give a mass which crystallises on cooling.

It is important in trying fusibility to direct the flame upon a fine point of the fragment under examination, so as to induce fusion as quickly as possible, otherwise an oxidising action, interfering with the result, may be set up. Thus the double sulphides of copper and iron are fusible, but if heated for some time in air below their melting point, they lose sulphur, and give a residue of oxide of copper and magnetic oxide of iron, which is practically infusible. In the same way, silicates containing ferrous oxide, although fusible, may by calcination be partly resolved into silica and ferric oxide, both of which are infusible.

When trying the fusibility of easily reducible metallic minerals, care must be taken that the points of the platinum forceps in contact with the as ay are not strongly heated, as an alloy of the platinum with the more fusible metal may result. It is safer in such cases to try the fusibility upon charcoal.

Flame colour tests. These may often be observed simultaneously with the trial of fusibility, the splinter of mineral used for the latter purpose in some instances giving a characteristic colour to the flame when first exposed to the point of the oxidising flame. It is more satisfactory, however, to make a special trial. Strontium, Lithium, and Calcium minerals give a red colour, which is best seen when the mineral, after being previously heated to redness in the reducing flame, is moistened with hydrochloric acid, and exposed in the outer blue envelope of the flame, without blowing. This is especially the case with the sulphates of barium and strontium, which may be partially reduced to sulphides in the reducing flame upon charcoal; the sulphides being converted to chlorides, by moistening with hydrochloric acid, give compounds which are eminently volatile at the high temperature of the blowpipe flame. Lithium gives a brilliant crimson; calcium a yellowish red; and strontium a purplish red tint. When the two latter substances are together, the colour due to lime is seen first, and then that of strontia. When the red flame is examined by a dark blue glass screen, the light due to lime and lithia is almost entirely extinguished, while that of strontia is but little altered. Sodium compounds, even in very minute quantity, colour the flame a deep yellow, which completely effaces the light due to other volatile bodies; the yellow colour may, however, be cut off by the blue glass.

Minerals containing barium give a feeble yellowish green colour to the flame when very strongly heated, especially when moistened with hydrochloric acid. Copper gives a bright emerald green, except in the state of chloride, when the flame is blue with a purple border. Phosphates, when moistened with sulphuric acid and exposed to the outer part of the flame, show a momentary coloration of a pale, bluish green; and borates, when similarly heated, give a brighter green colour.

Potassium gives a very characteristic reddish violet colour to the flame, which is completely hidden by even a very small proportion of sodium. If, however, the yellow light of the latter is cut off by the blue glass, which allows the violet rays to pass freely, a small quantity of potash may be detected even in the presence of a relatively large amount of soda.

In order to determine the presence of alkalies in silicates which are not decomposable by acids, they should be previously heated with fluoride of ammonium in the platinum spoon to decompose them, when the bulk of the silica is volatilised, and the residue can be examined in the flame with the blue glass. Chloride of calcium may be used for the same purpose, as the lime coloration is but slightly transmitted through the glass.

The alkaline metals may be determined with more certainty by the spectroscope,[1] the characteristic bright lines being, for their oxides—

Sodium, a bright yellow line.

Calcium, one green and one red line.

Lithium, a red line, which is at a greater distance from the soda (D line) than that of lime. This is cut off by the blue glass.

Potassium, a dull red line beyond that of Lithia; this passes through blue glass.

Strontium, an orange line near the D line, several red lines, and a blue line.

Barium, a group of several green lines close to each other.

The rare alkaline metals, Cæsium and Rubidium, are also readily detected by their spectra; while with the ordinary tests they may be confounded with Potassium and Lithium. For the detection of boracic acid in silicates, such as Axinite and Tourmaline, the mineral is heated on a platinum wire, with a mixture of Fluorspar and Bisulphate of Potash, when the characteristic green line is produced.

Heating on charcoal. For this purpose a rectangular prism, cut from a piece of straight-grained charcoal, having the rings of growth perpendicular to the ends and parallel to the length on two faces, is preferable to a moulded charcoal

[1] Browning's small direct-vision spectroscope with photographed micrometer, mounted on an upright pillar, is a convenient form of instrument for this purpose.

block. A slight depression being made by scraping the surface with the point of a knife near one end, the assay fragment is placed in it, unless the mineral decrepitates by heat, when it must be previously powdered, and the flame is directed downwards upon this point, the charcoal being held with a slight upward inclination, with its length parallel to the direction of the flame. The points to be observed are fusibility or infusibility, the production of phosphorous or arsenical odours in the same manner as in the open tube, change of colour, or the production of an alkaline mass. The latter reaction is characteristic of Calcite and Aragonite, which, when strongly heated, leave an infusible residue of caustic lime, giving an alkaline reaction with moistened turmeric paper. Many sulphides and other compounds containing iron are converted into infusible masses, which show magnetic properties. The principal special reaction upon charcoal, however, is the production of coloured incrustations of the oxides of volatile metals, which are produced from their combinations, either with or without an actual metallic residue. These incrustations cover the surface of the charcoal, commencing at a distance of an inch or less from the assay, according to the volatility of the metal. The deposit produced by Zinc minerals is yellow when hot and turns white in cooling ; that of Cadmium of a brownish yellow usually more or less irised, neither giving a globule of metal. Antimony and Bismuth compounds yield brittle metallic globules, the former with a thick white incrustation edged with blue, and the latter one of lemon yellow colour. Lead minerals are easily reduced, giving a malleable globule of metal, and a yellow incrustation ; but when notably argentiferous, the incrustation is more or less crimson at the inner edge. Minerals containing Tin, Copper, or Silver as principal constituents are reduced to the metallic state without producing an incrustation on the charcoal.

Tests with soda upon charcoal. The whole of the reactions depending on the reduction of metallic minerals and the formation of coloured sublimates on charcoal may be facilitated by adding a small quantity of carbonate of soda to the substance under examination, which forms a slag with the infusible constituents ; and with minerals containing Tin a small quantity of cyanide of potassium should be used to facilitate the reduction and to protect the reduced metal from oxidation. The most convenient way of applying this flux is to mix the assay in a finely powdered state into a paste with about its own volume of dried carbonate of soda upon the palm of the hand by adding a few drops of water. The mixture is made and removed by a small iron spatula or the blade of a knife, and the paste is spread over the charcoal, care being taken to heat it up gradually, so as to dry it before exposing it to the full strength of the flame. When sufficiently large, the metallic globules produced are tested as to their brittle or malleable character by the hammer and anvil ; but when, as generally happens, they are small and interspersed through a mass of slag, the assay with the adjacent portion of the charcoal must be removed and pulverised in the agate mortar. The charcoal is then removed by carefully washing the contents of the mortar in a gentle current of water, leaving the reduced metal, which, if malleable, will be found in flattened scales or spangles.

Other special uses of soda are in the decomposition of infusible silicates, the detection of Manganese, and of Sulphur in insoluble sulphates, or the lower sulphides that do not give a sulphur sublimate when heated. When Quartz or an infusible silicate, powdered and mixed with carbonate of soda, is heated on charcoal, the mixture effervesces from the escape of carbonic acid, and a fused alkaline silicate is formed, which is soluble in water, and may be decomposed by mineral acids with the production of silica in the gelatinous or soluble form, which may be rendered insoluble by heating to redness. This is the ordinary method adopted for rendering these minerals soluble for analysis in the laboratory, and the method may be sometimes adopted with advantage on the small scale by the blowpipe. Minerals containing manganese, when fused with soda in a full oxidising flame, give a green enamel-like mass of manganate of sodium. This test may be made upon charcoal, but generally some nitre is added to the mixture, and the fusion is effected on platinum foil. The test for sulphur consists in fusing minerals such as the sulphates of barium, strontium, or calcium with soda upon charcoal in the reducing flame until the bulk of the melted flux is absorbed by the coal, the slaggy matter remaining, and then the adjacent portions of the charcoal are cut out and placed upon a bright coin or plate of silver, with the addition of a few drops of water, in which the alkaline sulphide formed dissolves, producing a dark brown or black stain of sulphide of silver upon the metal. This is a very simple and delicate test; care must, however, be taken that the soda used is free from sulphates, which point must be previously determined by testing it upon the metal alone.

Fritting tests on charcoal. Some infusible metallic oxides which under ordinary circumstances are colourless, become coloured when moistened with a solution of nitrate of cobalt and strongly heated upon charcoal. The colours are blue with alumina, flesh red or pink with magnesia, and green with oxide of zinc. The first and last of these are in fact the ordinary Cobalt blues and greens used as water colours. It is essential for this test that the substance tried should be infusible, or otherwise a cobalt blue glass or enamel will be formed.

Tests with vitrifiable fluxes. The salts used for this purpose are Borax (di-sodic borate) and microcosmic salt or salt of phosphorus (ammonio-sodic phosphate), which melt into a clear glass at a red heat. The active agents are in either case the combined equivalent of boracic and phosphoric acids respectively, borax being an acid salt, and phosphorus salt, though a neutral or bibasic phosphate, loses its ammonia when heated, becoming a bibasic sodic phosphate with one atom of phosphoric acid free. The acids themselves may therefore be, and are sometimes, used instead of their sodium salts, but they are inconvenient from their excessively hygroscopic properties.

The sodium salts of boracic and phosphoric acids have, when melted, the power of dissolving up nearly all metallic oxides, producing glasses which are characteristically coloured by even very minute quantities of such oxides as are of strong colouring power. The test with borax is one of the most useful in the whole series of blowpipe operations, and is performed by melting in the loop of a platinum wire a bead of borax, which should be perfectly colourless both hot and cold. A small quantity of the mineral to be tried, preferably in a fine powder, is then added, the bead is remelted in one or other of the blowpipe flames until the substance is completely dissolved, when the bead is allowed to cool, and the colour due to the particular oxide and flame employed will be recognised. The following are the most characteristic reactions obtained in this way with compounds of different metals and borax :—

Iron : In reducing flame dark (bottle) green ; in oxidising flame, yellow while the bead is hot, but becomes nearly colourless when cooled.

Manganese : Colourless in reducing flame, deep violet or amethyst in oxidising flame.

Chromium : Grass green in both oxidising and reducing flame.

Uranium : Green in reducing, yellow in oxidising flame.

Cobalt : Deep blue in both flames. This colouration is produced by an exceedingly minute quantity of oxide of cobalt, and is probably the most delicate of all the tests for this metal.

Nickel : Reddish to brown hot, yellowish to dark red cold in oxidising flame, which is rendered blue by an addition of nitre. In the reducing flame the colour disappears, the bead becoming grey with finely divided metallic nickel.

These reactions refer to chemically pure preparations in minerals; they are generally obscured by the presence of cobalt.

Copper: Green while hot, turning blue on cooling in oxidising, and sealing-wax red in reducing flame. There is a great difference between the colouring power of the two oxides of this metal, the bead, which is perfectly transparent in the oxidising flame, being coloured with cupric oxide, becomes opaque from the formation of cuprous oxide or suboxide of copper in the reducing flame. This result is most easily obtained by heating the green bead on charcoal or touching it when melted with a piece of tinfoil so as to detach a minute globule of tin, which has an energetic reductive action upon the cupric oxide. In this way a small trace of copper may be recognised in the presence of iron, in spite of the strong colouring power of the latter metal in the reducing flame.

The colours obtained with beads of salt of phosphorus are generally similar to those of borax, but there are some characteristic differences, especially in the reducing flame. Thus Iron gives a yellow or reddish tint instead of the dark green obtained with borax; Vanadium gives a yellowish brown in the oxidising and green in the reducing flame, both being green with borax; Uranium gives green in the oxidising flame.

It is often of more importance for determinative purposes to know the colours given by minerals containing more than one element of strong colouring power than that of the components taken separately in a pure state, as in such cases very characteristic reactions are given by the combination. This is specially the case with the iron compounds of certain metallic oxides; thus, Tungsten in the form of tungstic acid, gives, with salt of phosphorus, a colourless or yellow bead in the oxidising, and a blue one, which is green while hot, in the reducing flame; but if Iron is present, as in Wolfram (natural tungstate of iron), the latter colour is changed to a brownish red.

Another example is afforded by titanic acid, which, when pure as in Rutile, Anatase and Brookite, gives, with salt of phosphorus, a bead colourless or cloudy in the oxidising, and red, passing into violet when cold, in the reducing flame. The latter colour is, however, changed to brownish red when iron is present, as in titaniferous iron ores, and the violet colour can only be brought out by the addition of tin or zinc to the bead.

The presence of titanic acid in iron ores may be rendered apparent by a method devised by Gustav Rose, which is of great interest. If the dark-coloured clear bead when completely saturated in the reducing flame be transferred to the oxidising flame, it loses colour, but becomes opaque from the separation of titanic acid TiO_2, which is much less soluble in melted phosphate of soda than the lower oxide Ti_2O_3. By flattening the cloudy bead between the forceps while still hot it may be rendered translucent, and when examined by the microscope is seen to contain minute crystals, having the characteristic square-based form of Anatase. These are perhaps more readily seen if a drop of water is added to the bead on the glass slide, when the vitrified sodium phosphate dissolves, releasing the crystals, which remain suspended in water, and are more visible.

Quartz and many silicates, when heated in sodic phosphate, do not dissolve, but leave an opaque mass usually called a silicious skeleton. This, according to Gustav Rose, is actually crystallised silica of the variety of low specific gravity known as Tridymite, a mineral found only in rocks of volcanic origin. The subject of the microscopic character of crystals formed by other oxides has recently received considerable attention from Sorby and other investigators, whose researches should be consulted by the student.

Tests by the wet way. The use of the methods of qualitative analysis by the wet way, in the determination of minerals, is necessarily restricted to those that can be applied without requiring any great amount of apparatus or laboratory appliances, and, as a rule, should only be used when the blowpipe tests are insufficient. They are therefore of most value in the examination of silicates and other insoluble compounds of the earthy and alkaline metals which are not readily recognisable by the dry way.

The most useful reagents are hydrochloric, nitric, and sulphuric acids, ammonia, sulphide of ammonium, * caustic potash or soda, * phosphate of soda, * nitrate or chloride of barium, * oxalate of ammonia, * nitrate of silver, * molybdate of ammonia. These, with the exception of those marked *, are liquids, and must be kept in stoppered bottles, while the others may be kept either as solutions or in the dry state, the former being most convenient when they are often used, but when they are only required occasionally it is better to make the test solutions when wanted. In every case distilled water, or such as is free from sulphates and chlorides, is to be used.

The principal pieces of apparatus, in addition to those already mentioned as requisite for use with the blowpipe, are a few porcelain capsules, the largest about 2 inches across, test tubes, a few small beakers and funnels, paper filters up to 2½ inches diameter, or sheets of filter paper, a wire filter stand, and, if possible, a small platinum crucible.

The course of examination followed should be similar to that adopted in systematic analysis with substances of unknown composition, and in all cases the mineral should be finely powdered, the test for solubility in water being first applied. Among the minerals readily soluble are the different alkaline Chlorides and Sulphates, the Alums, the Sulphates of Magnesium, Zinc, Copper, Iron, Nickel, &c.; less so are Sulphate of Calcium (Gypsum), and Arsenious acid. Hydrochloric acid is next used, first without and then with the aid of heat, and subsequently the same acid in a more concentrated form when necessary. Minerals may be completely, partially, or not at all soluble by this treatment. The solution is attended with effervescence in the case of carbonates and sulphides, carbonic acid and sulphuretted hydrogen being respectively evolved. The latter may be recognised by its action upon a slip of test paper made with acetate of lead, which is blackened when exposed to the current of gas at the mouth of the test tube or beaker used for making the solution. The different carbonates belonging to the Calcite-Aragonite group vary considerably in their solubility in acids. Thus, calcic carbonate in both forms effervesces readily with very weak hydrochloric or even acetic acid, while those of magnesium and iron, as well as the more complex varieties containing two or more of these metals, are but slowly acted upon even by strong acid in the cold. Compounds containing the higher oxides of manganese dissolve in warm hydrochloric acid with evolution of chlorine. Silicates belonging to the Zeolite group are decomposed with a separation of silica in the gelatinous form, while others, such as Labradorite, leave a pulverulent or granular residue of silica, the metallic bases passing into solution as chlorides. In doubtful cases the effect of the acid may be established by filtering the insoluble residue, and testing the clear solution with ammonia and phosphate of soda; if these reagents produce no precipitates, the mineral has not been acted upon.

Nitric acid is used for the decomposition and solution of the higher sulphides and arsenides, and of native metals, most of which are attacked with the evolution of red fumes of peroxide of nitrogen, the production of nitrates of the metals, and, in the case of sulphides of sulphates, with a separation of sulphur. Nitrates being, however, inconvenient for most analytical determinations, it is customary to convert them into chlorides by adding hydrochloric acid, and evaporating to dryness previously to commencing the systematic examination of the solution.

Gold, Platinum, and the allied metals, being insoluble in either nitric or hydrochloric acid alone, are brought into the soluble form as chlorides by the use of aqua regia, a mixture of nitric and hydrochloric acids, or any similar combination producing free chlorine. Sulphuric acid is principally useful in decomposing fluorides, producing sulphates and hydrofluoric acid. Experiments of this kind must be performed in a leaden or platinum vessel, the finely powdered minerals being heated with strong sulphuric acid: the mouth of the crucible is covered with a glass plate protected by an etching ground of wax, upon which a design is marked by removing the wax with a steel point. Wherever the glass is laid bare, it is corroded by hydrofluoric acid vapour, reproducing the lines drawn upon the glass plate. Liquid, or rather a strong watery solution of hydrofluoric acid, is also an exceedingly convenient reagent for the decomposition of silicates where it is desired to obtain the metallic bases, especially those of the alkalies, the whole of the silica being removed as hydrofluosilicic acid gas; but it cannot be used safely except in laboratories provided with good ventilation, owing to the corrosive action of its vapour on all glass and most metallic objects.

The usual method of decomposing silicates insoluble in acids—by fusion with carbonate of sodium in a platinum crucible over a spirit lamp or Bunsen gas burner, as adopted in quantitative analyses—may be imitated on a small scale before the blowpipe, the fusion being effected upon charcoal instead of on platinum. In such cases a small quantity of borax should be added to the alkaline flux to prevent the latter being absorbed by the pores of the charcoal, and to form a well-defined bead of the fused mass that can be removed from the charcoal without loss. The bead so obtained is dissolved in dilute hydrochloric acid, the solution evaporated to dryness, and the residue moderately heated, which renders the silica insoluble, while Lime, Magnesia, Alumina, and other bases may be removed by diges ion with dilute hydrochloric acid or soluble chlorides. The following are among the most characteristic reactions obtained by the wet way with the more abundant metallic oxides and mineral acids, and, as tests that can be used without requiring much refinement of apparatus or manipulation, are useful to the mineralogist.

Alumina. This base, when contained in the hydrochloric acid solution of a silicate after the silica has been separated by evaporation and heating, gives a white gelatinous precipitate with ammonia, which is soluble in solution of caustic potash. When both Iron and Alumina are present, as is the case with many silicates, potash must be used to precipitate the former as ferric hydrate, and after filtration the Alumina is thrown down by adding carbonate of ammonia to the solution, the excess of potash having been previously neutralised by hydrochloric acid. The precipitate, when dried and ignited, is tested with cobalt solution in the manner described on p. 309.

Iron. The solution of any iron mineral, when brought to the condition of a ferric salt by heating with a few drops of nitric acid, gives a dark blue precipitate with a drop of solution of yellow prussiate of potash. The ammonia precipitate is rusty brown with ferric, and green with ferrous salts, the latter passing into the former when exposed to the air.

Lime. In concentrated solutions a white precipitate of sulphate is produced by sulphuric acid, which is, however, sensibly soluble in water, so that this reaction cannot be used with weak solutions; in such cases a precipitate may be produced by the addition of alcohol. Oxalate of ammonia is the most sensitive test, giving a white precipitate in very dilute solutions, after other bases separable by ammonia have been removed.

Magnesia. This base may be recognised in the ammoniacal solution remaining after the removal of Alumina, Iron, Lime, and other bases by ammonia and oxalate of ammonia, by adding solution of phosphate of soda, which produces a crystalline precipitate of ammonio-magnesium phosphate. The formation of the precipitate is facilitated by well stirring the solution with a glass rod, and allowing it to stand for some hours.

In the systematic examination of minerals containing Silica, Alumina and Iron Oxides, Lime and Magnesia, when the tests are applied in the order given, the whole number of bases may be verified in the same solution. It is essential when Magnesia is present to have a considerable quantity of chloride of ammonia in the liquid. in order to keep it dissolved until the other bases are separated, otherwise it may be precipitated as hydrate of magnesia by ammonia.

Baryta. This base is recognised by the insolubility of its sulphate in water, the smallest trace of it in a solution producing, with a drop of sulphuric acid or of solution of a soluble sulphate, a cloudy precipitate, which, however, may require some considerable time to form when the quantity is very small. Strontia behaves similarly, but the sulphate being more soluble takes a longer time to precipitate. When a solution containing both of these bases in hydrochloric acid is evaporated to dryness, ignited, and digested with alcohol, chloride of strontium dissolves, leaving a residue of chloride of barium, and the properties of both can be established by their flame reactions.

Most of the so-called heavy metals are remarkable for giving dark-coloured precipitates when treated with sulphuretted hydrogen or an alkaline sulphide, and they are divisible into groups according to the condition of the solution necessary to ensure precipitation. Thus, Gold, Silver, Lead, Copper, Bismuth, Arsenic, and Antimony can be separated as sulphides from an acid solution, while with Iron, Nickel, Cobalt, Manganese, and Zinc, the solution must be alkaline, as the sulphides of these metals are decomposed by acids. Aluminium and Chromium are also members of the latter group, but the precipitates formed with sulphide of ammonium are not sulphides, but hydrates of alumina and chromic oxide. Whenever sulphuretted hydrogen is used, the solution of the metallic bases must not contain much free nitric acid, as in that case no precipitation of sulphides will be effected until the oxidising action of the acid upon the sulphuretted hydrogen, which is attended with a separation of sulphur, is exhausted.

The colours of the precipitates produced by sulphuretted hydrogen with the more abundant metals are as follows :—

Copper . .	Brownish black.
Lead . . .	Bluish-black.
Antimony . .	Orange red.
Arsenic . .	Bright yellow.
Aluminium . .	White gelatinous hydrate of alumina.
Chromium . .	Green hydrate of chromic oxide.
Zinc . . .	White.
Manganese . .	Flesh colour.
Nickel, Cobalt . .	Black.

The steps to be taken for the separation of the different metals existing in a mixture of precipitated sulphides will be found in the treatise on chemical analysis previously mentioned, and it will not be necessary to go further in detail upon this point here, as for most purposes required by the mineralogist the metals contained in such mixtures may be more readily determined by the dry way with the blowpipe.

Silver, when in solution, may be detected, even when in very minute quantity, by means of hydrochloric acid or any chloride, which produce a white curdy precipitate of chloride of silver soluble in ammonia, and becoming grey or violet, and ultimately black when exposed to sunlight. The chloride is readily reduced to the metallic state when mixed with carbonate of soda and heated on charcoal.

Lead. Solutions of this metal in nitric acid give a white

precipitate of sulphate of lead with sulphuric acid, and a crystalline deposit of chloride of lead with hydrochloric acid, insoluble in the cold, but dissolving readily in boiling water. The presence of silver in lead can be best determined by the method of cupellation on bone-ash, a special modification of the process for use with the blowpipe having been contrived by Plattner. In this way very minute silver beads having the characteristic lustre and colour of the metal may be obtained, but it is a useful precaution to verify their properties by dissolving them in nitric acid and testing them with salt. If the silver should contain gold, it will remain as a black, insoluble speck in the acid solution.

Gold may generally be best recognised by the dry way, but when in a moderately concentrated solution, it is precipitated as a brown metallic powder by sulphurous and oxalic acid or ferrous sulphate. In a weak solution, chloride of tin produces a purple or ruby-red coloration, or when sufficiently concentrated, a substance known as purple of Cassius, whose colour is due to finely-divided metallic gold.

Copper. Solutions containing this metal as chloride or sulphate, when slightly acid, are decomposed by metallic iron with a deposition of copper. Thus a knife-blade is coppered in a few minutes when plunged into such a solution. Ammonia gives a deep blue colour to the solution when the proportion of copper is very small, and yellow prussiate of potash is a still more delicate test, producing a brown precipitate, or, when only the minutest trace of copper is present, a reddish-brown tint to the liquid.

Sulphuric acid. The test for this acid is the inverse form of that for baryta, a solution of nitrate or chloride of barium producing a white precipitate of sulphate insoluble in acids.

Hydrochloric acid is detected by means of nitrate of silver solution, which produces chloride of silver, as described under the head of Silver.

Phosphoric acid. When in minute quantity, this may be best detected by adding molybdate of ammonia to the solution, which must be acidified with nitric acid and heated for a short time, when a yellow precipitate containing about 6 per cent. of phosphoric acid is produced, and subsides slowly. When the proportion is larger, it is separated as ammonio-magnesian phosphate, in the same way as described for magnesia, except that some soluble magnesian salt, as sulphate or chloride of magnesium, is added instead of phosphate of soda. This test may be performed in the presence of ferric oxide and alumina if a small quantity of citric acid is present, as these oxides are not precipitated by ammonia when organic matter is present.

Quantitative analysis. Although the methods of qualitative testing are sufficient in the greater number of instances for the identification of well-established minerals, there is a large remainder even of these which cannot be so identified, a knowledge of their actual composition being required in addition; and the same holds good with those of doubtful or unknown constitution, especially when the latter are from new localities. For those, therefore, who may desire to extend the field of mineralogical knowledge, a practical acquaintance with the methods of quantitative analysis as applied to minerals, or, at any rate, to the more abundant and simply constituted species, is essential, but the larger number of students, who may require only a good sight knowledge of known minerals, may be content to accept the results furnished by analytical chemists without further inquiry. In this case, however, a knowledge of the method of calculating the results of analyses, or the deduction of formulæ from percentage quantities, will often be found of great use.

Constitution of Minerals. Among minerals are included not only elementary substances, but combinations of two or more elements forming the classes of compounds known as oxides, sulphides, acids, bases, haloid salts, oxysalts, sulpho-salts, double salts, anhydrides, and hydrates.

Such combinations, being fixed, have a definite constitution. Thus, quartz invariably contains 7 parts of silicon, and 4 of oxygen; calcite, 10 of calcium, 3 of carbon, and 12 of oxygen; fluorspar, 20 of calcium, and 19 of fluorine; and so on. The proportion of the different components being constant in those forms that are normally constituted, deviations from the normal types can be shown to be caused either by foreign substances existing as mechanical impurities, or by partial substitution of one or more of the component elements by others of analogous properties, according to the laws of isomorphism. In the former case, the proportion between the essential constituents, notwithstanding their absolute diminution in quantity, is unchanged; while in the latter, the change, though less simple, consists in the substitution of one element for another in the proportions of their chemical equivalents. Thus, a specimen of calcite containing 10 per cent. of silica, clay, or other constituents insoluble in hydrochloric acid, cannot be considered as differing essentially from the normal composition, if the remaining 90 per cent. is so constituted as to make up the proportion 10 : 3 : 12, or 9 : 2·7 : 10·8, which are the ratios of calcium, carbon, and oxygen in calcite when in a pure state.

Variation in composition, due to the second of the above causes, or the partial substitution of analogous elements, is commonly observed in the same mineral, as almost all specimens of calcite show a deviation from the typical composition by containing small quantities of the elements magnesium, manganese, iron, or zinc. These, however, are held to be in partial substitution of the normal amount of calcium, and the analyses, when interpreted according to the theory of equivalent proportions, are found to be in accordance with the normal constitution.

This theory supposes every element to have a combining

value, or quantity peculiar to itself, and that its compounds with other elements are formed either in the ratio of that quantity, or of one or more simple multiples of it. Thus, supposing A and B to be two elements, they may form compounds: $2A+B$, $A+B$, $2A+3B$, $A+2B$, &c.

It is further supposed that analogous compounds may be made with another element C: $2A+C$, $A+C$, $2A+3C$, $A+2C$, in which the values of A remain unaltered; and in like manner, other elements, D, F, &c., may be substituted for A, giving a series of compounds with B and C, in which the latter are unchanged. The special quantities of the elements so substituted, or the combinations, are said to be equivalents; and if the weight of any one be known or assumed, the others may be referred to it, whereby a series of constant numbers expressing the combining proportions of the different elements is obtained. These are called chemical equivalents.

In forming such a series it is necessary to fix upon some one element as a basis, and this choice is necessarily an arbitrary one. For this purpose the elements oxygen and hydrogen have been chosen: the former as being the most abundant element in nature, and the latter as being the lightest. In the oxygen series of equivalents, introduced by Berzelius, oxygen was taken at 100, with the result of giving inconveniently large values to most of the elements, especially to such metals as silver, gold, antimony, &c. In spite of this drawback it was for a long time current in France and Germany, and is used in the greater number of works upon mineral chemistry of the first half of the present century, a period which has been more prolific in discovery in this branch of science than those immediately preceding or following. In England it has been customary to use Dalton's scale, upon the basis of which hydrogen is assumed as unity, being the lightest of all the elements, from which, assuming its combination with oxygen in water to be in the proportion of single equivalents by weight, the equivalent of the latter is found to be 8, that of water HO, or protoxide of hydrogen, 9, and so on. Latterly, however, a modification of Dalton's original hypothesis, founded upon the weight of equal volumes of the elements when in the gaseous form, has come into general use among chemists, and the older schemes, founded upon considerations of weight alone, have been practically abandoned. This is founded upon the proposition known as Avogadro's law: namely that equal volumes of all gases contain an equal number of molecules. By the term *molecule* is meant a quantity of an element, or compound of elements, capable of independent existence, but so small as to be incapable of further division. The molecule of a compound is a complex of still smaller portions of the component elements, which are known as *atoms*, an atom being defined as the smallest combining proportion of an element. A molecule must therefore contain at least two atoms which, in the case of an element, are both of the same kind, but in that of a compound are of dissimilar kinds (or those of the constituent elements).

The atomic weight of an element is the weight of a volume of its vapour expressed in terms of the weight of a similar volume of hydrogen; the unit weight adopted being that of one cubic centimetre of hydrogen under the normal pressure of 760 mm. of mercury at 0° Centigrade. The molecular volume of an element or compound is that corresponding to two volumes of hydrogen; the molecular weight of an element is therefore usually double its atomic weight.

When elements or compounds can be obtained in the state of gases, their molecular weights may be determined by direct experiment, otherwise a vapour density must be assumed upon considerations founded upon analogies drawn from known compounds supposed to be similarly constituted. Such determinations must necessarily be doubtful.

The atomic weight of an element is determined from the analysis of some one of its best defined and most stable compounds, the molecular constitution of the latter being assumed. The researches of Dulong and Petit have shown that the atomic weights of elements are to each other inversely as their specific heats.

The following table contains the atomic weights of the elements as far as they have been accurately determined. The symbol prefixed to each is held to signify an atomic unit, when used in combination, in the construction of molecular formulæ.

TABLE OF THE ATOMIC WEIGHTS OF ELEMENTS.

Name	Symbol	Class	Atomic weight
Aluminium	Al.	II. IV. VI.	27·3
Antimony (Stibium) .	Sb.	III. V.	122
Arsenic	As.	I. III. V.	75
Barium	Ba.	II. IV.	137
Beryllium (Glucinum) .	Be.	II.	9·33
Bismuth	Bi.	V.	208
Boron	B.	III.	11
Bromine . . .	Br.	I. III. V. VII.	80
Cadmium . . .	Cd.	II.	112
Calcium	Ca.	II. IV.	40
Carbon	C.	II. IV.	12
Cerium	Ce.	IV.	92
Cœsium	Cs.	I.	133
Chlorine . . .	Cl.	I. III. V. VII.	35·5
Chromium . . .	Cr.	II. IV. VI.	52
Cobalt	Co.	II. IV.	59
Copper (Cuprum) . .	Cu.	II.	63·4
Didymium . . .	Di.	II.	96
Erbium	Er.	II.	112·6
Fluorine	Fl.	I.	19
Gallium	Ga.	VI.	69·8
Gold (Aurum) . .	Au.	I. III.	196
Hydrogen . . .	H.	I.	1
Indium	In.	III.	113·7
Iridium	Ir.	II. IV. VI.	198
Iron (Ferrum) . .	Fe.	II. IV. VI.	56
Iodine	I.	I. III. V. VII.	127
Lanthanum . . .	La.	II.	93
Lead (Plumbum) . .	Pb.	II. IV.	207
Lithium	Li.	I.	7
Magnesium . . .	Mg.	II.	24

Name	Symbol	Class	Atomic weight
Manganese	Mn.	II. IV. VI.	55
Mercury (Hydrargyrum) .	Hg.	II.	200
Molybdenum . . .	Mb.	II. IV. VI.	92
Nickel	Ni.	II. IV.	59
Nitrogen . . .	N.	I. III. V.	14
Niobium . . .	Nb.	V.	94
Oxygen	O.	II.	16
Osmium. . . .	Os.	II. IV. VI.	108
Palladium . . .	Pd.	II. IV.	106
Phosphorus . . .	P.	I. III. V.	31
Platinum . . .	Pt.	II. IV.	198
Potassium (Kalium) .	K.	I. III. V.	39
Rhodium . . .	Rd.	II. IV. VI.	104
Rubidium . . .	Rb.	I.	105·5
Ruthenium . . .	Ru.	II. IV. VI.	104
Sulphur	S.	II. IV. VI.	32
Selenium . . .	Se.	II. IV. VI.	79
Silver (Argentum) . .	Ag.	I. III.	108
Silicon (Silicium) . .	Si.	IV.	28
Sodium (Natrium) . .	Na.	I. III.	23
Strontium . . .	Sr.	II. IV.	88
Tantalum . . .	Ta.	V.	182
Tellurium . . .	Te.	II. IV. VI.	128
Thallium . . .	Tl.	I. III.	204
Thorium . . .	Th.	IV.	234
Tin (Stannum) . .	Sn.	II. IV.	118
Titanium . . .	Ti.	II. IV.	248
Tungsten . . .	Tu. }	IV. VI.	184
Wolfram . . .	W. }		
Uranium . . .	U.	II. IV. VI.	240
Vanadium . . .	V.	III. V.	51·4
Yttrium	Y.	II.	61·7
Zinc	Zn.	II.	65
Zirconium . . .	Zr.	IV.	90

[Elements marked I. are monads; II. dyads; III. triads; IV. tetrads; V. pentads; VI. hexads; and VII. heptads.]

The elements are classified according to their atomicities, or combining power, as measured by the number of hydrogen atoms with which they combine to form definite compounds. Thus hydrogen unites in the following manner, with—

Chlorine,	1 equivalent	to form	hydrochloric acid	HCl.
Oxygen,	2 „	„	water	H_2O.
Nitrogen,	3 „	„	ammonia	H_3N.
Carbon,	4 „	„	marsh gas	H_4C.

—in which instances the quantities 1, 2, 3, 4 express the combining power, or quantivalence, of the respective elements, which are said to be monad (I.), dyad (II.), triad (III.), or tetrad (IV.), according to the special number of hydrogen atoms attached to them. Besides these, there are higher ratios of quantivalence, known as pentads (V.), hexads (VI.), and heptads (VII.). In the table the different elements are classified by these numbers, and it will be seen that one element may have several distinguishing atomicities, or that it may form several different types of compounds. Thus, sulphur is dyad in sulphuretted hydrogen, H_2S; tetrad in sulphurous anhydride, O_2S (oxygen being dyad); hexad in sulphuric anhydride, O_3S.

The hexad forms of chromium, aluminium, manganese, iron, nickel, and cobalt are typified in the compounds known as sesquioxides, or those containing two equivalents of metal to three of oxygen, or 1 : 1½. In such cases the double equivalent of the metal is usually considered as a unit, and is represented by a barred symbol, whose atomic weight is double that of the ordinary atom. Thus ~~Al~~ represents Al_2; ~~Fe~~, Fe_2; ~~Al~~ O_3, $Al_2\,O_3$; ~~Fe~~ Cl_3, $Fe_2\,Cl_3$, &c. This arrangement is specially convenient in representing the composition of minerals where the same metal occurs in two states of combination, as is often the case.

Construction of chemical formulæ. The number of atoms of the different elements entering into the constitution of a mineral, when its composition has been determined by analysis, is found by dividing the percentage proportion of each element by its atomic weight, and, subsequently, dividing the quotients so obtained by the smallest among them, which gives a series of numbers standing in a simple relation to each other, which, when reduced to whole numbers, give the number of atoms required. For example, Baryte, or Heavy Spar, gives the following analysis :—

Barium	58·80	137	$\frac{58·80}{137}$	= 0·429 = 1
Sulphur	13·73	32	$\frac{13·73}{32}$	= 0·428 = 1
Oxygen	27·47	16	$\frac{27·47}{16}$	= 1·717 = 4
	100 00			

The numbers in the last column being in the ratio 1 : 1 : 4, the formula required is $BaSO_4$.

When three or more elements are contained in a mineral the formula obtained by writing down the number of atoms of the constituents side by side is known as an elementary or empirical formula. This is merely the simplest numerical expression obtainable, not expressing any opinion on the probable arrangement of the components. In most cases, however, it is necessary to obtain some idea of the grouping of the constituents, for which purpose a knowledge of the principles of chemical classification is requisite ; and the formulæ constructed by these means are known as rational or constitutional formulæ. The range of compounds occurring in minerals is, however, comparatively small. It will only be necessary here to consider the principal types of composition known as acids, bases and salts. According to modern views, an acid is a compound of hydrogen with an electronegative element or combination of elements known as a compound radical.

Hydrochloric acid HCl, hydrobromic acid HBr, and hydrofluoric acid HFl are examples of the first kind, or acids with simple radicals. The acids of compound radicals may contain either oxygen or sulphur in the radical, but they are otherwise analogous in constitution ; the former are called oxygen acids, and the latter sulphur acids. They are further distinguished according to the number of atoms of hydrogen, as monohydric with one, dihydric with two, or trihydric with three atoms.

The constitutional formulæ of acids are constructed on the hypothesis of containing one, two, or three atoms of oxygen or sulphur, one half of which is united with an equal number of atoms of hydrogen, and the other half with an acid radical, which may be either mono- di- or tri-valent, according to the character of the third element.

Thus, nitric acid has the following formula :—

	Elementary.	Constitutional.
	HNO_3	$H-O-(NO_2)'$
Sulphuric acid . .	H_2SO_4	$H_2=O_2=(SO_2)''$
Phosphoric acid .	H_3PO_4	$H_3\equiv O_3\equiv(PO)'''$
Sulpho-carbonic acid .	H_2CS_3	$H_2=S=(CS)''$

where the accents represent the equivalency of the radical.

When the molecule of an acid is broken up by the removal of the whole amount of hydrogen, and the corresponding quantity of oxygen required to form water, H^2O, or one atom of the latter to two of the former, or in the case of a sulphur acid of sulphur to form sulphuretted hydrogen, H^2S, a compound is obtained known as the anhydride of the acid, which is, in fact, an oxide of the radical. In the case of mono- and tri-hydric acids two molecules are required in order to express the results in whole numbers of atoms. Thus :—

Nitric acid—

(2 mol.) or 2 (HNO_3) less 1 eq. H_2O giving $\binom{NO_2}{NO_2} O = N_2O_5$

Sulphuric acid—

(1 mol.) H_2SO_4 ,, 1 H_2O ,, $(SO_2)\,O = SO_3$

Phosphoric acid—

(2 mol.) 2 (H_3PO_4) ,, 3 H_2O ,, $\binom{PO}{PO} O_3 = P_2O_5$

Carbonic acid—

(1 mol.) H_2CO_3 ,, 1 H_2O ,, $(CO)\,O = CO_2$

Sulpho-carbonic acid—

(1 mol.) H_2CS_3 ,, 1 H_2S ,, $(CS)\,S = CS_2$

The compounds in the last column are therefore called anhydrides of the corresponding acids, or generally but improperly, anhydrides, without further qualification. In the older works on chemistry they are called anhydrous acids, the acids defined as above being considered as hydrated acids.

A base is defined to be a combination of hydrogen with a compound radical of an electro-positive character, consisting of an electro-positive element or metal united with oxygen or sulphur, the former being called an oxygen, and the latter a sulphur, base.

The constitution of bases is represented similarly to that of acids, the constituent atoms of oxygen being considered as combined to the extent of one-half with an equal number of hydrogen atoms, and the other half with an equivalent atom of the metal, producing, as in the case of the acids, monohydric and polyhydric bases, or hydroxides, or with sulphur as hydrosulphides, thus :—

$H-O-\overset{I}{K}$	is	Hydropotassic oxide,	or of the type	$(HO)\overset{I}{R}$
$H_2=O_2=\overset{II}{Ba}$	,,	Hydrobaric oxide	,,	$(HO)_2\overset{II}{R}$
$H_3\equiv O_3\equiv\overset{III}{Bi}$	,,	Hydrobismuthic oxide	,,	$(HO)_3\overset{III}{R}$

The Roman figures represent the equivalency of the metallic element, and in the type formulæ in the last column R stands for any metal of corresponding equivalency.

And of the analogous sulphur compounds :—

$H-S-\overset{I}{K}$	is	Hydropotassic sulphide,	or of the type	$(HS)\overset{I}{R}$
$H_2=S_2=\overset{II}{Ba}$	,,	Hydrobaric sulphide	,,	$(HS)_2\overset{II}{R}$
$H_3\equiv S_3\equiv\overset{III}{Bi}$	,,	Hydrobismuthic sulphide	,,	$(HS)_3\overset{III}{R}$

Basic anhydrides are produced in the same manner as the acids by the removal of the hydrogen, with sufficient oxygen or sulphur to form water or sulphuretted hydrogen from the hydrometallic oxides or sulphides respectively. Thus :—

$2\ HOK - H_2O = K_2O$ known as potassic oxide or potash.
$H_2O_2Ba - H_2O = BaO$ „ baric oxide or baryta.
$2\ H_3O_3Bi - 3\ H_2O = Bi_2O_3$ „ { bismuth oxide, or strictly di-bismuthic trioxide.

And from the analogous sulphur compounds—

$2\ HSK - H_2S = K_2S$ or dipotassic sulphide.
$H_2S_2Ba - H_2S = BaS$ „ baric sulphide.
$2\ H_3S_3Bi - 3H_2S = Bi_2S_3$ „ dibismuthic trisulphide.

The last of these is the mineral known as Bismuth Glance. The anhydrides of bases are therefore the oxides and sulphides of their respective metals.

A salt is considered to be a combination formed by the action of equivalent quantities of an acid and a base upon each other when the whole of the hydrogen and one of the atoms of oxygen are removed, as water. Thus :—

Nitric acid . .	$H-O-(NO_2)$	=	$K-O-(NO_2)$ or KNO_3 potassic nitrate, and
Hydropotassic oxide .	$H-O-K$		H_2O water.
Sulphuric acid . .	$H_2=O_2=(SO_2)$	=	$Ba=O_2=(SO_2)$ or $BaSO_4$ baric sulphate, and
Hydrobaric oxide .	$H_2=O_2=Ba$		$2\ H_2O$ water

Acids and bases containing like amounts of hydrogen combine in equal molecules, but when they are unlike the molecular relation of acid to base in a salt is dissimilar.

Salts formed by acids of a simple radical, such as hydrochloric acid, HCl, and hydrofluoric acid, HFl, are called Haloid Salts, and those with acids of compound radicals Oxy-salts and Sulpho-salts. A neutral salt is that resulting from equivalent quantities of acid and base ; it is also called a normal salt. An acid salt is a combination of a molecule of normal salt with one or more molecules of acid ; while a basic salt is similarly a normal salt combined with one or more molecules of base. Bisulphate of potassium $\left\{ \begin{matrix} K_2SO_4 \\ H_2SO_4 \end{matrix} \right\}$ is an example of the former, and Malachite or basic carbonate of Copper $\left\{\begin{matrix} CuCO_3 \\ H_2CuO_2 \end{matrix}\right\}$ of the latter class.

Acid and basic salts may in some cases be free from hydrogen, that is, they may consist of a normal salt combined with the anhydrides of the acid or base respectively. Of this character are acid bisulphate of Potassium $\left.\begin{matrix} K_2SO_4 \\ SO_3 \end{matrix}\right\}$ which is obtained by heating the salt $\left.\begin{matrix} K_2SO_4 \\ H_2SO_4 \end{matrix}\right\}$, basic chromate of Lead $\left.\begin{matrix} 2PbCrO_4 \\ PbO \end{matrix}\right\}$, and the oxychlorides of Lead $\left.\begin{matrix} PbCl_2 \\ PbO \end{matrix}\right\}$ and $\left.\begin{matrix} PbCl_2 \\ 2PbO \end{matrix}\right\}$ forming the rare minerals Matlockite and Mendipite.

Double salts are compounds of two different salts, which may be either similar or dissimilar in class or constitution. Thus, Carnallite $\left\{\begin{matrix} KCl \\ MgCl_2 \end{matrix}\right.$ is a compound of two similar haloid salts, the chlorides of Potassium and Magnesium: Cryolite $\left\{\begin{matrix} 6NaFl \\ Al_2Fl6 \end{matrix}\right.$ of the two haloids of dissimilar constitution; Blödite $\left\{\begin{matrix} Na2SO_4 \\ MgSO_4 \end{matrix}\right.$ of two similar sulphates; Potash Alum $\left.\begin{matrix} K_2SO_4 \\ Al_2S_3O_{12} \end{matrix}\right\}$ of two dissimilar oxysalts (sulphates) and Chlorapatite $\left.\begin{matrix} CaCl_2 \\ 3Ca_3P_2O_8 \end{matrix}\right\}$ of an oxysalt and a haloid, phosphate and chloride of calcium. The acid sulphate of potassium $\left.\begin{matrix} K_2SO_4 \\ H_2SO_4 \end{matrix}\right\}$ might also be regarded as a double sulphate of potassium and hydrogen, except for the special signification attached to the term acid.

Many minerals, especially alkaline sulphates and other easily soluble salts, give off water with more or less readiness when the crystals are exposed to the air, in some cases without, but more readily by, heat. Such water is usually regarded as not essential to the constitution, or as water of crystallisation, when it is given off at the boiling point of water or a little above it, and when the salt so dehydrated takes the same amount again when dissolved and recrystallised. The water so combined is often distinguished by the symbol Aq. Thus, Potash Alum, containing 24 equivalents of water, is represented by $\left.\begin{matrix} K_2SO_4 \\ Al_2S_3O_{12} \end{matrix}\right\} + 24Aq$; but when a high temperature is requisite to drive off the water, the latter is to be regarded as the result of decomposition of an actual hydrogen compound essential to the constitution of the mineral. For example, the silicate known as Dioptase, which yields by analysis oxide of copper, water, and silica in equivalent proportions, is not decomposed below a red heat, so that it should be regarded as consisting of H^2CuSiO^4 rather than as a hydrated silicate of the form $CuSiO_3 + Aq$. Such cases, however, are often of doubtful interpretation, depending upon the consideration of compounds of analogous constitution or other more or less assumed data, so that in most cases it is simpler to accept the result of analysis, which at any rate expresses the fact that water has been found.

The class of minerals known as hydrated oxides are similarly of uncertain constitution, as they may be regarded either as consisting of anhydrous oxides combined with one or more molecules of water, or as compounds of bases (hydroxides) with their anhydrides, and as such forming a class of compounds intermediate between acids and bases. For most mineralogical purposes, however, the former is the more convenient view.

CHAPTER XVI.

RELATION OF FORM TO CHEMICAL CONSTITUTION.

Heteromorphism.—When a mineral is of well-defined chemical constitution, its physical and crystallographical characters are, as a rule, similarly well defined; while, on the other hand, substances that are only known in the amorphous condition are very generally variable in composition. In addition to these, many instances are known where a substance of particular chemical constitution appears in forms belonging to different crystallographic systems, and also with differences in physical character. This property is called heteromorphism, and the substances possessing it are said to be di- or tri-morphic, according as they appear in two or three different crystalline systems, the latter being the largest number yet observed in any heteromorphic mineral.

These different varieties, when found in nature, are spoken of as different minerals; but when they are artificially produced, no such distinction is made.

The following are among the more remarkable cases of natural heteromorphous substances:—

Sulphur.—This element is known in three conditions—1. As an amorphous plastic substance, produced when melted sulphur heated to 200° Cent. is poured into water, having the sp. gr. 1·92, and insoluble in bisulphide of carbon; 2. in crystals belonging to the rhombic system, combinations of rhombic pyramids, of sp. gr. 2·06, which are deposited by spontaneous evaporation from a solution of sulphur in bisulphide of carbon; and 3. in prismatic combinations belonging to the oblique system, of sp. gr. 1·97, which are the common forms of sulphur crystallised from fusion. Both of the crystalline varieties are of the characteristic yellow colour of sulphur, while the amorphous one is dark brown. The second or rhombic form is the only one found as a natural mineral.

Carbon.—Also in three conditions—1. amorphous in coal as the deposit produced from the imperfect combustion of a volatile carbon compound, as soot, lamp black, &c.; 2. crystallised in the cubical system in Diamond, a non-conductor of electricity, having the sp. gr. 3·55 and hardness 10; and 3. crystallised in the hexagonal system, as Graphite, having the sp. gr. 2·3, and hardness 1·5, and a considerable degree of electric conductivity.

The allied elements, *Boron* and *Silicon*, have been shown by Deville to occur in dimorphous modifications analogous to those of Diamond and Graphite in carbon, which are distinguished as the adamantine and graphitoidal forms of the elements respectively. They do not, however, occur as natural minerals.

Antimonic oxide—Sb_2O_3, is both cubical and rhombic, the former being the mineral called Senarmontite and the latter Valentinite. *Silica*, SiO_2, is known in four different states. When fused before the oxyhydrogen blowpipe it forms an amorphous glass of sp. gr. 2·22, which also occurs in the minerals opal or hyalite. The crystallised varieties are—1. Quartz, which is hexagonal (tetartohedral) and sp. gr. 2·66; 2. Tridymite, also hexagonal, but holohedral, sp. gr. 2·3; and 3. Asmanite, rhombic, and of sp. gr. 2·24.

Titanic acid, TiO_2, is another example of a trimorphous oxide; the commonest variety, Rutile, belonging to the tetragonal system, has the fundamental parameter $a : c = 1 : 0·6442$, and sp. gr. 4.25; the second form, Anatase, is also tetragonal, but the crystals are of pyramidal habit, having $a : c = 1 : 1·778$, and sp. gr. 3·9. The third form, known as Brookite, is rhombic, and has the sp. gr. 4·15. *Carbonate of Calcium*, $CaCO_3$ is perhaps the most familiar example of a dimorphous substance being rhombic, and of sp. gr. 2·9, in Aragonite, and rhombohedral in Calcite, whose sp. gr. is 2·7. The cause of these differences in form was for a long time sought to be explained by slight differences in chemical composition, Aragonite being supposed to owe its rhombic character to the presence of a small proportion of carbonate of strontium; but the fact of their essential similarity in composition was demonstrated by G. Rose, who proved that carbonate of calcium, when precipitated from a cold solution, consists of microscopic rhombohedra of calcite; but when the precipitate takes place from a boiling solution, minute prisms of arragonite are obtained. The original establishment of the phenomena of dimorphism is due to Mitscherlich, who in 1823 first demonstrated the differences between the form of native sulphur crystals and that obtained artificially by crystallisation from fusion. Subsequent researches have proved that numerous other substances possess the same property, and their number may be considerably enlarged when the comparison is extended to substances not absolutely identical in composition, but represented by similar formulæ.

Isomorphism.—The fact that small variations of the proportions of particular components are possible in minerals without changing their crystalline form was known to the earlier crystallographers, and an explanation was propounded by Haüy, upon the assumption of the dependence of the former upon the quality of the constituents alone. Thus, in the series of rhombohedral carbonates calcite, magnesite, dolomite, siderite, ankerite, calamine, &c., the similarity in crystallographic characters was supposed to be caused by the presence of a small proportion of calcium in each of the different members, which exerted a dominating formative influence over the other constituents, and gave a general resemblance to the type species of the series calcite or carbonate of calcium. This view was shown to be erroneous by Mitscherlich, who demonstrated in 1819 that the relation is essentially based upon quantity, and applied to it the name of Isomorphism (from ἴσος, 'equal,' and μορφή, form). Minerals of analogous constitution, that is, containing the same number of atoms of like kinds, are isomorphous. Thus in the series in question, that of calcite, the whole of the members are constituted upon the common type, $R^{II}CO_3$, whence, by substituting for R^{II} equivalents of the analogous metals, calcium, magnesium, iron, manganese, and zinc successively are obtained—Calcite $CaCO_3$, Magnesite $MgCO_3$, Siderite $FeCO_3$, Manganese spar $MnCO_3$, and Calamine $ZnCO_3$; and, in addition to these, there are others, in which two or more metals are present, such as Dolomite, containing both calcium and magnesium; Ankerite, calcium, magnesium, and iron, &c.—all of which are rhombohedral in form and similar in constitution.

In the strict literal sense the term 'isomorphous' is only applicable to such substances as are cubic in crystallisation, as in the other systems, the relation between the members of isomorphous series is one of close analogy, but not of identity. Thus in the above series the polar angle of the rhombohedron varies from 105° 5′ in calcite to 107° 40′ in calamine; the carbonates of iron, manganese, and magnesium giving intermediate values. These differences, though small, are sufficient to establish the crystallographic independence of the different species, and therefore the term 'homeomorphous' is used by some authors to express the relation of such minerals; but the practice is not followed to any very great extent.

In addition to the example already given, the following are among the more important isomorphous groups:—

Corundum group. Hexagonal rhombohedral.

Corundum $\overline{Al}O_3$. Hematite $\overline{Fe}O_3$. Chromic oxide $\overline{Cr}O_3$.

Apatite group. Hexagonal pyramidal hemihedral.

Fluor-Apatite $3(Ca_3P_2O_8)\ CaFl_2$
Chlor-Apatite $3(Ca_3P_2O_8)\ CaCl_2$
Pyromorphite $3(Pb_3P_2O_8)\ PbCl_2$
Mimetesite . $3(Pb_3As_2O_8)\ PbCl_2$
Vanadinite . $3(Pb_3V_2O_8)\ PbCl_2$

These occur in closely allied forms, in spite of the qualitative differences in composition; the isomorphous relations of these elements being:—calcium and lead; phosphorus, arsenic and vanadium; and chlorine and fluorine.

Spinel group. Cubic.

Spinel . .	$MgAlO_4$
Magnetite .	$FeFeO_4$
Chromite .	$FeCrO_4$
Franklinite .	$(ZnFeMn)(FeMn)O_4$

Alum group. Cubic.

Potash alum	$K_2AlS_4O_{16}$ 24Aq
Ammonia alum . . .	$(NH_4)_2AlS_4O_{16}$ 24Aq
Potash-iron alum . .	$K_2FeS_4O_{16}$ 24Aq
Ammonia-iron alum .	$(NH_4)_2FeS_4O_{16}$ 24Aq
Potash-chrome alum .	$K_2CrS_4O_{16}$ 24Aq
Ammonia-chrome alum .	$(NH_4)_2CrS_4O_{16}$ 24Aq

Only the first two members of the above series occur as natural minerals, the others having been prepared artificially.

Most of the isomorphous groups in the rhombic system are dimorphous with those of other systems. For instance, the rhombic form of carbonate of calcium or Aragonite is the type of a series nearly as extensive as that of the rhombohedral form of the same substance, or Calcite, including the following species:—

Aragonite, $CaCO_3$	Strontianite, . $SrCO_3$
Witherite, $BaCO_3$	White-lead ore, $PbCO_3$

The Diaspore group, also rhombic and dimorphous with that of Spinel, includes—

Chrysoberyl, $BeAlO_4$	Göthite, H_2FeO_4
Diaspore, H_2AlO_4	Manganite, H_2MnO_4

The artificial compound produced by fusing ferric oxide and lime together, described by Percy, to which the name Calciferrite may be applied, probably belongs to this series, being very similar in character to Göthite; its crystallographic characters have not, however, been exactly determined.

By an extension of the idea of isomorphism, the dimorphism of many substances may be indirectly established. Thus the carbonates of the type RCO_3 are dimorphous; but only one of them, carbonate of calcium, is actually known in both systems as Aragonite and Calcite. Isomorphous mixtures of rhombohedral carbonates, as might be expected, assume the calcite form, and rhombic ones those of aragonite. Carbonate of lead, $PbCO_3$ in the species whitelead ore, belongs to the latter group; but, in combination with carbonate of calcium, it forms Plumbo-calcite $(PbCa)CO_3$, which is rhombohedral, and cannot therefore be supposed to be derived from the aragonite series but from calcite and a dimorphous rhombohedral variety of carbonate of lead, not known independently.

The isomorphous mixture of the carbonates of calcium and barium, $\left.\begin{matrix}BaCO_3\\CaCO_3\end{matrix}\right\}$, known as Alstonite, has the same symmetry as its constituents Witherite and Aragonite; but Baryto-Calcite, which is of similar constitution, belongs to the oblique system, proving the type RCO_3 to be actually trimorphous, although no carbonate of a single base is known to crystallise in the latter system. Similar cases are presented in the following series, which establish isomorphous relations between the dyad sulphates and carbonates and the combination of both:—

	Rhombohedral.		Rhombic.		Oblique	
RCO_3	Calcite	$CaCO_3$	Aragonite	$CaCO_3$	Baryto-Calcite	$\left.\begin{matrix}BaCO_3\\CaCO_3\end{matrix}\right\}$
RSO_4	Dreelite	$\left.\begin{matrix}CaSO_4\\BaSO_4\end{matrix}\right\}$	Anglesite	$PbSO_4$	Glauberite	$\left.\begin{matrix}Na_2SO_4\\Ca\ SO_4\end{matrix}\right\}$
$\left.\begin{matrix}RSO_4\\nRCO_3\end{matrix}\right\}$	Susannite	$\left.\begin{matrix}PbSO_4\\3PbCO_3\end{matrix}\right\}$	Leadhillite	$\left.\begin{matrix}Pb\,SO_4\\3PbCO_3\end{matrix}\right\}$	Lanarkite	$\left.\begin{matrix}PbSO_4\\PbCO_3\end{matrix}\right\}$

Substances of the above kind that are both isomorphous and heteromorphous are said to be *isodimorphous* or *isotrimorphous*, according to the number of different crystalline systems in which they occur.

Polysymmetry.—One of the most important series of minerals, known as the Hornblende-Augite group, is represented by the general formula $R^{II}SiO_3$, where R^{II}=Ca, Mg, Fe, or Mn. The members of this group are not isomorphous in the sense of having the same crystallographic symmetry, as some of them occur in the rhombic, others in the oblique, and others in the triclinic system; but a general similarity in the geometrical elements of the four is observed. Thus, Diopside or Augite, of the type $\left.\begin{matrix}CaSiO_3\\MgSiO_3\end{matrix}\right\}$ crystallising in the oblique system, has the fundamental parameters $a : b : c = 1·094 : 1 : 0·591$, and $\beta = 74°$; while those of Tremolite or Hornblende $\left.\begin{matrix}CaSiO_3\\3MgSiO_3\end{matrix}\right\}$ are $a : b : c = 0·544 : 1 : 0·294$, and $\beta = 75° 15'$. The parameters of the axes *a* and *c* in Augite are therefore approximately double those of Hornblende, while the angle β is nearly the same in both species.

A second group, $\left.\begin{matrix}mMgSiO_3\\nFeSiO_3\end{matrix}\right\}$, represented by the species Bronzite, Hypersthene and Enstatite, is rhombic, with the parameters $a : b : c = 1·031 : 1 : 1·177$. These may be compared with those of Augite, if the latter be referred to a system of axes that are rectangular or nearly so, which is done by considering the face (001) as (101), which gives the parameters $a : b : c = 1·052 : 1 : 0·296$, and $\beta = 89°40'$. If the same face be further noted as (104), the following close approximation between these new oblique parameters and the rhombic ones becomes apparent:—

	a	:	*b*	:	*c*	β
Bronzite .	1·031	:	1	:	1·177.	90°
Augite .	1·052	:	1	:	1·182.	89° 40'.

Similar approximations between the elements of substances of analogous or identical composition, but crystallising in different systems, are observed in the rhombic and hexagonal varieties of Sulphate of Potassium and in Albite and Orthoclase. They are included by Rammelsberg under the general head of isomorphism, but the special term polysymmetry has been applied to them by Scacchi.

When a mineral contains both dyad and hexad bases, it may, by the progressive substitution of one metal for another of the same class, vary considerably both in composition and physical characters without change of form. One of the best examples is afforded by Garnet, which occurs in many varieties, differing considerably both as regards colour and density, but all crystallising in the cubic system—the rhombic dodecahedron being the dominant form; the observed range of the four principal bases being as follows:

Alumina .	(AlO_3)	o to 22 per cent.
Ferric Oxide .	(FeO_3)	o to 30 „
Lime . .	CaO	o to 37 „
Magnesia .	MgO	o to 22 „

When, however, the proportions of the bases in a lime-alumina and a lime-iron garnet respectively are reduced to the above values, it is found that in the first case Ca : Al = 3 : 1, and Al : Si = 1 : 3, and in the second Ca : Fe = 3 : 1 and Fe : Si = 1 : 3; while in both Ca : Si = 1 : 1. Whence it appears that the two compounds are of the analogous composition—

$$Ca_3AlSi_3O_{12} \text{ and } Ca_3FeSi_3O_{12},$$

and that those containing both alumina and ferric oxide are isomorphous mixtures of both types in varying proportions.

In addition to these, other varieties are known containing the following silicates—

$Mg_3AlSi_3O_{12}$	$Mg_3FeSi_3O_{12}$
$Fe_3AlSi_3O_{12}$	$Fe_3FeSi_3O_{12}$
$Mn_3AlSi_3O_{12}$	$Mn_3FeSi_3O_{12}$

either independently or in combination with the calcium silicates given above.

The term Garnet, therefore, is not special to any one of these compounds in particular, but distinguishes a group of isomorphous silicates, which, however much they may differ qualitatively, have the above ratio, 3 : 1 : 3, for their dyad and hexad metals and silicon respectively common to all, or may be represented by the generalised formula, $\overset{II}{R}_3\overset{VI}{R}Si_3O_{12}$, which covers every possible variety of composition indicated by the above special types.

The isomorphism of compounds, not containing the same number of elementary atoms, supposes the substitution of the elements to take place in the proportion of their equivalence, two atoms of a monad replacing one of a dyad element, &c. This is seen in the Diaspore group, where H_2 in Göthite and Manganite represents Be in Chrysoberyl, and Fe, Mn, or Zn, in the analogous dimorphous species, Magnetite and Franklinite of the spinel series. Another example is afforded by Oxygen and Fluorine, O replacing Fl_2, or $R_2O = RFl$, and $RO = RFl_2$. This is seen in Topaz $\left.\begin{matrix}5AlSiO_5\\ Al\ SiFl_{10}\end{matrix}\right\}$, which is rhombic and isomorphic with Andalusite $AlSiO_5$.

The isomorphism of analogous compounds of monad $(\overset{I}{R})$, dyad $(\overset{II}{R})$, and hexad $(\overset{VI}{R})$ elements is apparent in the Augite group of silicates which, in addition to the varieties already mentioned as represented by the constitution $RSiO_3$, contains others both in the augite and hornblende series, in whose composition sodium, aluminium, and ferric silicates form part, in addition to the dyad metals Ca, Mg, Fe, &c. Of these, the following, Babingtonite $\left.\begin{matrix}9RSiO_3\\ FeSi_3O_9\end{matrix}\right\}$, Achmite $\left.\begin{matrix}3Na_2SiO_3\\ FeSiO_3\\ 2FeSi_3O_9\end{matrix}\right\}$, and Aegirite $\left.\begin{matrix}Na_2SiO_3\\ 2RSiO_3\\ FeSi_3O_9\end{matrix}\right\}$, appear in the augite form, while Arfwedsonite $\left.\begin{matrix} Na_2SiO_3 \\ RSiO_3 \\ \cancel{Fe}Si_3O_9 \end{matrix}\right\}$ assumes that of horn-blende. These formulæ suppose Na_2SiO_3 and $\overset{II}{R}SiO_3$ to be equivalent molecules, three of which correspond to one of ferric silicate $\cancel{Fe}Si_3O_9$. In one instance, in the augite group, the $\overset{II}{R}$ elements are completely absent: this is in Spodumene $\left.\begin{matrix} 3R_2SiO_3 \\ 4\cancel{Al}Si_3O_9 \end{matrix}\right\}$, where $\overset{II}{R}$ is replaced by $2\overset{I}{R}$ $=$Li, Na, and $\cancel{Fe}$ by $\cancel{Al}$. Here $2Al_2$ is equivalent to $3\overset{I}{R}$.

The isomorphism of $\overset{II}{R}\overset{IV}{R}O_3$ and $\overset{VI}{R_2}O_3$ is illustrated by the case of Titanic iron ore, a term applied to several minerals of varying composition, containing Iron, Titanium, and Oxygen, and usually some Magnesium, but which, according to Rammelsberg, can be represented by the general expression $\left.\begin{matrix} mFeTiO_3 \\ nFe_2O_3 \end{matrix}\right\}$, all having the crystalline form of Hematite, or Fe_2O_3. This view is not universally adopted, as another hypothesis supposes them to contain Ti_2O_3, the blue oxide of titanium. The aluminous varieties of augite and hornblende may be similarly represented by $\left.\begin{matrix} nRSiO_3 \\ Al_2O_3 \end{matrix}\right\}$.

The most remarkable examples of isomorphism combined with dissimilarity of constitution are afforded by the alkaline nitrates; potash-nitre, or saltpetre, KNO_3, being crystallographically almost identical with Aragonite, while nitrate of sodium, $NaNO_3$, is equally close in form to Calcite.

There are two cases of isomorphism of minerals not of analogous constitution among the class of silicates. These are Spodumene and Petalite, and Anorthite and Albite. The former are both oblique and closely allied in form, but completely dissimilar in composition—

Spodumene, being $\overset{I}{R_6}\cancel{Al}_4Si_{15}O_{45}$, or a bisilicate; and

Petalite $\overset{I}{R_6}\cancel{Al}_4Si_{30}O_{75}$, or a quadrisilicate.

Similarly, in the second case, both minerals being triclinic and isomorphous members of the lime-soda felspar group—

Anorthite, $CaAlSi_2O_8$, is a monosilicate and
Albite, $Na_2AlSi_6O_{16}$, a trisilicate.

When a crystal of a salt is placed in a solution of some other similar salt of an isomorphous metal brought to the crystallising point, it will increase in size by the addition of layers of the new salt, which will be symmetrically disposed about the planes of the nucleus, exactly in the same manner as would have happened had the growth been contained in the original solution. The form will therefore be preserved, but the crystal will obviously be only a mixture of heterogeneous substances, and its composite nature will be apparent if there is any marked difference in physical characters between the different constituents. One of the best examples of this kind of structure is furnished by the double sulphates of the alum series $\overset{I}{R}_2\overset{VI}{R}_2S_4O_{16}24Aq$, where $\overset{I}{R}_2$ may be either ammonium, potassium, or some other monad metal, and $\overset{VI}{R}_2$ either Chromium, Iron, or Aluminium. Two of these, the ammonia-aluminium and potassium-aluminium salts are colourless, while the chromium and iron salts are strongly coloured, the former being dark green and the latter violet, so that crystals formed from the solutions of two or more of them present strongly contrasted alternating bands of colour upon a cross section; or, if one of the colourless salts is from the outer layer, they may appear as transparent octahedra with coloured centres. Crystals of this kind, although illustrating the phenomena of isomorphism in a graphic manner, are obviously only mechanical mixtures whose heterogeneous character is plainly visible, and cannot therefore be represented as compounds of isomorphous bases in the same sense as those of dolomite, pearl spar, and other minerals are, where the combination extends to the individual crystalline molecules. They are, however, of considerable importance as illustrations of facts which occur in nature tolerably frequently. Thus crystals of Vanadinite, $3PbV_2O_8 PbCl_2$, from Russia, are occasionally found to contain a nucleus of the isomorphous species Pyromorphite, $3PbP_2O_8PbCl_2$, which is of the same hexagonal form. Crystals of Tourmaline when transparent are also commonly observed to be banded in different colours which correspond to differences of composition in the $\overset{II}{R}$ bases. In the felspar group, apparently homogeneous crystals are often made of alternations of the isomorphous minerals Albite and Orthoclase, and these being colourless, it is often difficult to distinguish one from the other, the use of optical tests being necessary in such cases. The same thing probably occurs in many other minerals, and is the cause of the discrepancies between the theoretical composition as required by the formula, and the results obtained by analysis. In this respect minerals differ essentially from crystallised salts artificially prepared, which may, by particular manipulation, be obtained in a state of almost absolute purity, while the former almost invariably contain some matters foreign to their essential constituents.

Another interesting case of crystals made up of alternating layers of isomorphous compounds of different composition is occasionally seen in the arsenides of Nickel and Cobalt ($NiAs_2$, $CoAs_2$). These are both cubical, and found in large lead-grey crystals apparently perfectly uniform in composition, but, when exposed to damp air, become oxidised with the formation of basic arseniates of the respective metals—that of cobalt being pink and that of nickel pale green, so that the crystals when broken across often weather in layers which are alternately coated with pink and green incrustations, according as one or other metal predominates in the particular layer.

It is probable that the universal presence of gold in minute quantities in such minerals as galena, PbS, and iron pyrites, FeS_2, may be due to a mechanical isomorphous intermixture of this kind, as all these species are cubical in form, and there is no reason to suppose that the gold is in chemical combination as it may often be extracted by the process of solution in mercury known as amalgamation.

CHAPTER XVII.

ASSOCIATION AND DISTRIBUTION OF MINERALS.

MINERALS, when exposed to the action of air, water, carbonic acid, and other agents of a similar kind tending to produce alteration in chemical composition, show very unequal degrees of stability; some species, such as gold, diamond, and graphite, the different forms of carbon, quartz and tin ore being practically unalterable, as they are neither susceptible of change by oxidation, nor reduction in air at the ordinary temperature, and almost, if not quite, insoluble in meteoric or ordinary spring waters; while, on the other hand, soluble and hydrated salts, especially those containing the alkaline metals, and the dyad forms, iron, manganese, and calcium, are, in a greater or less degree, liable to change either in form or composition under ordinary atmospheric vicissitudes. The following are some of the most general cases :—

1. *Alteration by loss of water.* This is called *efflorescence*, and is characteristic of minerals containing water of crystallisation which may in dry air be given off either entirely or in part. Crystals of Laumonite, $\left\{ \begin{array}{l} CaSiO_3 \\ Al_2Si_3O_9 \end{array} \right\} + 4Aq$, lose water by exposure and fall to pieces, although the change may be very gradually effected.

Crystals of gypsum, $CaSO_4 2Aq$, which are perfectly transparent when fresh, are often found in dry countries to become opaque either wholly or in part from a partial dehydration when exposed to the air.

2. *Solution and absorption of water.* Minerals whose crystals lose their form by absorption of atmospheric moisture, and are ultimately converted into solutions are said to be *deliquescent*; this property is common to many of the soluble alkaline salts, such as nitrate of soda, common salt, sal ammoniac, &c.

3. *Change by oxidation of one or more constituents.* This is a very common occurrence in ferrous or manganous compounds, and is generally known as *rusting*. It is most rapidly developed in the soluble salts of these metals. Thus, Ferrous sulphate ($\overset{II}{Fe}SO_4 7Aq$) is of a well-defined constitution and form, but the crystals can only be preserved in absolutely dry air, or in the vapour of a hydrocarbon, as under ordinary conditions of exposure they become dull and rusted through the production of ferric salts, a very large series of which are known in nature, and are produced by progressive oxidation, the ultimate product of such alteration being a Ferric hydrate ($H_6Fe_4O_9$) and acid ferric sulphate. In the same way, carbonates containing the same base, such as Siderite, Pearl spar, Dolomite, become invariably rusted externally when exposed to the air, even when the proportion of iron present is but small. Manganous compounds which, when fresh, are of a delicate rose red, are even more susceptible, as they turn brown by exposure to sunlight, and ultimately become brown or black by the formation of manganic-oxide MnO_2, the brown oxide Mn_3O_4, or their hydrates. For this reason, specimens of minerals such as Rhodonite $MnSiO_3$ and Diallogite $MnSO_4$ are generally kept in the dark, or the cases containing them in museums are screened from direct light.

Most metallic sulphides and arsenides are similarly liable to change by oxidation in damp air, with the formation of oxysalts of their constituents. Thus, Galena PbS gives rise to Anglesite $PbSO_4$, Zinc blende ZnS to Zinc vitriol $ZnSO_4 7Aq$, Cobalt speiss $CoAs_2$ to the hydrated arseniate

$Co_3As_2O_8 . + 8Aq$, known as Cobalt bloom, and Nickel speiss $NiAs_2$ to the corresponding Nickel salt or nickel bloom. Bisulphide of iron FeS_2, which is among the commonest of minerals, forming the dimorphous species Pyrites (cubical), and Marcasite (rhombic), besides being found in various isomorphous mixtures with other metallic sulphides and arsenides, yields, by the simultaneous oxidation of both constituents, Ferrous sulphate or Green vitriol $FeSO_4$, and Sulphuric acid, the former salt going through the changes previously noticed, the ultimate product being Ferric hydrate and basic ferric sulphate; but when aluminium or sodium compounds are within reach, the iron salts may be completely destroyed with the formation of the sulphates of these metals, Glauberite, Gypsum, and Alum. This group of reactions is one of the most important in the whole range of natural chemistry, as it is concerned in the production of the quantity of soluble sulphates present in most terrestrial waters, and which, in the case of mineral springs, often amounts to a considerable percentage. This change, which is generally known as *vitriolescence*, goes on more rapidly in the rhombic and granular varieties of pyrites than in the crystallised cubes; and when specimens of such minerals are kept in cabinets they often develop the unpleasant property of 'eating up their labels,' that is, the labels are rotted and destroyed by the sulphuric acid formed, the change being accompanied with the formation of capillary crystals of ferrous sulphate.

Sulphuric acid is also formed by the oxidation of sulphurous acid in the steam jets of volcanoes or fumaroles. In such localities it is common to find the felspathic component of the rocks reduced to a mass of clay variegated with parti-coloured patches, which are essentially alum and ferric sulphates.

4. *Change by action of carbonic acid.* All natural waters, whether terrestrial or atmospheric, hold more or less carbonic acid in solution, and, as such, are a cause of alteration,

which, though less energetic than sulphuric acid in any special case, is, as a whole, more important from the universality of its action. The effects produced are of two kinds: 1st, the solution of carbonate of calcium, magnesium, and analogous metals, and of their phosphates and fluorides, is promoted, as these salts are not sensibly soluble in pure water, but dissolve more or less readily in water saturated with carbonic acid; and 2nd, the double silicates containing alkaline metals and aluminium, typified by the felspar group, are attacked and decomposed with the separation of the alkaline silicate, which dissolves and is ultimately decomposed with the production of alkaline carbonates and soluble silica, while the insoluble aluminium silicate becomes hydrated, forming the mineral Kaolin or China clay. This change, known as *kaolinisation*, affects many minerals besides felspars, and is probably concerned in the production of most clay deposits. Silicate of lime $CaSiO_3$ is also decomposed by carbonic acid water, as are also ferrous silicates, but less readily than the former. Waters containing alkaline carbonates, especially when concentrated as in hot springs, also have a marked solvent effect upon silica, which is afterwards deposited as opal, hyalite, or chalcedony, or even quartz. The sulphides of the heavy metals, lead, zinc, or copper, are slowly converted into carbonates under the action of atmospheric waters containing carbonic acid, and are therefore commonly found to have been so changed in mineral veins at or near the surface.

5. *Change by reducing agents.*—These are, to some extent, inverse reactions to those involving oxidation. Ferrous sulphate, when kept in contact with decomposing organic matter, whether animal or vegetable, is converted into sulphide, crystals of iron pyrites having been produced in this way both naturally and artificially. Hydrated ferric oxide is in like manner reduced by decaying vegetable substances and carbonic acid, producing ferrous carbonate. Water containing alkaline sulphates may, in contact with

organic matter, produce sulphides of copper, lead, &c., by direct action upon those metals, as is shown by Daubrée to have taken place in the basins of certain thermal alkaline springs at Plombières, where copper pyrites and antimonial grey copper ore have been produced from Roman coins imbedded in the mud of the spring.

6. *Alteration by chlorides.*—The mineral Atacamite, or oxychloride of copper, $\left.\begin{matrix}CuCl_2\\H_2CuO_2\end{matrix}\right\}$, is a tolerably common product of the alteration of copper pyrites in the mining districts of Chili, Peru, and South Australia. It is readily produced when sulphuretted copper ores are exposed to the joint action of air and sea-water, and is probably due to a similar action in the mines in question which are situated in hot, dry countries, with little or no rainfall, and where, in consequence, the alkaline chlorides in the rocks have not been completely washed out. Probably most of the chloride of silver found in mineral veins is to be attributed to the same cause as a product of the alteration of sulphide of silver.

Evidence of alteration.—Pseudomorphism.—The transformation of a mineral by any of the methods previously described may be more or less evident, according to the nature of the change and the extent to which it has progressed. The earliest stages of alteration are marked principally by change of colour or lustre in the faces of the crystals, which become superficially altered while preserving their original structure and composition within; while, on the other hand, the alteration may be so great, and the development of new minerals so completely effected, that the derivation of the latter can only be inferred by generalisation upon evidence obtained in other cases. Such evidence may in some cases be obtained by direct experiment; but in the larger number of instances it is furnished by what are known as pseudomorphs, i.e. minerals that appear in crystalline forms not compatible with their chemical constitution, and must therefore have been altered by a partial or complete modification of their constituent elements, while retaining the forms proper to their original composition. The study of this most interesting branch of mineralogy has been systematised by Haidinger, Blum, Volger, and other observers, and the observed cases have been classified under the following heads :—

1. *Pseudomorphism by substitution.*—This implies a gradual replacement of the original substance by another, by means of simultaneous solution and deposition, without involving chemical action. The pseudomorphs of Quartz, or other varieties of Silica, after Calcite, Fluor Spar, Barytes, and similar minerals, are of this kind : the siliceous matter having been deposited from solution, *pari passu*, as the original crystal was attacked and dissolved ; the general order observed in such cases being that the replacing substance is less soluble than that forming the original crystal. The fossilisation of the remains of plants and animals by Silica and other minerals is also to be referred to this kind of action.

2. *Pseudomorphism by incrustation.*—In this case one mineral having been deposited upon another, and the older one subsequently removed, evidence of former existence of the latter is supplied by the hollow impression of its crystals retained in the second or incrusting species. As instances of this kind may be mentioned : Quartz upon Fluor Spar, Iron Pyrites upon Barytes, Quartz upon Barytes, Chlorite upon Dolomite, and Siderite upon Barytes. In all these instances, the free or outer surface of the incrusting mineral is generally developed according to its own form, while the under side of the layer forms an empty mould of the crystal of the mineral upon which it was originally deposited, but which has been since removed. It often happens, however, that the hollow space so produced is subsequently filled either with the incrusting substance or some third mineral, with the production of a substitution-pseudomorph of a more complex kind than those of the previous case.

3. *Pseudomorphism by alteration.*—Under this general head are included the following particular cases :—

(*a*) By loss or diminution of constituents.

(*b*) By gain or increase of one or more constituents.

(*c*) By interchange or substitution of one or more constituents.

The first of these cases is exemplified by pseudomorphs of Anhydrite after Gypsum, where the change is a simple dehydration ; Calcite after Gaylussite $\left\{ \begin{matrix} Na_2CO_3 \\ CaCO_3 \end{matrix} \right\}$, where the carbonate of sodium is removed ; and native copper after Cuprite (Cu_2O), where there is reduction or removal of oxygen.

The following are examples of the second case : Gypsum after Anhydrite, involving the addition of water; Malachite $\left\{ \begin{matrix} CuCO_3 \\ H_2CuO_2 \end{matrix} \right\}$ after Cuprite (Cu_2O), by addition of oxygen, carbonic acid, and water ; and Anglesite ($PbSO_4$) after Galena (PbS), by simultaneous oxidation of both lead and sulphur.

The third and last case is wider in scope than either of the preceding, and includes the larger number of observed pseudomorphs, some of the most prominent being the following : Limonite ($H_6Fe_4O_9$) after Iron Pyrites (FeS_2), and Siderite ($FeCO_3$), by loss of sulphur in the first, of carbon dioxide in the second, and addition of water in both instances ; White Lead ore ($PbCO_3$) after Galena (PbS), by loss of sulphur and gain of carbonic acid ; and Kaolin ($Al_2Si_2O_7 2Aq$) after Felspar ($\overset{I}{R}_2Al_2Si_6O_{16}$), by the loss of an alkaline silicate and the addition of water.

The term *paramorphism* is applied to a particular class of pseudomorphism where a mineral occurs in a form proper to its composition, but having the structure proper to a dimorphous mineral of the same composition. Examples of this are furnished by the change of oblique, prismatic crystals of sulphur into an aggregate of rhombic crystals without change of exterior form, aragonite with calcite structure, and augite with hornblende structure in Uralite.

The derivative character of pseudomorphs is evident from the absence of the structural peculiarities, such as cleavage, lustre, &c., proper to the forms imitated, and as a rule they are dull, made up of amorphous material, very generally hydrated, such as Kaolin and Steatite, dull or waxy on a fractured surface, and usually spongy or hollow, especially when produced by total replacement; but in some cases of pseudomorphism by alteration, the change may be effected without alteration of volume, so that the pseudomorph may be as compact as the original crystal. This is especially well seen in the common case of Limonite pseudomorphs after cubes of Iron Pyrites, where the latter, or original, substance is completely transformed into a compact mass of the former, while preserving all the external characters of the crystals, even to the twin striations upon the faces. Here the proportion of the unaltered constituent Iron (46·7 per cent.) in the molecule of Pyrites, is to that in the molecule of Limonite (60 per cent.) as 1 to 1·3, while their specific gravities are in the inverse ratio of 1·4 to 1, or 5·0 for Iron Pyrites and 3·6 for Limonite.

In the parallel case of pseudomorphism of Limonite after Siderite, the change is attended with diminution of volume, as both are of nearly the same density, 3·5, but Siderite contains only 45 per cent. of iron against 60 per cent. in Limonite, so that the volume of the pseudomorphs can only be about $\frac{3}{4}$ths of that of the original crystal. Actually, however, the volume is considerably less, as Siderite invariably contains more or less of the isomorphous bases—Lime, Magnesia, and Manganous Oxide, which are either removed in solution as carbonates or separate as Pyrolusite, or other manganese ores, by oxidation. These differences do, however, actually correspond to structural differences in the resulting minerals on the large scale, as brown Iron Ores (Limonite) produced by the alteration of masses of Pyrites are usually dense or compact in structure, while those produced similarly from Spathic Carbonates (Siderite) are, as a rule, spongy or cellular.

In many cases the alteration of minerals is attended with very considerable increase of volume, as, for example, in the conversion of metallic Copper into Malachite, or Iron into Limonite. In the latter instance the increase of volume is from eight- to tenfold, so that the change may be attended with considerable mechanical action when effected in a confined space. Numerous examples of this action may be seen in old wrought-iron work, which has been long exposed to the weather, where rusting has gone on between surfaces originally in contact, but which have been thrust apart by the rust formed between them. An analogous action may be assumed as taking place in the change of complex silicates into alkaline carbonates, silica and clay, owing to the greatly increased volume of these products as compared with that of the original mineral.

Origin of minerals. Questions as to the probable origin and method of formation of minerals are among the most interesting in the whole field of mineralogy, but the material available for their solution is comparatively small. In many instances the evidence of pseudomorphs is sufficient to show a secondary origin, or transformation from a pre-existing combination. The researches of Daubrée upon deposits formed in the conduits of the thermal springs supplying mineral baths in France and Algiers, which have been in use since the Roman conquest of Gaul, have shown conclusively that minerals of the class of Zeolites may be readily produced by the action of slightly alkaline heated waters upon the rocks they traverse when continued for a period of many centuries; and similarly, the condition of the bronze objects found in the remains of Assyrian and other ancient cities, by their conversion into Ruby Copper ore and Malachite afford not only a proof of the essentially secondary character of these minerals, if it be needed, but also, in some degree, a measure of the time required for their formation.

With minerals of a more complex character, especially among Silicates, direct evidence can rarely be obtained, and in such cases, therefore, the undesigned production of compounds similar in form and composition to natural minerals in slags and other furnace products, or what are usually known as artificial minerals, is of special significance.

The following are amongst the minerals that have been most frequently observed in furnace products definitely crystallised:—

Lime Augite ($CaSiO_3$) in the slags of several blast furnaces smelting iron ore. *Manganese Augite* ($MnSiO_3$), or approximating to that composition, which resembles the natural mineral Babingtonite, but is not exactly similar. This is found in slags produced in the Bessemer process of steel-making. *Potash Felspar* ($KAlSi_3O_{12}$), in minute crystals in the walls of furnaces smelting the copper-schist of Mansfeld, exactly similar to the natural mineral, but of rare occurrence. *Humboldtilite*, in modified square prisms, very much larger than those of the natural mineral which occurs in the lavas of Vesuvius: these are commonly seen in the older blast-furnace slags of South Staffordshire. *Iron Chrysolite*, or olivine (Fe_2SiO_4), corresponding in composition to the doubtful mineral species Fayalite. This occurs very commonly in the slags of puddling furnaces, and also at times in those produced in smelting lead or copper ores, or generally where slags that are essentially ferrous silicates are formed. The crystals have the form of the isomorphous mineral chrysolite or olivine (Mg_2SiO_4), but the latter is not found in furnace products, for the obvious reason that slags consisting mainly of magnesian silicates are not susceptible of formation under the ordinary conditions of working, owing to the refractory character of such compounds, and the use of magnesia in fluxes is therefore carefully avoided. *Magnetite* (Fe_3O_4) is common in octahedral crystals in the slags produced in the later stages of the puddling process, when the amount of iron taken up is in excess of that required to form a definite silicate with the silica present. It is also produced when steam is passed at a red heat over ferrous sulphide, which has probably been the mode of formation of the brilliant, artificial crystals found occasionally in the deposits formed in furnaces smelting pyritic silver ores at Freiberg.

Galena, PbS, in brilliant, cubical crystals and columnar aggregates, is tolerably common in deposits apparently the result of sublimation in the throats of blast furnaces smelting lead ore, and where the ore contains zinc, Blende (sulphide of zinc) and oxide of zinc (the latter not occurring as a natural mineral) may be deposited in a similar manner.

When metallic sulphides are roasted by burning in heaps in the open air, numerous minerals are formed by partial oxidation of the more volatile constituents, especially sulphur and arsenic, and deposit in the cooler portions of the heap in a manner analogous to that observed in solfataras and other volcanic emanations. Among these are *Sulphur*, *Arsenious acid* (As_2O_3), *Realgar* (AsS), and *Orpiment* (As_2S_3). This association of realgar crystals with sulphur is common at the solfatara of Naples. *Sal Ammoniac* (NH_4Cl) is occasionally deposited in the same manner upon waste heaps over burning coal slack. *Specular hematite* (FeO_3) in minute, brilliant crystals, is occasionally found on surfaces of salt-glazed pottery (such as drain-pipes, &c.); these are a consequence of the action of steam upon ferric chloride (Fe_2Cl_6), which results in the production of ferric oxide and hydrochloric acid. In this process, common salt is thrown into a kiln when the clay goods are brought up to a bright red heat, and in the presence of water vapour is decomposed with the formation of a glaze (silicate of soda) upon the heated surface of the ware, hydrochloric acid and some ferric chloride from the iron contained in the clay being volatilised. This reaction explains the formation of the very brilliant crystals of specular iron ore found upon the lavas of Vesuvius, Ascension Island, and other volcanic centres, hydrochloric acid and ferric chloride being commonly found in the steam emitted from fumaroles during and after periods of eruption.

Graphite in the form of crystalline scales, and occasionally in masses of considerable size, and closely resembling the natural mineral, is a common product of blast furnaces smelting iron ores, but the method of its formation, namely, separation from solution in molten cast iron, cannot be considered as analogous to any process likely to produce it in nature.

Water at high temperatures, when it is made to act under conditions by which the formation of vapour is prevented, or when under considerable pressure, has a powerful solvent action upon many substances which are not affected by it under ordinary temperatures and pressures, and may in such a case give rise to considerable danger to substances submitted to its action. The principal experiments upon this point are due to Daubrée, who found that hard glass, a homogeneous silicate of lime and soda, may be converted into crystallised quartz and pyroxene at a comparatively low temperature in this way. It is probable that an action of this kind may be concerned in the production of crystalline rocks containing quartz, with orthoclase and other silicates, by the slow rearrangement of masses originally homogeneous, when solidified by the process which is generally known as devitrification, or the transformation of a glassy substance into an opaque mass containing crystals.

Another agent of considerable importance in the production of minerals is probably boracic acid, which, though practically fixed when exposed alone to a very high temperature, is sensibly volatile even at the boiling point of water in an atmosphere of steam. The method of occurrence of minerals containing borax, especially tourmaline, in veins in granite and other rocks, seems to indicate a formation by a process analogous to sublimation; but more direct evidence is afforded by the presence of boracic acid in the condensed steam issuing from volcanic vents and furnaces in Tuscany and other places, and which are the principal source of supply of this mineral for commercial purposes. By the solvent action of boracic acid, or borax, at very high temperatures, many refractory or ordinarily infusible substances may be made to combine and crystallise from fusion. In this way, Ebelmen produced such minerals as Spinel ($Mg\overline{Al}O_4$) and Chrysoberyl ($Be\overline{Al}O_4$) from mixtures of magnesia and alumina, and glucina and alumina respectively, boracic acid being used as a solvent. The colours of the natural minerals were imitated by the addition of oxide of chromium for red, iron for black, and cobalt for blue spinel. Alumina was also converted into crystals of corundum and ruby by heating with borax. The analogous method of producing crystallised titanic acid by the use of salt of phosphorus, due to Gustav Rose, has already been noticed at page 312. Hydrofluoric acid and fluoride of silicon have also been used to induce combination between silica and metallic oxides. In this way Staurolite has been formed by passing hydrofluoric acid through alternating layers of silica and alumina and the analogous silicate Topaz, which contains fluoride of aluminium, by the action of fluoride of silicon upon alumina.

For further information upon this most interesting class of subjects the reader is referred to the various memoirs published by Ebelmen, Daubrée, Senarmont, and others. A compendious notice of these will be found in Percy's 'Swiney Lectures on Geology,' published in the 'Chemical News,' vol. xxiv., and Daubrée, 'Etudes de Géologie Synthétique,' Paris, 1879.

Speaking generally, two great and contrasted groups of causes may be said to be concerned in the production and modification of terrestrial minerals. These are, 1. The production by heat within the crust of the earth of homogeneous silicates of the alkaline and other light metals, analogous to glass. Silicates either appear at the surface by eruption, or remain, as deep-seated molten masses, to undergo more or less complete devitrification and differentiation into aggregates of quartz, felspars, and other silicates, phosphate of calcium, magnetic iron ore, metallic sulphides; 2. The conversion of these aggregates by the action of air and carbonic acid, atmosphere and thermal waters, into the various classes of hydrated silicates and metallic oxides, alkaline and earthy carbonates, metallic sulphates, &c. These actions are essentially compensatory, the tendency of the second being towards the production of quartz, soluble silica, clay, brown iron ore, alkaline carbonates, and carbonate of lime; the latter, being removed in solution, can only be returned to the general circulation by being brought within the range of the earth's internal heat. That causes of this kind have been in action from a very early period of the earth's existence as a solid body is evident from the identity of the minerals found in the oldest rocks with those recurring in other places in similar rocks, which can be shown to have been formed at very different geological periods.

Daubrée, from his researches on the synthesis of meteorites, has suggested the probability of an earlier period of universal scorification, when heavy silicates, such as olivine and masses of the heavier metals, may have been formed within the crust of the earth and depressed below the range of our immediate observation, but of whose existence evidence is furnished by their presence in meteorites and volcanic masses. This view supposes the formation of basic silicates by the action of heat alone to have preceded that of quartz, and the silicates of the felspar group, where the intervention of water is assumed to be essential.

Association and grouping of minerals. This subject is intimately connected with the preceding, as it is only by a study of the relative positions of dissimilar minerals in the same aggregate or mass that their order of succession or relative ages can be made out. The branch of mineralogy devoted to this class of investigation, and which stands in very close relation with geology, is usually known as *Paragenesis*, a term first applied by Breithaupt in 1849, in a work upon the associations of minerals observed in the veins worked in different mining districts. The further development of the same class of observation consequent upon the application of the microscope to mineralogical investigation, and more particularly to the constitution of rock masses, has given rise to another special branch known as Petrology, which forms the subject of a companion volume in this series. The nature of these associations can best be considered in the description of individual minerals, but some of the more prominent groups may be mentioned here.

Quartz occurs in association with almost every other mineral, but is more commonly found together with orthoclase, felspar, and other so-called acid silicates, than with those containing less silica. It is also very frequently found with mica, tourmaline, rutile, tin-stone, topaz, and the richer silver ores. Together with the amorphous varieties of silica (agate, chalcedony, &c.), it accompanies hydrated silicates of the zeolitic group in basalt and vesicular lavas, where it is obviously of secondary origin, as is also the case when it occurs in mineral veins traversing limestone strata.

Labradorite (soda-lime felspar) is almost invariably found with pyroxene, hypersthene, and titaniferous iron ore, forming the rocks known as norite, basalt, &c.

Of the different minerals containing iron, *Magnetite* is commonly associated with all rocks containing ferrous and magnesium silicates, and less so with quartz or micaceous schists, where hematite and titaniferous iron ores are more generally found. Magnetite occurring in large masses worked for iron ores, usually contains iron pyrites, chlorite, garnet, hornblende, and apatite in small quantities. Hematite deposits in stratified rocks as a rule contain as associates quartz, barytes, fluorspar, calcite, and aragonite.

Spathic iron ore is generally associated with sulphides, such as iron pyrites, copper pyrites, galena, &c., but very unequally in different localities, and also with the various isomorphous carbonates of calcium, zinc, and manganese, and the products of the alteration of the latter such as pyrolusite (MnO_2).

Iron Pyrites is the most abundant of all the metallic sulphides, and is widely diffused through rocks and mineral deposits of all kinds. In small quantities it is found in clays and other rocks impermeable to water, and in coal, and other carbonaceous deposits, where it is protected against oxidation. When in large masses, it is commonly associated with copper pyrites, arsenical pyrites, and the various sulphides and arsenides of nickel and cobalt, gold and silver. It also forms a general constituent of mineral veins containing the ores of tin, copper, and lead.

Tinstone and its associates constitute a very special group of minerals, which, though restricted to a small number of areas, are often very abundantly developed in particular localities within these areas. In this group are included Tinstone (Stannic oxide), tourmaline, topaz, wolfram, scheelite, iron pyrites, mispickel, copper pyrites, chalcedony, fluorspar, hematite, pitchblende, and occasionally bismuth ores.

Galena, or sulphide of lead, the principal ore of that metal, forms part of numerous well-marked groups of minerals, which as a rule are characteristic of particular districts. Among its more usual associates are the products of its own alteration, sulphate, carbonate, and phosphate of lead, and zincblende (ZnS), calamine (carbonate and silicate of zinc), iron and copper pyrites, and the 'waste' or earthy minerals, calcite, fluorspar, aragonite, dolomite, and barytes, when in slate and limestone districts; in addition to which quartz, and occasionally zeolites, are found when the veins are in siliceous rocks, such as granite, gneiss, &c. When in company with antimonial minerals, gold and silver ores usually enter into the group, as in the Hartz and Hungary.

Nickel and cobalt ores, when found in quantity, i.e. as rich arsenides, and not merely as mixtures with iron pyrites, severally accompany native arsenic and various arsenides, the products of their oxidation (pharmacolite, cobalt bloom, nickel bloom), and the different ores of silver and bismuth.

Copper ores are very commonly found in connection with minerals containing magnesia, such as hornblende, chlorite, serpentine, and dolomite, and also with quartz. Copper pyrites, the most abundant ore of this metal, has two principal lines of association, the first being with iron pyrites in more or less intimate mixture, forming the so-called coppery pyrites, and the second with copper-glance (Cu_2S) and erubescite ($FeCu_3S_3$), forming a series richer in copper. Native copper, and the various oxides and oxy-salts, carbonates, sulphates, and phosphates of this metal, are common products of the alteration of the sulphuretted minerals; but in the district of Lake Superior the metal occurs exceptionally and in enormous quantities over a very large area practically without other copper ores, and in association with quartz, calcite, and zeolitic minerals.

Rock Salt (NaCl), though often found in very large masses in a perfectly pure state, is generally associated with salts of calcium and the alkaline metals, especially gypsum and anhydrite. Where the deposits have been isolated in such a manner that the more soluble salts contained in the original salt water have been preserved, a numerous series of double sulphates and chlorides of potassium, magnesium, &c., are developed. Among these are Kainite, Carnallite, Kaluszite, Boracite, Polyhalite, &c., besides bromide of magnesium. This is only seen on a very large scale at Stassfurt in Prussia, and Kalusz in Gallicia.

As regards the frequency or scarcity of the occurrence of minerals, it may be useful to remember that a substance may be rare in two different ways—being either widely distributed, but in minute or even invisible quantities in other minerals. or restricted to a few localities, where, however, it may be found isolated in quantity. Sulphide of cadmium may be taken as an instance of the first kind, having only been found in a single locality as an independent mineral (Greenockite), and in a few minute examples; but as an isomorphous associate with the corresponding sulphide of zinc, it is present in the larger number of samples of zinc-blende, so that some tons of the metal are made annually by fractional distillation of the first deposits of zinc oxide obtained in the zinc works, which are usually found to be cadmiferous. Other examples are afforded by the rare alkaline metals cœsium and rubidium, common in certain mineral waters, but almost unknown in individual minerals; also by thallium, gallium, and similar elements existing in spectroscopic traces in common minerals such as pyrites, zincblende, &c. Cryolite (6NaFl+ AlFl6) is an example of the second kind of rarity, it being almost entirely restricted to one spot on the coast of Greenland, where, however, it is found in such masses that it can be utilised as a commercial source of soda and alum.

INDEX.

PRINTED BY
SPOTTISWOODE AND CO., NEW-STREET SQUARE
LONDON

Made in the USA
Monee, IL
04 November 2026

67387317R10213